Günther Anders' Philosophy
of Technology

Also Available from Bloomsbury

Hans Jonas: Life, Technology and the Horizons of Responsibility, Lewis Coyne
The Bloomsbury Companion to Arendt, ed. Peter Gratton and Yasemin Sari
Chance, Phenomenology and Aesthetics: Heidegger, Derrida and Contingency in Twentieth Century Art, Ian Andrews
Adorno and the Ban on Images, Sebastian Truskolaski
The Dialectics of Music: Adorno, Benjamin, and Deleuze, Joseph Weiss

Günther Anders' Philosophy of Technology

From Phenomenology to Critical Theory

Babette Babich

BLOOMSBURY ACADEMIC
LONDON • NEW YORK • OXFORD • NEW DELHI • SYDNEY

BLOOMSBURY ACADEMIC
Bloomsbury Publishing Plc
50 Bedford Square, London, WC1B 3DP, UK
1385 Broadway, New York, NY 10018, USA
29 Earlsfort Terrace, Dublin 2, Ireland

BLOOMSBURY, BLOOMSBURY ACADEMIC and the Diana logo are trademarks of
Bloomsbury Publishing Plc

First published in Great Britain 2022
This paperback edition published in 2023

Copyright © Babette Babich, 2022

Babette Babich has asserted her right under the Copyright, Designs and
Patents Act, 1988, to be identified as Author of this work.

For legal purposes the Acknowledgements on p. xii constitute an extension
of this copyright page.

Cover design by Charlotte Daniels
Cover image © Babette Babich

All rights reserved. No part of this publication may be reproduced or transmitted
in any form or by any means, electronic or mechanical, including photocopying,
recording, or any information storage or retrieval system, without prior
permission in writing from the publishers.

Bloomsbury Publishing Plc does not have any control over, or responsibility for, any
third-party websites referred to or in this book. All internet addresses given in this
book were correct at the time of going to press. The author and publisher regret any
inconvenience caused if addresses have changed or sites have ceased to exist,
but can accept no responsibility for any such changes.

A catalogue record for this book is available from the British Library.

Library of Congress Cataloging-in-Publication Data

Names: Babich, Babette E., 1956-author.
Title: Günther Anders' philosophy of technology: from phenomenology to
critical theory / Babette Babich.
Description: London, U; New Yor: Bloomsbury Academic,
Bloomsbury Publishing Plc, [2022] |
Includes bibliographical references and index. |
Identifiers: LCCN 2021014332 (print) | LCCN 2021014333 (ebook) |
ISBN 9781350228580 (HB) | ISBN 9781350228597 (ePDF) |
ISBN 9781350228603 (eBook)
Subjects: LCSH: Technology–Philosophy. | Technology–Social aspects. |
Anders, Günther, 1902-1992.
Classification: LCC T14.B29 2022 (print) |
LCC T14 (ebook) | DDC 601–dc23
LC record available at https://lccn.loc.gov/2021014332
LC ebook record available at https://lccn.loc.gov/2021014333

ISBN: HB: 978-1-3502-2858-0
PB: 978-1-3502-2862-7
ePDF: 978-1-3502-2859-7
eBook: 978-1-3502-2860-3

Typeset by Deanta Global Publishing Services, Chennai, India

To find out more about our authors and books visit www.bloomsbury.com and
sign up for our newsletters.

Contents

List of Figures	viii
Preface	ix
Gratitude and a Prefatory Note on Acknowledgements	xi
Acknowledgements	xii
Abbreviations	xiii

Introduction	1
Black Stars	1
On the "We" of the "With-World" and Social Media	4
Life and Legacy	6
A Star Among Other Stars	7
Suppression: Positivity and Neutrality	10
Consumers and "Situations"	12
Nomen est omen: "Other" Reflections	14

Part One A Critical Theory of Technology   19

1 Criticizing Technology	21
Questioning Being-in-the-World, the Work of Art, and Technology	21
On the Intersection of Sociology and Anthropology, Critical Theory, and Technology: Günther Anders and Power	21
Insiders and their "Discounted" Others	25
Oblivion	26
Whither?	36
Whose Critical Theory?	38
Media Branding: How to do Politics with Programming and iGadgets	40
2 Heidegger's Authenticity and Günther Anders' Neg-Anthropology	42
Heidegger's Authenticity and Günther Anders' "Humanism"	42
Jemeinigkeit: On Becoming a Question to Oneself	44
Anthropologism and Psychologism	45
Heidegger on Dasein and the *Seinsfrage*	47
"Who," Then, is Dasein?	48
On Animal *Dasein*: Anders and "Other" Others	49
Sein and *Dasein* without God	53

3	Günther Anders and Hannah Arendt: Love, Triangles, and the Political	58
	Philosophical Triangles: Students, Scholars, Lovers	58
	Triangulating the Triangle: Arendt–Heidegger–Anders	63
4	Anders and Arendt Reading Rilke: Love Songs to God	69
	Kafka	69
	Rilke	71
	Angels	74
	Elegies: Poetry and Hearing/Poetry and the *Schma*	76
	The Cherry Slaughter: Back to Love	82
5	Between the Lines: Benjamin's Angels of History and Anders' Apocalypse	88
	Having Been	88
	Angels of History	89
	Time	92
	Heidegger and *Time:* New Rules	94
	Time/Space	101
	Whose Holocaust? Which Genocide(s)?	106
	Once More, With Feeling	107
6	Anders and Adorno: Genocide	110
	On What We Have Done	110
	Singulare Tantum: Whose Genocide? Which Genocide?	114
	Thinking the Holocaust	115
	Colonialism and the Exploitation of the Globe	118
7	From Anders' Sexless Capuchin to Virilio's Chimeras	123
	Economy, Power and Possibility, Impotence and Sexuation	123
	From Consumption to Biotech	125
	Manufacture and Art: *Homo Materia*	129

Part Two Anders, Media, Music		135
8	Radio Ghosts	137
	Ghosts	137
	Radio Transforms: Politics and Music	138
	Phenomenology's Ghosts: Anders' Phenomenology of Radio Listening	141
	Media-Induced, Collective, "Autism"	148
9	Being-in-Music	152
	Music Critique and Musical "Situation"	152
	Situational Phenomenology: Underway to a Hermeneutics of Music	154

	Positive Attunement: Sociological Reflections on the Musical "Situation"	160
	Transformation and Transfiguration	163
	Critical Sociology of Music	165
	Being-In	166
10	Transistor Radios and Media "Überveillance": From Anders' "Radio Leash" to Tracing	168
	"Ground Control to Major Tom"	168
	Political Philosophies of Technologies	169
	The Skies Down to Earth: Being Without Time	176
11	Pop Music (and Jazz), and Covers (and Copies)	181
	Transitioning: From Anders' Radio "Leash" to Cohen's *Hallelujah*	181
	Death and Taxes	183
	Pop as Contemporary Music: Fortunes and Futures	184
	Cover Culture	187
	Darker Stars: Death and Silence	193

Part Three Schizotopic Thought: Planetarism and Apocalypse Blindness — 195

12	Political Media Theory, Hiroshima, and Nuclear Power Plants	197
	From the Holocaust to Hiroshima: "Chernobyl is Everywhere"	197
	Other Than, One More Time	197
	"Seit ein Gespräch wir sind"	202
	Violence Contra Violence	205
	What about Violence: Yes? or No?	207
13	"The Devil's New Apartment"	209
	Apocalyptic Thinking	209
	Nightmare	210
	Counting Industrial Revolutions	213
	Weather Talk: How to Do Things with Clouds	215
	The "Modern Prometheus"	221
	Geoforming, Masks, and Us	221
	"Cosmic Parvenus"	223

Notes	227
Bibliography	278
Name Index	312
Subject Index	318

Figures

1. Film scene of *Metropolis*, 1927, directed by Fritz Lang. Photo credit: bpk Bildagentur / Horst von Harbou / Art Resource, NY — 31
2. Film scene of *Metropolis*, 1927, directed by Fritz Lang. Alfred Abel, Brigitte Helm, Rudolf Klein-Rogge. Photo credit: Adoc-photos/ Art Resource, NY — 31
3. Hans Baldung Grien, "Weibermacht." Woodcut of Phyllis seated on Aristotle on all fours, with Alexander peering from the parapet. 1513. Wikicommons. Public domain — 60
4. Bethesda Fountain, Central Park, New York City. Photograph, Babette Babich, December 12, 2018 — 73
5. Günther Stern and Hannah Arendt, January 1, 1929. Photo credit: Courtesy of the Hannah Arendt Bluecher Literary Trust/Art Resource, NY — 80
6. Paul Klee, Angelus Novus, 1920. Gift of Fania and Gershom Scholem; Courtesy of the Israel Museum, Jerusalem. Photo credit: HIP/Art Resource, NY — 90
7. Paul Gustave Doré, Dante, *Paradiso*, Canto 31. The Saintly Throng in the Form of a Rose. Wikicommons. Public domain — 104
8. Trinity atomic bomb explosion, Rapatronic image. July 16, 1945, 05:29:45, Mountain Wartime. Alamogordo Test Range, Jornada del Muerto Desert. Courtesy of US government Defense Threat Reduction Agency. Wikicommons — 104
9. Ferdinand Barth (1842–92), *Der Zauberlehrling* (1797). *Goethe's Werke*, 1882, Ink drawing. Wikicommons. Public domain — 173

Preface

This book was begun before the Covid-19/Coronavirus crisis of 2020 (as of this writing, still ongoing), and yet Günther Anders (1902–92), the philosopher of modern media technology, is perhaps more significant than most philosophers today not least because his was explicitly a philosophy of isolation, what Anders called "home-working," a kind of labour distinguished by being mediated and effected by the media. Anders argued that such homework focused on the homeworker, was conducted alone, and potentially without limit. In addition, this homework was often self-elected, without any direct connection between the worker and the industry that profited from that labour and that that 'homework' was often self-financed to boot: one paid in order to have the privilege to provide such work on behalf of corporate benefit, a kind of sharecropping practice in the age of what Anders' named the "second industrial revolution" and which we know as the age of social media. For this reason, Anders' argument seems uncannily relevant to the conditions of life as we live today, even before recent social distancing restrictions, as many of us 'lived' online, by way of and through the screen, the cell phone, the tablet, and laptop/computer.

This makes Anders worth reading, and yet, like other scholars of his era, other thinkers in the tradition of philosophy called, for want of a better designation, "continental" as opposed to "mainstream" or "analytic" style university philosophy, simply in order to begin to read Anders, one must have a sense of his antecedents, his teachers, his tradition. It adds complexity to note that Anders, as one of the founders of the early Frankfurt School, having been present from the start along with his first wife, Hannah Arendt, would preserve his allegiance to critical theory throughout his thinking and his life.

But philosophers and cultural and critical theorists whether in sociology or ethnography or political theory have their habitual references, and Anders is not quite included among such references. When a proper philosophical biography is available—and this must await a full understanding of his thought—one will be able to argue that Anders was obscured from a light that, by any number of rights, he should have shared. The reason for this is not a matter of accidental or incidental inattention but at times calculatedly academic: scholars refused to read him, experts asked to vet translation proposals refused, presses that commission such expert judgement declined to publish his work, both his creative efforts and his theoretical writings or else, especially in the case of the former, delayed publishing until such time as it ceased to matter for the vital communicative energy essential for the life of scholarship. I have for years sought to draw attention to what is missed via silencing, failure to mention, failure to note. There is no way I can make up for this in this study, but in general I try to footnote references to alternate interpretations as well as further reading.

To be sure, this seems to be changing, and Anders is increasingly read. At the same time, access is often advanced in a narrow fashion that ignores or else dismisses other readings, other perspectives. A blinkered approach, if common in academia, given 'received views,' insider circles and publication cartels and such like, is not the best way to approach Anders, if only owing to the critical question of breadth: reading Anders requires all the background indicated earlier. One needs the resources of phenomenology, hermeneutics, including both theological and literary hermeneutics, sociology, including sociology of music and literature, media studies especially but not only radio and television, film and print media, all in addition to taking the politics of Anders' own life, his own 'situation' very seriously and very much in his own way, read according to his own very particular and stylistically distinctive slant.

In addition, though this text can only graze this theme, Anders enjoyed a gnostic view of the world, even more so perhaps than Jacob Taubes or Hans Jonas, both of whom, unlike Anders, enjoyed a wide readership. The text to follow will only obliquely point to this dimension, fitting for the esoteric. At the same time, this text is predominantly philosophical, focused largely on technology. For this reason, it cannot encompass all arenas of Anders' thought, though it seeks to do more than most as this is dictated throughout by Anders' own focus as he mixed his own themes. There is always more to be done than has been done by other authors, and this book offers another perspective on Anders than can otherwise be found. There will, of course, be more to say.

Gratitude and a Prefatory Note on Acknowledgements

In Anders' own spirit, it is hoped that there will someday be a broad and vital community of interlocutors for the sake of scholarship and critique. Essential for such hoped-for community cannot but be mutual engagement, citation, and cross reading: scholar to scholar.

I gratefully acknowledge the contributions of a key webpage initiated and, most importantly, *maintained* by Harold Marcuse on Anders' life and work, chronicling research on and translations of his work.

I am grateful to Günther Anders' nephew, David Michaelis, the son of Anders' sister, Eva Stern, for engagement and support via email and, indeed, social media: his encouraging words were sustaining.

I thank those who made it possible to give lectures on or to publish work on Günther Anders, whether directly or indirectly. Thanks are owed to Andreas Beinsteiner, Roger Berkowitz, Dmitri Ginev, Annette Hornbacher, Christopher John Müller, Vallori Rassini, David Rasmussen, Rüdiger Schmidt-Grépaly, Georg Zenkert, and, most recently, for a kind invitation to Galway, Felix Ó Murchadha as well as his colleague, Rod Stoneman.

Here, too, I acknowledge, with gratitude, Bloomsbury's Lucy Russell and Liza Thompson. I also thank Joseph Gautham and Giles Herman. Thanks are owed to Michaela Latini along with Alessandra Sannella for inviting me to contribute a lecture for a conference on "Violence" in Cassino, Italy. The opportunity, even after so many years, to visit a landscape and architectural sites still marked by the contour of death was important for this book. And I thank Teodosio Orlando.

The New School philosopher, Jerry Kohn, has my admiration for his scholarship and gratitude for his kindness in writing to me on some of the work relating to the current project.

I thank Tracy Burr Strong for his insights on the array of complicated topics that make up the current study.

And, despite his efforts against Anders, or perhaps because of these, I thank my Stony Brook teacher, Don Ihde.

Acknowledgements

Books grow out of an author's earlier essays, scholarly debates and exchanges, retrospective reflections, collegial conversations. I thank my indirect interlocutors: these are those numerous authors who have written on Anders in German as in French and in English. Some of their names are found in the bibliography. I have also noted conversational partners earlier, including scholars and not less those editors who have published my writing on Anders. And, although in every case, substantially revised for the present book, I draw here on my published work developed over many decades on Friedrich Nietzsche, Martin Heidegger, Theodor Adorno, Hannah Arendt, Ivan Illich, Jean Baudrillard, and not less on Günther Anders himself. Because of the changes made and updates needed for the sake of the constellation that is the current book, it would be misleading to list previously published chapters but I do draw on all the work listed under my name in the bibliography.

All translations, unless otherwise indicated, are the author's own.

Abbreviations

AM I	Anders, *Die Antiquiertheit des Menschen 1: Über die Seele im Zeitalter der zweiten industriellen Revolution.*
AM II	Anders, *Die Antiquiertheit des Menschen; Zweiter Band. Über die Zerstörung des Lebens im Zeitalter der dritten Industriellen Revolution*
EH	Nietzsche, *Ecce Homo*
FW	Nietzsche, *Die fröhliche Wissenschaft*
GT	Nietzsche, *Die Geburt der Tragödie*
KS	Anders, *Die Kirschenschlacht. Dialoge mit Hannah Arendt*
MS	Anders, *Musikphilosophische Schriften*
PC	Stern/Anders, "On the Pseudo-Concreteness of Heidegger's Philosophy"
PM	Anders, "The World as Phantom and Matrix."
QCT	Heidegger, *The Question Concerning Technology*
SZ	Heidegger, *Sein und Zeit*
TA	Sloterdijk, *Terror from the Air*

Introduction

Black Stars

The *OED* tells us that a "schlimazel" is a "consistently unlucky accident-prone person."[1] The meaning I grew up with (between Brooklyn and Manhattan) references bad luck inherited, as it were, from birth: no amount of decency, hard work, or effort can shift the odds in such a person's favour. Thus, for the etymology, a *slimazel* is a Yiddish term combining the old High German *slim*—in today's German that is *schlimm*—with the Hebrew, *mazzāl*, meaning luck. If Günther Anders might seem to fit such a description, he also utterly lacked any sense of fated *ressentiment* all the while being fully aware that favours were often denied him.

To this extent, this book seeks to avoid the typical Anders-cliché that one can find in many publications emphasizing that Anders was, as indeed he was, one of the most persistently ignored thinkers of the twentieth century. Instead, attention to the complexities of ill-aspected fortune or luck can underscore that it is arguably the extent of Anders' thinking that complicates his reception, as his thought ranges over Frankfurt School–style critical theory as well as phenomenology and hermeneutics via Edmund Husserl, Max Scheler, and Martin Heidegger, but also over sociology and literature and music, including literary theory and the sociology of music as articulated across the spectrum of music history and theory, composition, and performance practice, and also popular reception and musical critique, expression, and reception, quite in addition to a philosophical anthropology in the lineage of Scheler and Walter Benjamin, who was Anders' cousin, and Theodor Adorno.

And then there is media criticism.

And then there is technology.[2]

Anders undertook to raise the question of the human being in an age of technologically intensified reproducibility, much along the lines that Benjamin had written about this technical capacity/promise. Rather than addressing the work of art as such, despite Anders' interest in art as such, including music and literature, as well as, although critical reflection on this exceeds the limits of this study, Anders' own creative writing as essayist and novelist, thinking the working of the literary work as art, and reflecting on the performative dimension of actively composing–, actively playing–, and actively *listening-to*-music, Anders was also interested in the *soul*.

Such was the topic of his first book on the outdated/antiquated human being whose "soul" Anders understood by way of a certain attuned or musical sense with reference to the Judaic tradition and its acoustics as also with reference to St Augustine and his musical examples intertwined with a reflection on time. This includes allusion to

Kant and Husserl and Heidegger as a gnostic dimension, although as I have already noted, fully tracing out this last will be beyond the current focus, except in the most liminal way.

In a series of iterations, Anders' book on "the second industrial revolution" is addressed to readers including both atheists and theologians in the era of not the Hegelian but the Nietzschean 'death of God.' For Anders, this is characteristic not of religious experience but its "frustration" at least as expressed variously by Rilke, Kafka, and Beckett, authors Anders counts as those—notice the formula—who "*do not experience God.* [*daß sie Gott nicht erfahren*]"[3] Spanning two world wars, the instauration of the industrial turn in the twentieth century installs the human being, for Anders, in express counterpoise and, hence, note thereby Anders' reference to shame and inadequacy, to a different kind of genesis, illuminated as deficient by contrast with the manufactured object, the *Gerät*, the gadget, the tech fetish or consumer product in a serried scheme and component part of the same industrial assembly Heidegger named *Ge-Stell*. Although a critic, Anders assumes and presupposes the Heideggerian *Ge-Stell* as the assemblage essential to the perfection of the *Gerät*, as the "must have" as this variable object changes in a social-historical constellation, materially speaking, from moment to moment. Our technological absorptions are thus articulated as a series of metonymic sediments, via the machine, as Kant epitomized the schema inherited from Descartes and Spinoza and, above all, Leibniz. Not a clock, not a computer (and today one can add genomic and nanotech technologies to the mix), in an era confident of the machine's perfectibility and its progressing-accession, what is key is the next "new" thing. Different from Gilbert Simondon's more popular object concerns, Anders' soul musing in his "anthropological phenomenology"[4] explores the fate of the human in a nihilistic era quite in place of deity and by phantasmatic contrast with a technological product of a system of technology and its products—Anders never forgets that this is a capitalist constellation, even in post-war cultures that claim the flag of communism or socialism—for the members of that technological society. There are parallels to be made with the work of the sociologists Jacques Ellul, Jean Baudrillard, and Paul Virilio in terms of symbol and sign and not less fetishized design *qua* mark of "distinction" and thereby also parallels with Pierre Bourdieu and Michel de Certeau and others. Thus, one can speak of sociology and anthropology or ethnography, as of psychology, but always in a philosophical and, for Anders, expressly *phenomenological* context.

Characteristic of Anders—and here perhaps the best parallels would be with Ivan Illich and, in a more contemporary voice, with Giorgio Agamben and, again, Virilio—is a clear focus on the postlapsarian condition. This all-too-human condition was expressed via the language of shame: the shame of having been born as such but not as contrasted with deity in its eclipse, much as Heidegger also emphasized following a long theological tradition of reflection on silence as speech *and* as medium of encounter. Anders' focus on the soul was also a focus on technology, and to this extent he undertook a theological turn in critical theory, raising questions that were not always received and were often otherwise unasked with respect to "the world," as Anders spoke of it, as "phantom and matrix," adumbrated through the spectral echoing of radio broadcasts, ghosts of television and film, and doubling and displacing humans in what had been a different situatedness in the world, now directed by a set of programmes to which the human being himself or herself actively attended, directed, and curated and, ultimately, in which they lived their lives. For Anders, this is all about copies in the age, as Benjamin spoke of it, of technological reproducibility and digital reproducibility. This is our world today, and Anders can help us to understand it.

Anders' two main works correspond less to conventional academic *monographs*— Anders lacked an academic position which also meant that he was free from the tribal tithes required of the same—than collections of expositions, deliberately arrayed variations in the spirit of the phenomenological tradition in which he was trained. Thus, the aforementioned focus on the soul [*die Seele*] could be allied to musical sociological and phenomenological anthropological reflections articulating nothing less than the world around the human being, the human in *medias res* of the world, music, radio, film, television, literature, and theatre but also war and in particular the atomic bomb that made wartime a never-ending affair. To this extent, Anders offers us a philosophical reflection on what may also be named a phenomenological anthropology of the human being, and to this same extent he tends to steer clear of the Hegelianism that had in his era so dominated the twentieth century and that still continues to reign, typically associated with other names, in the twenty-first century.

If Anders argues in his first book that the human soul had undergone a series of transformations in the *second* industrial, technological epoch, in his second book (collection) he reflected on the transforms effected by the *third* industrial age, inaugurated, as he wrote of the succession of such eras and well in place by the mid-twentieth century. In each case, the human being was antiquated/outdated by no less effective means than that of his own projections, against which that same human being found himself or herself lacking. In his first book, the shame was in being a creature rather than a precision product manufactured to specs, corrigible and thus perfectible to infinity. Today's GMO/mRNA transforms fit this as well, if more concerned with 'gain of function' (again to spec) and trackability.

Anders' phenomenological anthropology is thus articulated as a critical theory of technology; its concerns include broadcast media and, given his own cultural theoretical interests, music. In addition, and this addition presented for some readers a certain dissonance, Anders includes a reflection on violence, personalized and reflected and unremitting. These reflections range from not only the First World War, which he observed in France as a member of a youth corps, but also the Second World War, including the violence in the Pacific as well as, on its cusp, the atomic bombing of Hiroshima and Nagasaki, through to Korea and Vietnam and a range of wars that followed, including, at the end of Anders' life, Grenada and the Gulf War itself, mediated via television and embedded or circumscribed reportage. Anders' philosophy takes us beyond his century into our own, as we meet the new face of war, increasingly conducted from the sky and with the sky, and now given air pollution and socially mandated masks, through the air we breathe (or do not breathe), all the way to the force multipliers of climate manipulation, with scalar and sonic warfare, uncontrolled forest fires, earthquakes, and geo-forming, along with drone warfare at a distance, literal weather warfare, and geoengineering.

On the "We" of the "With-World" and Social Media

It is a characteristic of reflections on Anders that these typically begin by focusing on the fact that Anders' name is not his name. Thus the name Anders is often read as a mark of distinction. *Anders* means "other," or "otherwise," and to this extent, underscores the character of an outsider, *Aussenseiter*, a *Fremdling*, a word Anders also uses to describe himself.[5] But it is characteristic of human nature that those who are other, those excluded, are often regarded as an object of some envy by the groups that excluded them. This dynamic is certainly part of what Hegel analyses as his master–slave dialectic, and yet we need more, if only because the Hegelian dialectic as such will not suffice, not after Nietzsche,[6] nor, even more specifically, after Anders himself as not after Adorno (given his *Negative Dialectics*),[7] nor after Horkheimer, or Paul Virilio or Jean Baudrillard. "Etc., etc.," just to quote Anders' own tribute to the excess that belongs to any such listing in the recollected course of his recollection of his conversations with Hannah Arendt.

The dynamic between exclusion–persecution and resentment–envy is complicated. Thus, Anders observes that *everyone* in a technological age can take himself or herself to be an outsider, to be as Moses once found himself, *a stranger in a strange land*, a designation taken over by a popular science fiction author for a crucial and crucially successful novel in the genre. Anders himself would not have success with his own novels, but he nonetheless characterized today's very human condition: today we are all outsiders, no one more than anyone else; all of us are onlookers, spectators, consumers captivated by programmes (today that can be no more than a social media feed) advertising the newest news along with the newest consumer products. Here the parallel is with Nietzsche's language of the "all-too-human" rather than the numinous constellations that underly "posthuman" ambitions. Anders thus advanced the consequences of the claim that we are alienated from ourselves, "shamed" by means of nothing less "concrete" than radio programmes and television broadcasts, transmissions, and mediations. These are the vapours Marx and Engels allude to when they write in 1848 in their *Communist Manifesto*:

> All fixed, fast-frozen relations, with their train of ancient and venerable prejudices and opinions, are swept away, all new-formed ones become antiquated before they can ossify. All that is solid melts into air, all that is holy is profaned, and man is at last compelled to face with sober senses his real conditions of life, and his relations with his kind.[8]

Marx's allusion to what *becomes* "antiquated" resonates in the title of Anders' constellation of intersecting reflections on human antiquatedness, *Die Antiquiertheit des Menschen*.[9] At the same time that Marx foregrounds, quite as Anders does, the "constant revolutionizing of the instruments of production,"[10] ageing before time (think of new products and new ages, space-, computer-, information-, digital-, Anthropocene-), including for Marx the transformation of guilds and traditional mastery, Anders highlights the becoming-antiquated effected by ghosts, shadows, on and with radio, the projections of film, programmed television broadcast, permanently returned in closed circuits, cable or wired access, wirelessly networked, now streaming, live or on

YouTube or social media videos on demand, with- or without-your-leave in public squares, on mass transit, in taxis, in sports bars.

Pocket convenience turns out to be the mode for Dick Tracy-style wristwatch walkie-talkies, Star Trek-era communicators not at all unlike (depends on the model) our cell phones. Today's smartphones and tablets do the same job: we are become-cyborg via the same transportable means Anders describes invoking transistor radios, and connected or meaning to be connected at all times, we carry our instruments of alienation on our person, or even embedded as digital hacks or via injection, more conventionally, at all times. In a literal sense, we are all Americans-by-proxy, living as global citizens, quite as Nietzsche described the new world ethos of American ambition, driven by speed, "One thinks with a watch in one's hand, the one eats one's midday meal, with one eye on the financial papers, one lives as if one way continually might 'miss out on something'" (FW, §329).

Anders' term seems to be Dick Tracy dated, "gadget-envy," but it is an enduring condition to this day. Hence, by updating Goethe's 1797 tale, *Der Zauberlehrling*, "The Sorcerer's Apprentice,"[11] Anders appropriates the second-century (CE) Lucian's morality tale, complete with mystery cults and roses and the seemingly as if made for TV, transmogrifications that could turn a boy into an ass, and Shakespeare draws on the same magic, transmogrifications that trap him in that new condition—the Pinocchio story simply tells this account in reverse, ditto the mechanical mystery dolls of the *Tales of Hoffmann*. Anders tells the same cautionary tale by way of Goethe, who himself took over Wieland's translation of Lucian's account, in order in the middle of the 1950s to tell us ourselves in the age of the gadget quite up to the current day as we await not the second coming but the next iPhone, apple computer, new camera, new car, new booster shot. But by now we are talking about transhumanism.

Today, this is mediated by pocket electronics, "pocket robots,"[12] as such things are the hacks that rather than jetpacks turn out to be the essence of the transhuman. For some, especially among the young, but also among the disaffected, this complex dynamic of voyeurism and trolling characterizes all of us. Richard Seymour writes in his book titled with reference to Paul Klee's 1922 painting of the same name, *The Twittering Machine*,[13] that "we are all addicts" and "we are all trolls"—whereby both addiction and trolling are effected, and this is the beginning point of my *The Hallelujah Effect*, by nothing more than the miniscule feedback of a *click*—liking or sharing or retweeting something.

It was to analyse media programming on social media, Facebook and Twitter, that I began the reflections that led to *The Hallelujah Effect*, musical observations initially published in an online music magazine, with the glorious name *PerfectSoundForever*. The essay focussed on YouTube music videos, specifically Leonard Cohen's *Hallelujah* (and k.d. lang's interpretation and gender norms and expectations and so on), on the face of it a vastly more innocuous phenomenon by contrast with the attention shock of Seymour's example, as he begins by describing a series of self-curated suicide videos.[14]

Here we note the ongoing consequences of homeworking *in extremis*, as Anders had already analysed in "The World as Phantom and Matrix,"[15] already published in English more than half a century ago as the world-estranged "situation" of the human being in a technologically mediatized epoch, not metaphorically but even before Lockdown:

"Only after the human being has closed the door behind him, does the outside world become visible to him; only after we have been transformed into windowless monads, does the universe reflect itself in us."[16]

The mix of distracted fascination, as Adorno and Anders (and so I would argue Nietzsche likewise) tell us, using the first-person plural form for emphatic inclusion— "we" like what we recognize—combined with the transient urgency of responding to this in real time, the sooner the better — for Seymour, "we are all celebrities," or, as I remark in a lighter mode, regarding our affective lives: "we are all royals."[17]

This can shed further light on the "outsider" condition that is a consequence of sustained self-preoccupation. In his *Art & Fear*, Paul Virilio highlights some of the darker consequences of the streaming human condition that has for some time corresponded to video culture.[18] And there are compounding factors, as recently the analytic philosopher Kate Manne has described a phenomenon called "himpathy," a phenomenon of kipped sympathy, in this case a tendency to privilege the male offender over his (often) female victim, which is no exception but mainline.[19] This is also the dynamic of George Orwell's dystopian 1945 fantasy, *Animal Farm*, where some are more favoured—the current global "pandemic" only dramatizes this with greater urgency—and some are more "equal" than others.

The mix in these foregoing introductory reflections of themes (and more besides) is necessary to read Anders. He approached the problems of his day—problems that continue to be contemporary problems—with insights drawing upon and across classical philology and theology, anthropology, sociology, and musicological and musical critical reflections which made up his particular *stylistic* kind of continental philosophy (I have argued that continental philosophy is primarily a stylistic distinction,[20] as opposed to a geographic one or as opposed to subject matter or focus on any given historical figure).

Earlier I referenced Kate Manne's notion of "himpathy" because it may illustrate what can seem to be strangely characteristic of the privileged, who can tend to insist on the singular quality of their own sufferings, claiming exceptionality in all respects, including the "exceptionality" of exclusion, even persecution as one can see this in the phenomenon Manne also analyses of men's rights groups, including their self-designation as "incels." This last complex of privileged self-absorption is the stuff of psychology—about which Anders knew a great deal from his earliest years—his psychologist mother analysed and publicized his childhood—as it is also about competition. In our media-driven, exposed age, this can appear as the desire to claim exceptional status (the tag line of Garrison Keilor's radio comedy show, *Lake Woebegone*, "where all the children are above-average"), deserving of consideration, compensation, sympathy, and understanding and, above all, *more* regard than others, has never been more widespread.

Life and Legacy

Analyses of addictive self-absorption are not rare but Anders was a born *Aussenseiter*. Born as Günther Stern in Breslau, on July 12, 1902, he died on December 17, 1992, in Vienna. There have been a number of books on Anders in the decades of what nonetheless continues to be dampened reception, and one of these exemplifies what has become cliché in the interim, introducing him as "the most neglected German philosopher of the twentieth century."[21] In addition, Anders' scholarship tends to be insular (specialized scholarship is generally so); thus, the claim is less than accurate. Thus, Christopher John Müller and David Mellor can point to an explosion of interest in his work, counting some fifty books over the past two decades.[22] But this is not to say that today scholarly inattention has been overcome, and Anders' name retains its outsider qualities largely because his concerns, as I argue, are not reflected in today's mainstream philosophical canon. In addition, Christian Fuchs argues that Anders is not prominent in media studies.[23] To the disciplinary roster, as I have already-noted theology and political theory, anthropology/ethnography and sociology, hermeneutico-phenomenological approaches are rare enough in any case, and I hardly need to add music studies despite the fact that there are studies of Anders and music. That Anders also has vital contributions to critical cultural theory more broadly, including poetry and aesthetics, has had the effect of making it harder, and not easier, to receive him, even after all this time, quite in spite of the efforts of many scholars, generating a kaleidoscope of various claims to have "discovered" him.

To be sure, there are similar issues when it comes to other "outsider, outlier" thinkers like the reception/non-reception of the already-noted Jacques Ellul or Ivan Illich (parallels may also be found in the reception, as it comes and goes, of the more mainstream Michel de Certeau). But what can be done? It seems scholarly *habitus*, to echo Bourdieu, to constitute pockets of scholarship that turn on themselves. In this way, fiefdoms are established, and scholars exclude in their own turn consideration of other authors who work on the same thinker, while other scholars similarly tend (quite for the most part) themselves to exclude others in their turn as opposed to including other scholars on the same path. This is done with a good conscience as one from time to time reflects that one cannot read everything. The conundrum haunts the academy if it is, at the same time, the precisely normativizing essence of what Thomas Kuhn defined as "normal science." Indeed, outsiders conduct themselves quite as insiders in their turn.

The point here with respect to Anders goes beyond inadvertence or neglect. There *is* a legacy of scholarship on Anders, and yet, and at the same time, this scholarship is insular, excluding, unless one takes pains to do otherwise, attention to different voices. This exclusion plays out on many levels, ranging from analytic or mainstream philosophy to so-called "continental" philosophy with the result—the tactic is self-destructive—that such scholarship tends in turn to be ignored.[24] "*Nevertheless*"—to quote Katherine Hepburn's response to a hostile German captain in the 1951 film, *The African Queen*, challenging the fact that she and Bogart successfully traversed a putatively unnavigable swamp—Anders remains an essential thinker for our over-surveilled, digitally absorbed era, and this is a book about him.

A Star Among Other Stars

Anders' fame is of the metonymic variety as he is associated with the well-known. And Hannah Arendt begins her "Introduction" to the translation of the collection of Walter Benjamin's *Illuminations* by writing about fame, under the title of the Roman goddess *Fama*: observing of "posthumous fame" that it is hardly to be aspired to—"rarer and least desired"[25] (an odd contention given that Kant himself predicted philosophical fame as at best to be attained only after an interval of a century, and Nietzsche described himself as "posthumous," and Heidegger, her former lover, apparently dedicated himself, so it has been argued, to the pursuit of little else). Even more perplexing is Arendt's twofold contention, the first part of which is indisputable but the second (regarding the lucrative value of the work) precisely not so, whereby "The one who stood most to profit is dead and hence it is not for sale. Such posthumous fame, uncommercial and unprofitable, has now come in Germany to the name and work of Walter Benjamin." And all of this *is* true of Benjamin.

It is almost inevitable that I start with Arendt and Benjamin, as Anders is arguably the most famous *ex-husband* in philosophy—as Arendt's first husband—or else the most famous *relative*—Anders was Benjamin's cousin[26] and, in the same lineage, Anders was the brother of the Holocaust hero, Eva Michaelis-Stern.[27] He was also, as already noted, famous as a child for being a child: the subject of a certain intimate notoriety as mid-born son of famous experimental psychologists in a book authored by his father and mother, William and Clara Stern (complete with scenes from his childhood featuring in a popular case study account of child rearing) and a further *Tagebuch* published by his mother.[28] Name-wise, Anders was also his father's son, his father having been the famous William Stern, inventor of the IQ test.

Besides his ex-wife and his famous cousin—for many, Benjamin has more star value than both Adorno and Heidegger put together—and apart from serving as case material for his famous father's research in developmental child psychology, Anders was also a student of leading philosophical movements of the day, both the famous and the faded. He studied with Heinrich Wölfflin and Wolfgang Köhler, among others, as well as Husserl and Heidegger. In 1924, he wrote his doctoral dissertation on situational categories and logical propositions under Husserl.[29]

Anders went on to work (with less-than-concretely positive results) with Paul Tillich, and he was from the start a member of the original Frankfurt School—which one must say to distinguish it from what it became after Adorno's death from the 1970s onwards—with Adorno, Horkheimer, the elder Marcuse, and others. This proximity led some (Hannah Arendt) to blame Adorno for Anders' lack of advances, while others faulted the then-increasing Nazi politics in Frankfurt. Although Anders worked with Max Scheler (he was briefly his assistant) and with Helmut Plessner on philosophical anthropology,[30] he sought to habilitate himself with a project in the situational sociology of music under Tillich: *Philosophische Untersuchungen über musikalische Situationen*.[31] The habilitation project failed to come to fruition for a host of reasons—things of this kind are fairly overdetermined—but most decidedly because Tillich asserted his incompetence as reader.[32]

Anders published his doctoral dissertation, quite as one must, along with a monograph on the nature of "having,"[33] reflecting the influence of Husserlian phenomenology. The latter was to influence Gabriel Marcel, who took it over.[34] Importantly, given Anders' hostility to him, Heidegger's influence is patent both in

Anders' monograph on having/not having (published one year after *Being and Time*) and Anders' musical phenomenological reflections on "listening-to."[35]

In his musical phenomenology, beginning 1927 onwards and under the influence of Husserl and others, Anders, who played piano and violin, followed the experimental psychology of Carl Stumpf, the research framework being practically familiar to him from his own father's technical study (and invention) of the *tone variator* (a device used in psychological research to study tone perception).[36] Anders thus raised the question of the phenomenology of music from the experiential perspective of the musician, especially the composer but also the performer who lived, as it were, "in-music," to use Anders' lived experiential term. His existential-phenomenological orientation afforded Anders a perspective from which he might underscore the difference between lifeworlds, as he argued that the musician counted his "real life" *not* as the life of the everyday lifeworld but "in-music."[37]

In exile from his country and lacking the security of university employment enjoyed by many post-war émigrés, Anders, in his 1956 book, *Die Antiquiertheit des Menschen*,[38] wrote one of the most prescient accounts of the human relationship to technology in terms of the human relationship to the very media we today name "social media." Often rendered, doubtless owing to a Francophone influence, as *The Obsolescence of Humanity*, the rendering is inapt not least because Anders ascribes the term "obsolete" to Ernst Bloch, lamenting his resistance to "even considering the event of Hiroshima" (AM II, 20). Arguably, *The Antiquatedness of Humanity* would suit as "*Antiquiertheit*" includes many of the meanings associated with the English "antiquated." At issue is being "outdated": *périmée*.

It is relevant to note that Anders would use the same title for a second volume of his essays, subtitled *Concerning the Destruction of Life in the Age of the Third Industrial Revolution*, published in 1980.

I have suggested that Anders does not offer a systematic monograph but to say this is not to say that his work is not systematically executed. What will be important to note, and in what follows I will try to highlight this, is Anders' style together with his attention to his titles, systematically arrayed. Thus, the first volume of *The Antiquatedness of Humanity*, subtitled *Concerning the Soul in the Age of the Second Industrial Revolution*, begins with a reflection "On Promethean Shame,"[39] followed by "The World as Phantom and Matrix," a chapter published in the very same year in English. Indeed, adding Anders' intermezzo in his first book, "*Sein ohne Zeit*. Zu Becketts Stück '*En Attendant Godot*'"—available since 1965 as "Being Without Time: On Beckett's Play, *Waiting for Godot*"[40]—to his "Reflections on the H-Bomb," it is significant that as of the current writing, more than the half of Anders' first book is extant in English. But still there is inattention, and part of the reason may have to do with the entirety of the third section, published as it was more than a decade after Hiroshima and Nagasaki, dedicated to what will become a lifelong engagement with atomic weapons/power: "Concerning the Bomb and the Roots of our Apocalypse-Blindness." Although some of Anders' writings on atomic power are also available in English,[41] Anglophone philosophy can be inclined to distinguish such concerns from philosophy proper, regarding them as activist or politicized or part of a peace movement, or anti-war/anti-atomic power movement. This tendency testifies to the various ways that a thinker like Anders (and like Ellul, like Illich, like Virilio, and Baudrillard, etc.) can remain marginalized in philosophy proper and most perniciously in philosophy of technology.

Both of Anders' volumes on *The Antiquatedness of Humanity* have been available in French as of 2002 and 2011, respectively. And, as noted, the first portions of Anders' *The Antiquatedness of Humanity* have already been in English for more than sixty years, a point that should be coupled with the presence in English of Anders' Kafka book (as of 1960, with an early essay available in 1949), such that some part of Anders' enduring invisibility has to be ascribed to scholarly inattention rather than a simple lack of translation. Still, it matters to attend to Anders' own systematic articulation of his own endeavour and in the important and perhaps all-too-timely second volume on "life" in the era of the third industrial revolution, Anders, who was preternaturally fond of lists (and 'commandments'), lays out a systematic array of variations, as it were, on *Antiquatedness*. Thus, if I emphasize that the Francophone "obsolescence" does not quite convey the force of "*Antiquiertheit*," it is firstly because Anders himself uses the German equivalent of obsolescence but also because what is at stake is more ancient, more outmoded, and more out of date, than the simply obsolete. As Baudrillard would remind us, the latter is always part of, which is to say that the latter assumes, a "system" of objects.[42] And the problem for Anders is older and more complex than that—which is why he needs a reference to both Aeschylus and *Genesis*.

In the second 1980 collection, comprised of essays composed after 1956, Anders offers an account of not merely his "three industrial revolutions" (1979) but also of the outmodedness of appearance (1979), materialism (1978) and products (1958)—and it should be noted that these essays are themselves divided into significantly titled subsections, such as the first section here, crucial for the young Baudrillard: "Serial products are born for death" (AM II, 38ff),[43] is succeeded by the human world (1958/1961), the masses (1961), work (1977), machines (1969), philosophical anthropology as such (1979), the individual (1970), ideologies (1978), conformism (1958), limits — with reference to atomic weapons (1979), privacy (1958),[44] dying (a single page from 1979), reality (for a symposium on mass media, 1960), freedom, history (1978), fantasy (1955), of space and time (1959), seriousness, concerning "happenings" (1968), "meaning" (1972), application (1979), incapacity [*Nichtkönnens*] (1975), and signally, the antiquatedness of evil, with a first section on the transfigured sorcerer's apprentice (1966), along with a methodological retrospective (1979). I've added the dates to indicate the durability of Ander's thought but also because the constellation makes up the force of his book.

Suppression: Positivity and Neutrality

Anders' work remains muffled. This muffling has incidental and accidental components—many authors are unreceived, after all. But, quite by contrast with the French and Italian tradition on Anders (the German tradition is its own separate story), in philosophy of technology in the Anglophone world, inadvertence to Anders is certainly, at least in part, the direct result of a tactic of suppression. Anders was silenced by the Stony Brook philosopher who was also one of my teachers there,

Don Ihde (1934–) who made a name for himself initially owing to the simplicity or "accessibility" of his work as well as his explicit aggression contra any and all thinkers apart from himself. This initially meant that he would seek to exclude reference to past thinking; this is the rather literal force of 'post-phenomenology' as he conceived it, such that he would favour, and this made him wildly popular among contemporary scholars *only* contemporary scholars. Now at work here is a distinction between primary and secondary scholarship. Secondary literature comments on a primary author and, sometimes, on other secondary sources. By not commenting on other secondary authors, consider the case of Stanley Cavell, say, one has a greater chance of being considered, it worked in the case of Cavell, a primary author. More specific to Anders, here I draw on his own account of it, Ihde nixed a proposed translation of Anders' *Die Antiquiertheit des Menschen — The Antiquatedness of Humanity*. Ihde's reasons for this derived from his emphasis, in addition to excluding past voices, on avoiding any hint of luddism and thus the importance of approbative or expressly *positive* takes on technology. The result could not but cool Anders' reception. Even more than Heidegger and Marcuse, and more than Ellul and Illich, Anders was an uncompromising critic of technology.[45]

Here, I follow Anders thematically and stylistically. To this extent, I engage some of Anders' more controversial positions, in particular his critical perspective on technology. It is for this reason that I took care to note Ihde's view that philosophy of technology was best advanced by explicitly sidestepping critique: no Heidegger, or as little Heidegger as possible, no Ellul, no Illich, no Baudrillard, no Virilio, and no Anders. Ihde's techno-positive approach dominates philosophy of technology especially but not only in the Netherlands where Ihde's influence is marked—a tad ironic given that one of the first books on Anders and technology was written by the late Paul van Dijk.[46]

Anders himself writes of the presumption of techno-neutrality—this is the thesis that it is not the tool but the motivation or use that determines the effects of any given technology:

> Nothing is more misleading than the (to be sure rarely explicitly elaborated, but in both right and left wing politics presupposed as self-evident) "Philosophy of Technology," which claims that instruments [*Geräte*] are in the first instance "morally neutral," and are freely available at our disposal for whatever use; the sole issue that matters is only *how* we use them; what use we subsequently impose on them: be it moral or immoral, humane or inhumane, democratic or antidemocratic. (AM II, 216)

The idea here is that one is to have one's tools *and* use them too, that is, that through our good intentions we manage to avoid being, in Thoreau's sense, in Bert Dreyfus's sense, and most particularly as the popular Neil Postman propagated this, the "tools of our tools"[47]—being free to use them, quite as Heidegger argued as a high goal, the goal of *Gelassenheit*, quite as if already a given. But for Anders, as he emphasizes repeatedly, a tool, a technical object, is already its application or use.

If Ihde's pro-tech imperative continues to reign, if scholarship has found ways to advance positive assessments, if not quite "how to stop worrying and learn to love the bomb," from the subtitle of a film current in Ihde's era, pro-transhumanist accounts seem the order of the day as we may read in techno-salvific thinkers like Stefan Sorgner with respect to a certain take on Nietzsche and with Steve Fuller making common cause with Sorgner in sociology among to be sure many others writing on technology and nihilism and so on.[48]

Sherry Turkle's reflections on her own trajectory are instructive,[49] quite where she might have given Ihde some pro-tech competition at the start of her career but where the course of her research brought her to the spirit of her *Alone Together*,[50] a book in which, quite independently, Turkle repeats Anders' insights as he details these in "The World as Phantom and Matrix," the paradoxical means whereby screen media isolate us from one another,[51] creating a nation of what Anders called "homeworkers."[52]

"Homeworker" is, once again, the term Anders uses for the "work" the consumer does at home, in what was or would traditionally have counted as his free time, gratis for both the government and for commercial interests, working on himself or herself to produce himself or herself as a consumer.[53] The product thus created at home is fabricated by nothing more than favour, enthusiasm, interest, attention, "likes," and such. Anders and Adorno, and later Baudrillard, would emphasize this as in effect fabricating or manufacturing a product sold, that is: our work on ourselves, as many theorists tell us, as the content—'you are the product'—of so-called "big data."[54] What we have most recently (re)"discovered" in the phenomenon of "surveillance capitalism" is the same culture industry Adorno and Horkheimer write about along with Anders, but today this cannot be understood without the phenomenology of simulation and illusion arguably best articulated between Debord and Baudrillard as well as, for another tradition, Ellul and Illich and Postman.

Consumers and "Situations"

Anders maintained that consumers became "homeworkers" simply by consuming mass media. Anders initially refers to radio and to television and print media and only the internet in its earlier modalities and thus, some commentators may assume, apart from the interactive interface we today take for granted. Nevertheless, tracked and "surveilled" as we are, that interface remains unidirectional or programmatic, quite as Jean Baudrillard always took care to emphasize the lack of response constitutive of new media. This is the third point Anders emphasizes in his "The World as Phantom and Matrix": "Because the receiving sets speak in our place, they gradually deprive us of the power of speech, thus transforming us into passive dependents" (PM 17). This is the precondition for programming as much as it is the precondition for what we name "fake news." Here, despite his criticism of Heidegger, Anders emphasizes the consequences of this shift in terms of the loss of language itself, reflecting:

> For humanity's inward life, its richness and subtlety, cannot endure without the richness and subtlety of language; the human not only expresses himself through his speech, he is also the product of his language. (PM 19, citation slightly altered)

Elsewhere I emphasize the consequences of this same deficit with respect to music and the culture of its iterative variation-replication or "cover" phenomenon keyed as this is to recognition or what Anders analyses as "pseudo-familiarity." Non-response is trained via the illusion of instant response—"click here," as the social commentator Evgeny Morozov teases in a politicological context[55]—that makes such media engagements/non-engagements as addictive as they are.[56] Thus, simply by listening to radio, by following television programmes, clicking on Facebook videos, Instagram, etc., the consumer works tirelessly at this self-assigned, unpaid task. Nor was this homeworker housebound, as Anders emphasized the then already utter transportability of homeworked mobility: one could work as easily from home as on the road. Today, to be sure, we do know what he means as we dedicate ourselves to working on our online selves: living our lives in the private/public sphere dedicated to fabricating ourselves as mediatic consumers. Anders liked to highlight the fact that we are eager to do this homework all by ourselves, paying for the privilege of doing so.

The radio "homeworker," like the television homeworker, like the new smartphone or internet homeworker, paying as they go for a cellphone or an internet subscription, not to mention the cost of equipment, accessories and upgrades,[57] "works" from "home" as they likewise "work" on the job, on the road, wherever that worker happens to be. And it is *work*, even though we are not compensated as traditional homeworkers might have been, even though we pay for the privilege of fabricating ourselves into "consumers," masses consuming mass goods created in self-absorbed bubbles of isolation, all the more perfectly as the act of creation of ourselves in this image is tweaked and is consumed or enacted at the same moment in a virtual, digital, socially mediated age.

For Anders, reflecting in 1979 on observations he had already worked out in the mid-1950s:

> In such cases it is not actually the products themselves that serve as means of production, but our *acts of consumption*—a truly shameful circumstance, inasmuch as our human role here is reduced to guaranteeing, by way of the consumption of products (for which, in addition, we also must pay), that production continues.[58]

The other side of Promethean shame, as Anders speaks of this, is not merely our tech-anxiety, our desire to be the machine or to match the machine we have made (as if we had a choice, assuming we wish to *use* the machine), with all the perks we imagine that it has (thanks to advertising), but the attribution of responsibility. The "shame" in question corresponds to a certain techno-theological reflection, and our acts of consumption ensure the ongoing circumstance, be that matters of media or plastics in the ocean, industrial farming and fishing, or indeed the mining of rare earths and the extractions of fossil fuels flooding the ocean and the land with deadly pollution spills, and the explosive damage done by fracking: we make ourselves consumers, but we also make thereby the material conditions of many other problems that have for some time been wreaking planetary havoc. Greater than the damage done to our psyche and our health as consumers is what is done to the world by those same acts of consumption. This aspect of Anders' critique remains ongoing.

Nomen est omen: "Other" Reflections

Misses and near misses can seem to have characterized the life of a man named "star" [*Stern*], who, quite as if he declined the name Stern that was his own, also set himself under a different star, one that resolutely refused to support him. According to a popular story, owing to a surfeit of Stern surnames on a Berlin newspaper masthead, when asked by an editor for another byline, Stern casually suggested (there is an echo in German) "Anders" (something else, something other), which, if it is unlikely to be literally true, is surely *ben trovato*.

Indeed, in her introductory essay to her collection of Walter Benjamin's writings, Arendt invokes what she calls "the element of bad luck,"[59] citing Benjamin's fondness for the German fairy tale image of the "*buchlicht Mannlein*," a cripple or hunchback, who constantly interferes, a rather more literally destructive daemon than the cautionary negative influence Socrates invokes for his part, the unanticipated misfortune that seems to come along to add to one's troubles, like the phrase *ein Unglück kommt selten allein*, or as Arendt tells us that Benjamin's mother would say to him (and thus one cannot but imagine that Arendt herself heard the same from her own mother, not uniquely but quasi-universally as a theme inherent in the "mother tongue" that Arendt would underscore in a famous interview as missing in her own years as expatriate):

> like millions of other mothers in Germany, used to say, "Mr. Bungle sends his regards" (*Ungeschickt lässt grüssen*) whenever one of the countless little catastrophes of childhood had taken place.[60]

Arendt quotes the adult Benjamin reflecting, not unlike the disposition, in two senses of 'disposition,' of the angel of history constantly cited in the literature: "In consternation he stands before a pile of debris." And Arendt tracks this influence/perturbation/identification in Benjamin's life to Benjamin's commentaries on the letters of famous and not-so-famous figures: *Deutsche Menschen*, noting the array of subtitles, their importance, and so on. The words Arendt cites are not quite to be ascribed to Benjamin, deriving as they do from the publisher's suggestion for the title, and feature as subtitles on the cover, beautifully printed, all three lines:

Von Ehre ohne Ruhm
Von Grosse ohne Glanz
Von Wurde ohne Sold[61]

Certainly, Benjamin's appreciation for the "right word" holds for Arendt as it served as palimpsest for her introductory words.[62]

Arendt reads Benjamin reading Kafka as she could just as well have been reading Anders, although she does not say this:

> "an understanding of [Kafka's] production involves, among other things, the simple recognition that he was a failure" (*Briefe* II, 614). What Benjamin said of Kafka with such unique aptness applies to himself as well: "The circumstances of

this failure are multifarious. One is tempted to say: once he was certain of eventual failure, everything worked out for him *en route* as in a dream." (*Briefe* II, 764)[63]

The reference to hope, as we will see in the following text with respect to Anders as indeed Adorno, can be traced here, as Arendt reminds us that this is not taken from Kafka but Goethe's *Elective Affinities*: "Hope passed over their heads like a star that falls from the sky."[64] The language of the "bungler" is borrowed from Benjamin's essay on Kafka: "It is for them and their kind, the unfinished and the bunglers, that there is hope."[65]

With respect to bungling, as it were, little seems to have turned out well for Stern/Anders, be it his marriage(s) or unsuccessful efforts at habilitation or his literary efforts or with respect to the academic reception of his published works given the lack of resonance. On the other hand, success is its own trap for a scholar, and it may have been owing to his lack of reception that Anders was able to become the clear-sighted witness that he was to the force and violence of his century.

Who is who? Put the question differently, rather in the way Anders seeks to raise, phenomenologically as he conceives the question of music and "musicking" as this can be extended to academic presence: Who gets to be who? Who gets to be received? How so?

For Anders, as it turns out, the "who" question is crucial, quite as it was for Heidegger in *Being and Time*. Thus, Anders raises the same questions with respect to what he names "musicking," that is, music making. Who, Anders asks specifically, gets to musicize? Who gets the opportunity to make music, to play music, to arrange it, to listen to it? Related questions concern performers who are recorded: performers whose songs are taken up by contrast with those whose songs are not. In a later chapter we will see that in a "remix" culture, a culture of musical "covers,"[66] this is related to the tendency to credit some musician's names above others.[67] As Adorno asks, thematizing the culture industry as fait accompli, who gets plugged on radio (this is a paid arrangement on the part of the recording industry and not a matter of 'pop' reception) and consequently listed on a chart? Which performers are booked and celebrated, and which reviewers are "allowed" to "review" music? Who is received, mentioned, or heard?[68] In the case of art in general or what we count as "culture," the case of music is interesting because one can tend to assume both timelessness and absolute neutrality for the "artwork" in general: one does not assume it has a genesis, an "origin," as Heidegger reflects, or a "birth," as Nietzsche writes (or indeed, as this is Nietzsche's theme, a death).

How do names become canonic? What forces disperse these names, excluding them from scholarly attention, intellectual focus, collective memory? Who is Günther Anders? Who was Günther Stern? What themes capture and hold our interest? How do we pose the philosophical questions we regard as urgent in any age? What is the question of "having"—a question Anders made his own, influencing no one less than Gabriel Marcel who subsequently came to be associated with the same theme? And what of what Anders called the "situation," a term related to what Jaspers spoke of, what Adorno spoke of? What of his critical reflections on Rilke, Kafka, and Goethe himself? What of Anders' relation to Husserl? Scheler? Plessner, Heidegger? What about Adorno? And what about Nietzsche—just to the extent that attentive appropriation by mainstream thinkers often works to obscure a thinker's proper concerns, turning a recondite classicist, a specialist in Greek meter and rhythm,[69] into a bad-boy moralist.

In recent philosophy, there has long been attention to the "other" as such. This is largely filtered through Levinas and Buber. And to a certain extent such a focus has excluded not only Günther Anders but also Michael Theunissen, himself another student of Heidegger's towards the end of his life, a friend of William J. Richardson's, and author of an important study on the theme of *The Other: Studies in the Social Ontology of Husserl, Heidegger, Sartre, and Buber*.[70] The subtitle of Theunissen's study makes it plain that certain names of those concerned with the *other* can tend to be obscured—or 'othered'—on the terms of mainstream attention to "the" other. Such names not only include Anders, importantly, as well as Sartre and de Beauvoir, and Merleau-Ponty but also Michel de Certeau, just the names we know in addition to names that new attention to diversity has been recalling, those that are also long known as such, like Camus and Fanon, and so on. Thus, other "others" remain obscure: names like Illich to an extent and perhaps parallel to Anders to the extent that Illich's focus never meshed with that of the mainstream, not on education[71]—such that scholars to this day prefer Paulo Freire and the more apt reflections of Jacques Rançiere—and especially not regarding what Illich named "medical nemesis" where Illich remained as indomitable in his own way as Anders,[72] or with respect to technology where Illich offered his own 'convivial' take.[73]

In great measure, what we call "thinking," even including what Heideggerians call "thinking,"[74] is a matter of fashion when it comes to reception, engagement, discussion, insight. Thus, in a lecture presented in German in Normandy in 1955, *Was ist das die Philosophie?*, Heidegger reprises a point he had already raised in a 1944 Freiburg lecture, *Denken und Dichten, Thinking and Poetizing*, a key point echoing the forestructure and externality of the very transmission of, or thinking of, philosophy as such: what is presupposed is that those to whom one speaks of philosophy or thinking or poetry "stand initially outside of it."[75] As ever, Heidegger, as thinker of the question as such, wants to remind us that what is at stake in questioning is less to avoid the circle that is the hermeneutic circle but much rather how to get into it: "the aim of our question is to enter into philosophy, to tarry in it, to conduct ourselves in its manner, that is 'to philosophize.'"[76]

For a parallel illustration, Nietzsche wrote on *The Birth of Tragedy Out of the Spirit of Music*, thinking to connect both the scientific research he had done on metre and ancient lyric with the possibilities for a rebirth of culture. Little came of Nietzsche's philosophico-musical project apart from the erroneous certainty that all Nietzsche had ever been on about was a certain advocacy of Wagner's own dream of a transformation of art culture. To this day, in Nietzsche's case, his legacy is less his own philological project but more his incidental genealogical reflections on morality and a book and a figure—Zarathustra—written as a piece of dynamite to toss to the masses who otherwise are slow to pick up his thinking on the untimely, on the utility and detrimentality of history, on the all too human. Nothing else Nietzsche would write could change this reception, not even his *Gay Science* written in two instaurations to reprise the themes of his first book on tragedy, a text in which Nietzsche emphasizes that he was the first to raise the question of science as a question. But it is to this same extent that when Anders reflects on Nietzsche in the 1940s, Anders will draw not upon a scholar's specialist knowledge of Nietzsche but Anders' affinity with a life of non-reception as Anders seems to fulfil the destiny of his adopted name: *another, otherwise, other than*.

How do we, in a world framed by certain names and the thought associated with them, begin to read the neglected, the marginalized, the obscure? In cases where a thinker is completely overlooked, it can perhaps simply be enough to read their work, locate intersections, and so on. Where a thinker gradually enters the public eye, it is almost inevitable that some will seek to institute a canon to exclude divergent voices, and this is so in Anders' case.[77] In addition, there are intersections with the Frankfurt School, and Anders' obscurity seems to have gone hand in glove with the cohort with which he grows up—Benjamin, Arendt, Adorno—along with his teachers—Husserl and Heidegger as well as other less well-known names.

For his part, Anders would follow, after fleeing Berlin for Paris, a shared trajectory of German political emigres before/after the war, like Horkheimer, Adorno, Arendt. But part of the reason for reflecting on what might seem, as mentioned at the outset of this introduction, to have been a "black star" haunting Anders is the clear difference in reception and engagement at the same time that Anders was certainly part of a common cohort, namely the trajectory that was likewise shared by not only Leo Strauss, and Hans Jonas, but also Dietrich von Hildebrandt and including both Ludwig Marcuse and Herbert Marcuse, among others, some remembered, some less so, some not at all.

The guiding or unifying thread, and here we may also count in Carl Schmitt, is technology. The passion behind the ontic concern is a focus on broadcast sound transmission, thus a focus on radio for Beckett and Arnheim, Heidegger, and Anders and, of course, Adorno, foregrounding both social reception and the technical conditions of acoustic reproduction as this transfigures music above all. This element was key for Adorno but it was Anders who brought a critical and phenomenological reading to these two concerns and never forgot, as many managed to forget, the war that continued after it had ended, with two atomic bombs and a lifeworld altered forever, if invisibly, by the same.

Part One

A Critical Theory of Technology

1

Criticizing Technology

All the technical arts [τέχνες] have by Prometheus to mortals been given.
—Aeschylus, Προμηθεὺς Δεσμώτης (*Prometheus Bound*)

Questioning Being-in-the-World, the Work of Art, and Technology

Despite his early and unremitting criticisms, Günther Anders' techno-philosophy cannot be conceived without reference to Heidegger's phenomenological hermeneutics, including Heidegger's own reflections on tools, situated in a life of purposes and projects, being-with and -in-the-world in *Being and Time*. To this same extent, Anders' reflections on technology must be read in conjunction with Heidegger's lectures in the 1930s, given in three instaurations,[1] on "The Origin of the Work of Art" as well as his post-war lectures on technology, including "The Question Concerning Technology," lectures delivered in Bremen, Freiburg, and Munich.[2] And there is the constellation in thematic order of time between Heidegger's 1955 *Gelassenheit* lecture in Messkirch and Anders' 1956 *Die Antiquirtheit des Menschen/The Antiquatedness/Outdatedness of Humanity*.

Reviewing Anders' thinking in connection with power, including his reception on the issue of power and politics, this chapter explores the contributions of the original Frankfurt School of which Anders (along with Adorno, Horkheimer, Ludwig Marcuse, and so on) was an original member. Here we might also count, technically speaking and in ancillary way, Hannah Arendt, reading Anders' critique of technology together with Horkheimer and Adorno's *Dialectic of Enlightenment*, especially their reflections on what Adorno analyses as *The Culture Industry* for the sake of beginning to read Anders' *The Antiquatedness of Humanity*.

On the Intersection of Sociology and Anthropology, Critical Theory, and Technology: Günther Anders and Power

Anders' work is marked by both the depth of his insight and the sheer range of his thought. But if depth is a good thing, breadth is not, and it is the breadth of his concerns that makes Anders difficult to categorize as philosopher *and* political theorist who philosophized about anthropology, power, violence, logic, and what he calls, quite technically, the "situation," as well as music and media, specifically radio and television but especially in the lived context of society, specifically as an émigré observer, American post-war society. To be sure, at the time, these were common themes, especially for someone who was as Anders was, a student of both Karl Mannheim and Max Scheler in addition to Edmund Husserl and Martin Heidegger. Yet where even Scheler is only slowly gaining attention, and that too in an attenuated or mainstream sense, few political theorists are familiar with Anders such that Mannheim is better known in social studies of knowledge and Heidegger in political studies of technology along with Benjamin and Adorno. To be sure, gatekeeping scholars seeking to define the Frankfurt School in institutional terms (not the happiest for critical theory as such) exclude Anders as a "member," while many count in Walter Benjamin and Siegfried Kracauer. Most such demarcational efforts articulate the origins of their authors who tend to be students in the lineage of Jürgen Habermas or else Raymond Guess, and so on.[3] It is for this reason that it is not far off to regard Anders as "stepchild of the Frankfurt School."[4]

And then there is Hannah Arendt, Anders' first wife. From the start, then, Anders would be inscribed in the political, a thinker of critical theory. Yet readers tend to find Anders' reflections on political power somehow elusive precisely as reflections on power, despite the fact that he speaks about it and expressly thematizes war and violence.[5] Indeed, it can seem that even more than philosophers, intellectual historians, and media studies scholars, it is, perhaps, above all, political theorists, including the same thinkers who read Arendt, who have trouble engaging Anders' thought.[6]

Philosophy of technology has its own complex of allergies to critical theory inasmuch as the critical in Anders' case exceeds the meaning of critique in an academic and philosophical sense, for Anders and for Heidegger (and indeed, for Nietzsche in an earlier era) notably deriving from Kant. Making matters worse, *qua* theoretical terminus, "critique" is often taken conventionally such that "criticizing," holding a negative view of technology, for example, can seem to be no more than an affective orientation, quite as emphasized in the introduction, whereby conventional and mainstream scholars seemingly anxious to avoid being saddled with the term Luddite[7], or characterized as "negative," highlight only positive elements, dangerous for a host of reasons and not least, human as we are, forward thinking as we are, we are also very typically unable to project or anticipate (with any accuracy, this is a key limitation to judgement) downsides to any practical project.[8]

I've noted the mainstream distrust of critique and negativity in philosophy of technology.[9] Anders' writing bristles with both. In addition, the style of his writing style can be difficult to follow (he is not part of the mainstream and is thus under no compulsion to use the academic conventions and emphases mainstream thinkers take for granted). Arguably, this is less because Anders' style of writing is elusive—though it certainly follows its own rhythms and inheritances—than for substantive reasons. Anders' approach to questions of politics as to questions of media and musical experience has been systematically avoided in thinking about politics, about power as indeed about media and especially about music, which is not to say that there is no scholarly tradition of receiving his work. Matters are not improved by the fact that Anders wrote unsparingly and concentratedly about technology—quite as most political philosophers and, indeed, most political theorists, with few exceptions, do not. Thus, standard academic reflections on technology and sociology,[10] technology and anthropology,[11] technology and politics,[12] with certain exceptions, are more about the ancillary discipline than technology but, contra certain "empirical" or "material" claims, the solution is not to focus on technology alone. "To the things themselves" as a phenomenological orientation was never meant as tactic for dispensing with the lifeworld but as a means or method to thereby bring that world into view, arguably as Husserl showed, for the very first time. Heidegger—and Anders—would only take that phenomenological *turning* further, in quite distinct directions.

Hence, for example, George Kateb, the Princeton political theorist, writes about technology,[13] his themes fit certain paradigmatic schemata appropriate to political theory, and not quite parallel to Anders or Adorno, and to this same extent Kateb steps clear of a critical account of technology and does not quite as Anders does, conceive technology-as-such as "political."[14] John McCormick too, in the same broader spirit, writes on Carl Schmitt and technology, with the upshot that technology, politically speaking, can seem an adjunct to the argument.[15] By contrast, and he has been less received than perhaps he should have been, there is Langdon Winner, whose initial doctoral thesis, *Autonomous Technology: Technics-out-of-Control as a Theme in Political Thought*, won him the unwelcome sobriquet of Luddite as did his *The Whale and the Reactor*, though he excludes Anders and many others so decidedly that one is left to imagine this to be a conscious choice.[16]

"Luddite," "romantic," and "anti-technologist" are terms favoured by post-phenomenological writers on technology. This is the reason I noted Ihde's conscientious efforts (once again, quite as he himself maintained),[17] to "block" any effort to bring Anders to an English-reading public. At issue then is an exploration of some of the elements underlying what Gadamer called *Wirkungsgeschichte*—in this case negative—with respect to Anders' reflections on the "antiquatedness," typically translated (one supposes in homage to the language of fabricated obsolescence, planned and otherwise, characteristic of the era of the 1950s and 1960s): *the "obsolescence" of humanity*.

The same negating or "silencing" effect would be imposed on Dallas Smythe in the different field of political theory, a silencing which consigned Smythe to utter invisibility, and in sociology the term would be applied to Jacques Ellul.[18] Thus, and overwhelmingly, popular technology and media studies prefer popular accounts like Sherry Turkle or Jaron Lanier, or any of the mainstream reflections cited earlier or, for another generation, Marshall McLuhan or Neil Postman.[19] Heidegger's critical reading remained unassailable but also obscure enough that it seemed unclear to many what it was that he was saying, so much so that some Heidegger scholars undertook to claim, perhaps not altogether successfully, that Heidegger meant to argue that it would be technology itself that he proposed to advance as "the saving power."[20]

It is appealing to seek to make the case that things today are changing. However, calls for change tend to be calls to introduce novelty, and opposed to efforts to draw attention to older names, such as Anders. And the anxiety about critique remains as present as ever in discourse related to the philosophy of technology. Scholars who write on technology rarely introduce critical elements. Exceptions can include in the front line, thinkers such as Bruno Latour and Peter Sloterdijk as well as, to name younger names, Byung-Chul Han,[21] among important others like Yuk Hui, but I am only pointing now to those who are "received" or are well known, and I am (deliberately) leaving out the phantom of technological transcendentalism.[22] Thus, Sloterdijk, in particular, knows how to frame, politically as well as mediatically, a critique of technology, but such framing can leave one adrift when it comes to philosophy of technology as such. Can one think it? Can one frame a critique of technology? As Heidegger writes in the so-called Black Notebooks, which, as I argue, he intended for a posthumous publication (quite as if a kind of message in a bottle), reflecting on the complex conventionality of the Nazi view of technology, as if telegraphing one's alignment to the "spirit," "one continues to view 'technology' 'romantically,' as mere 'deviltry' ['*Teufelei*'] and romanticism."[23]

Ihde's apotropaic measures might strike one today, now after the fact of Anders' increased relevance, as simply high-handed, but the method he used is standard in the academic world, today arguably more than ever matching the bubble world of curated scholarship within which we live, complete with peer review exclusions and inclusions but increasingly complete with outright censorship. Importantly, other mavericks in the philosophy of technology (and it is not "good" to be such) like the already-noted Ellul[24] but also Illich and Virilio and Baudrillard raise challenges of their own only to be neglected by mainstream philosophy of technology.[25]

To this extent, what is needed is not a *re-reading* of Anders, as if he had been read, but and here I follow Heidegger's recommendations regarding Nietzsche in his own Nietzsche lectures given under Nazi rule in Germany, and even if Heidegger himself never seemingly did this, to follow as hermeneutic rule, a strategy of reading, in this case, namely to read Anders as one would read Anaximander. It is noteworthy that Anders read Aeschylus in this fashion. In addition, here there is the question of reading Anders and literature, which perforce involves literary scholars (and thus too their concerns and their values), including Anders and Rilke[26] and, more commonly, Anders and Kafka,[27] and perhaps more significantly, as I will seek to suggest, Anders and Goethe,[28] and so on. To this extent, Daglind Sonolet's discussion of Anders, Arendt, and Adorno as "Interpreters of Kafka" alludes to Bourdieu's *habitus* and, thereby, to the intersection of philosophy and sociology and the conjoining of "Literature and Modernity." And how should one read between the lineages (this is Bourdieu's *habitus*) of class and privilege and milieu with respect to literature and not less to art and philosophy and its questions, especially if one does not, as Anders did not, forget Heidegger?

In addition, and this adds a further layer of complexity, there is the question of style, that is, Anders' own contribution to literature as such, not only with his essays on literary figures—and a reflection on his style is already problematic as it bears on his philosophy—but also with his poetic and literary compositions including his absurdly delayed novel, *Die molussische Katakombe: Roman*, published, as if the publisher had quite intended to ensure mere posthumous impact, only upon Anders' death.[29] Thus, Astrid Nettling could apostrophize Anders as a "lonely herald in the desert," witness to nothing less consummately apocalyptic than "totalitarian threat" [*totalitären Bedrohung*] in a radio broadcast given, very fittingly as medium, upon his death, a few weeks before the publication of his novel.[30]

Insiders and their "Discounted" Others

As noted earlier, and one will have to continue to note it again and again, names counted in and names counted out are not incidental: the "insiders" constitute the history of academic reception, positively (and negatively) along with the value judgements intentionally (commonly) or accidently attached thereto.[31] We are, post Sartre, Fanon, Levinas, Derrida, Vattimo, and Virilio, not utterly without resources when it comes to taking the perspective of what Sartre called "the eyes of the least favoured."[32] But in the case of Anders, the question is different because to us it is also invisible: we do not see those we discount, devalue, and often literally so over time, because uncited it is almost as if these authors never wrote, never existed. How are we to recover a thinker passed over so thoroughly in the past that adverting to his lack of reception, as we began by noting, has become a sobriquet, and assuming, as one should now assume a certain reception, how to avoid the almost inevitable petrification in terms of the view that is called "received" because, quite on its own say so, it brooks no alternatives? At the very least, it makes for good copy. Thus, and the example can be varied, Anders can be the source for an "Undiscovered Critical Theory of Technology."[33] Nor do I disagree, except of course that Anders is not exactly 'undiscovered' *qua* critical theory both in German language reception and in Francophone theory of technology, as well as among those who had read him in his lifetime, and so on. Here what is at stake is the politics of the academy and that is both more trivial and more complicated.

What makes Anders "undiscovered," or better perhaps, as-yet-*unrecovered*, is the high or explicitly *critical* component many theorists avoid in practice, meaning to be optimistic rather than pessimistic concerning technology, quite in order to avoid being denounced as backwards, uninformed, insufficiently *au courant*, Luddite, or, as Heidegger wrote on the cusp of the 1940s in Nazi Germany, as we cited the concern then, all too "Romantic." For many, it seems, and certainly in Ihde's wake, this "backwards" status could only correspond to an inability to keep up with new developments, as if only a backwards disposition would lead one to raise any questions concerning, let alone criticize, technology. But the more one knows, the more critique one can—and must—afford. Critical thinkers on technology ought to, at the very least, include criticisms of technology. And such thinkers exist, for instance not only Ellul but also Baudrillard and Virilio, Illich, and, no less, Heidegger himself, as already noted earlier, and it is relevant in the current context that they too be reviewed a little. Today we also noted Sloterdijk along with, somewhat more complicatedly, Latour, crossing political theory from the side of Latour's Actor-Network theory.

But philosophers of technology like the Ihde enthusiast, Peter-Paul Verbeek, or like the late Bernard Stiegler, even those inclined to read their Deleuze with Simondon and Bachelard, tend to a certain optimism, if not quite at the level of the two Steves, the cheerfully mainstream and highly bankrolled level of a Pinker,[34] or, in sociology more conventionally academic, a Fuller.[35] Here what may be key to note, by way of difference, is that like the Vienna-born Illich, perhaps in the first measure, writing as Illich did from the perspective of history on related questions,[36] Anders wrote contrary to and thus against academic or intellectual expectations.

I have mentioned philosophy, as well as, latterly, sociology, and I have also mentioned media theory referring to David Berry and others like Baudrillard, but there is also the late Friedrich Kittler, who can seem to be a tech-enthusiast to the extent that readers can overread his critique, especially of war technology, including communications,[37] and who made his alliance with Heidegger clear but who often inspires readers less than familiar with Heidegger. I cited the more specifically playful dimension of Kittler's philosophy, as this is also a characteristic of Anders' work and his style.[38] In addition, there are exceptions among practical and moral philosophers who also engage with the Frankfurt School such as the contemporary critical theorist, Arne Johan Vetlesen.[39] There are other systems-theoretical names such as Arnold Gehlen, Niklas Lühmann, and Günther Ropohl, and there are mystical eschatological names, like Hans Jonas, already mentioned, for better or worse, and Jürgen Habermas.[40] I mention these last theorists, although, conspicuously given Anders' long-time precedence in just this regard, most of the aforenamed do not engage Anders. There is no way to properly bring all these themes and names together, but the one focus that continues to matter is the exclusion of Günther Anders, such that one can argue, seemingly again and again as many of the scholars who write on Anders as I cite them do tend to repeat, that one must attend to his work. Scholars who engage Anders can find themselves isolated from mainstream thinking especially on the themes that concerned Anders, especially Hiroshima studies, philosophy of technology, philosophical anthropology, and not least media studies. If Anders brought these themes together in his writing, the cost he suffered, namely exclusion from philosophical reception, continues.

Oblivion

I noted earlier that a kind of oblivion seems to follow Anders, possibly because of the metonymy pointed out earlier: Anders is better known by triangulation, via his relationships to his first wife, Hannah Arendt, or as Walter Benjamin's first cousin or as the son of his famous psychologist father and so on. And then there were the circles he seemed to have shared, to continue the resonance of inbred familiarity, including the original circle of scholars associated with Adorno and the Frankfurt School in addition to having been a student of Husserl's and of Heidegger's.

Anders made the question of technological mastery or excess along with the correspondent notion of human obsolescence his life's work. Not only that. Anders kept his observant powers throughout his long life, in this not unlike Kant's late-life productivity. Nor did Heidegger himself quite achieve this (as Arendt tells us and as Gadamer also attests). But what is still more significant, Anders kept his powers sharply attuned to the changing technological times.

Not that this mattered in terms of his lack of influence on the academy which then, as now, pays attention only to "important" names (and these are usually younger names we already know). And notwithstanding Anders' sustained philosophical focus, even as already mentioned, though it can hardly hurt to repeat that philosophers of technology such as Ihde and technoscience and social theorists such as C. Fred

Alford do not even mention, let alone engage, Anders; even Latour does not do so, despite Anders' reception in French technoscience. Surprisingly, even the activist scholar Stanley Aronowitz, himself very like Anders, and whose work is indispensable for a social and political theory of technology, does not refer to Anders, just as those interested in discussing crimes against humanity typically do not refer to Anders.

There may be good reasons for this in addition to the perennial scholarly desire to reinvent, all by oneself, whatever it is that one wishes to claim to be first to talk about or to mention. Thus, the point noted to begin with as this often appears at the start of most discussions of Anders, highlighting the rarity of said discussions, is that the claim that no one writes on him, or that he is the least "well-known," be it in general or in the Anglo-Saxon world, is overdone in Anders' case. Dozens of books (and still growing) demonstrate the contrary, as noted in the introduction. Yet the number of books on/about Anders does not seem to alter an ongoing lack of familiarity. Anders is not a name like Arendt or like Benjamin.

This is the quality of the outsider, and one cannot understand the outsider unless one understands the insider culture of academic cartels, peer groups, groups of friends, in-groups, and not less: ideologies. Thus, John Gray insists that the humanities "can't be saved."[41] Gray does not err when he points to the importance of ideology, and this is the point of mentioning citation—and hiring—cartels. Not only the Straussians engage in this. And to be sure, no one, today, gets a PhD or appointment, much less tenure (this is the reason Gray cites Kimball's dated study), *without* paying due fealty to prevailing ideology and the latest trends. As Gray, a political commentator, credits Kimball, the mechanism is identical with that of the academy itself and what is central; this is also the key insight to Kuhn's notion of normal [read: normativized] science:

> it is self-reproducing. Through their powers of patronage, the nomenklatura decide the prospects of new entrants, and exclude anyone who deviates from the party line. No young scholar who fails to genuflect to it has any prospect of a future in academic life.[42]

For most, this is an invisible tribute, unnoticed because they believe in it, and quite for the same reason, one cannot advert to it.

Beyond this question of academic inbreeding and self-absorption (these are the same), there is also the question of science as a problem as Nietzsche spoke of problematizing it. But with respect to Anders, the issue is more than the matter of the academy as such, although Anders was a dyed in the flesh academic, and beyond the question of science, is the question of method and substance as he himself would insist on this, especially in his own writing on music but also philosophical anthropology, as part of which discipline he himself would characterize his *The Antiquatedness of Humanity*. Withal, Anders followed his own course, and this made him difficult to boot. From some perspectives, this would mean that Anders, here not unlike Jacob Taubes, could be difficult to deal with, a bit like Ivan Illich, who was however, being a priest, the kinder sort of heretic. This means that anyone writing a biography of Anders for English readers, where such things do not go without the need for extensive explication, will need to explain the significance of his birth (in Breslau), and that he would, after returning from the United States, make Vienna his adopted home town with his second wife, Elisabeth Freundlich, and then his third wife, the pianist, Charlotte Zelka, who married him later in his life (Zelka performed as accompanist on a range of classical recordings) and who, to Anders' clear bafflement, left him perhaps for the same reasons that she married him—although they never divorced as such *and* although, in the aftermath, as if to complicate things, Anders reconciled with his second wife. If, as can be thought, Anders exemplified a "strong" personality, as the Germans speak of this, it can seem he was himself able to deal with comparably strong personalities. Thus, and significantly, he remained on speaking terms with Adorno.

Significantly, perhaps, Anders, like Adorno, was a teaser whose teasing, again as in Adorno's case, was unbearable for Americans as it tended to illuminate how much he knew and how much his interlocutors did not and which was inclined as a result to be perceived as mockery. Unlike the kind of "critical thinking" that involves thinking just and only what status quo science ("the" science) tells you to think, critical theory requires considerable breadth just to begin to be critical. Anders knew an enormous amount about both Greek *and* Hebraic cultures, including Augustine's Latin, as he knew about music, about art and literature, about Hegel and Marx, about Kant, and about Husserl and Heidegger. To this must be added the social sciences, and this last addition, which to be sure he shared with Adorno and other members of the Frankfurt School, distinguished him from Heidegger and company. Like Nietzsche but perhaps better said given the epoch of the twentieth century as such, and like Ivan Illich, the philosopher and social critic of education, medical science, and technology/ecology, Anders was also, and this is perhaps the most rare of all — I used the word above to speak of Illich, himself a believer — a *committed* heretic, that is, the sort of critical thinker who meant what he said and who acted on it at the expense of his career—and he did this from the start—and who suffered for this in terms of his reputation (he was for a long time not even mentioned) and his livelihood. Thus, Anders did what most social critics do not do and sometimes even suppose cannot be done: throughout his life Anders walked the talk.

What is more, the views Anders opted to champion were out of kilter, unpopular. Indeed, like Illich's political views, Anders's views were *anti-popular*. Thus, and instead of talking about the Holocaust as a Jew and as he might well have done (though he did this too, he did all kinds of things, including music and literary theory to boot), Anders made Americans (that would be the good guys in the Second World War from his perspective, and he should have been more grateful, etc.) uncomfortable by talking as incessantly as he did about Hiroshima.[43] Thus Anders would count off, almost kabbalistically, the dates of Hiroshima, where the bomb detonated on August 6, 1945, and of Nuremberg, where just two days later, on August 8, 1945, the legal rubric for defining crimes against humanity would be spelled out, followed by the bombing of Nagasaki the next day, on August 9, 1945.

Like many others of his era, Anders thought that the problem of evil was the bomb. And like his nemesis, Heidegger, he also insisted that the evolution of that same problem had to do with what, unlike Heidegger, he had seen from the start as the problem of humanity itself as standing reserve in Heidegger's terms, a bioresource that however would need, desperately need, 'improving.'

This Anders called the shame of being born human, a shame that carries with it its own mark or sign, that is, the shame of a navel. For the mark of creation, as a creation at the hand of god, which is (and here Anders concurs with Sartre) the perfected dream of modernity, is that we as human beings are the ones who, as Nietzsche's madman tells us, have "killed" God—"And *we* have killed him" (*FW* §125)—a deed done with our own hands, so that the sacred as Nietzsche puts it, bleeds to death as we watch (but then, what about the blood, and Nietzsche goes in for excessive realism: What about the stench? Gods, too, so he tells us, decompose!).

Much more than merely murdering God—this, after all, would be a piece of cake for Anders as a Jew, a secular Jew no less—we want to take his place. But that's the kicker.

The problem for us is that we are born and not made. Above all, we are born, and this is the Heideggerian point, as we are born, thrown as we are thrown, and we are not designed in accord—this is the anti-Cartesian impetus—with our preferences as we might have specified them (had anyone asked).

Anders' most dissonant insight—vying with anything Levinas argues about the Face as it also vies with anything Heidegger argues about death and thrownness, and with everything (and in the case of Anders, this is not by accident) that Arendt writes about natality—is that the whole of our problem with modernity begins and ends with our awful shame at having been born (oh gosh and now we begin to remember all the Theweleit anxieties about war, about Jews and others as very patent anxieties about women). What we much rather want to be instead, and there is always an instead, is the machine: perfectible, replaceable, immortal. Thus as Anders articulates the modern human fantasy today, the "dream" as he calls it, "was naturally to be like our gods, the apparatus, better said, to belong to these (mechanical) gods completely, to be to an extent co-substantial with these gods: *homologoumenōs zēn*" (AM I, 36). Our desire is to 'be' the machine or, as in the current era, to become one with the digital realm, hormonally augmented, virally enhanced, genetically tweaked.

Thus, and ultimately for Anders, our desire is to be manufactured, to be fabricated, to be a product, maybe one with serial numbers, perhaps an ISBN, quite such that we can market and upgrade ourselves: the point here would be interchangeable parts (AM I, 39). If something breaks, fix it; when something wears out, replace it.

Towards the end of his life, Anders would recollect his own collision with the spirit of the times after the First World War. No kind of poetic experience "on horseback," this was a direct confrontation with changes made by the medical technology of the day coupled with modern transport. The result of these technological transformations of human life at the very limit of everydayness, here retrospectively conceived as a Heideggerian everydayness, shattered that everydayness for him. Beyond anything so theoretically to the point of the ready-to-hand quotidian, more than Heidegger's misplaced or broken hammer, Anders recalled the dissonance of this vision, at the age of fifteen, as he was on his way home after the First World War:

> On my way back, at a train station, maybe it was in Liege, I saw a line of men, who strangely seemed as if they began at the hip. These were soldiers who had been set

on the platform on their stumps, leaning them against the wall. Thus they waited for the train that would take them home.[44]

These broken "fragments" of human beings, in Anders' foundational trauma as he encountered this—and we will have occasion to recall this again—*are* already transhuman. No one will ever need to tell them that their canes, their wheelchairs, and their prosthetic limbs *are* their extended selves. This they know.

To translate the little hymn Anders gives us for musically monotone Molossians, as this echoes the metric rhythm of the shuffling footsteps of the workers' change of shifts in Fritz Lang's *Metropolis*:

But if we ever succeed
in throwing off our burden
and stand as [iron] bars
fitted into [iron] bars

As prosthesis to prosthesis
in intimate conjoining,
and the flaw was what had been
and shame was yet unknown—[45]

The rhythm is one thing.

Fritz Lang's *Metropolis* (Figures 1 and 2) articulates the catastrophic juggernaut that is modernity and the capacity, thus the robot image that captures theatrical imagination as the playwright Karel Čapek first sees in his human fabrications in his 1920 play, *R.U.R*, quite as Lang realizes the same ideal on film using, this is Lang's genius, an iconically stylized iron maiden that automatically (of course) evolved/morphed into human form. The filmic magic is in the dissolve, the cuts that effect the glorious streamlined vision of the robot together with its subversive metamorphosis into a "gynoid" otherwise indistinguishable from the human. This is leagues away from the clunkiness of Čapek's rigid players, which certainly continued as B movie robot ideals up to the television "ninny," as Jonathan Smith named the "Robot" in the 1960 television series, *Lost in Space*. But beginning with Lang's filmmaker's device, the illusion of the ideal transhuman, the metamorphosis into a machine is that the actor who plays the *human* Maria is the actor who plays the *mechanoid* almost-Maria (Figure 2)—Lang famously insisted that Birgitte Helm had to be physically inside the heavy armour in all shots although, as she complained, one could not see her—and of course, Helm also plays the evil, *gynoid*, Maria (the humanoid version otherwise indistinguishable from the original).

These ancillary points are relevant here, as a "cover," to use this language, of *Pygmalion*. This includes the notion/detail that the first female robot would also exemplify the characteristics of both an industrial fetish *and* a sex doll, a reflection Anders includes; it is part of the language of shame as he speaks of it, if a full elaboration necessarily exceeds the present context.[46] For his part, Lang sought artistic redemption of the heart, which will include eros, and the hand or work. This vision illuminates

Criticizing Technology

Figure 1 Film scene of *Metropolis*, 1927, directed by Fritz Lang. Photo credit: bpk Bildagentur / Horst von Harbou / Art Resource, NY.

Figure 2 Film scene of *Metropolis*, 1927, directed by Fritz Lang. Alfred Abel, Brigitte Helm, Rudolf Klein-Rogge. Photo credit: Adoc-photos/Art Resource, NY.

the streamlined aero-tech city that is *Metropolis*, the bodies are those of the workers sacrificed in the bowels of the city hidden beneath the surface (cf. Figure 1).

And the Maria-robot, as the fully transformed, "gynoid" copy designed to function as a changeling under the command of the rulers/owners to replace the Maria-human, who agitated on behalf of the workers, now betrays them by leading them into destructive frenzy, in a dance that quite matches the phrase we read in Anders himself: "hammering the syncopated rhythms of the machine god into the body" (AM I, 84). Thus, Anders describes "Jazz as an Industrial Dionysus Cult" (AM I, 83). Thereby—and note that Anders' analysis helps to understand why just this would work to undermine any revolutionary impulse in Lang's film—such a mechanized, (pseudo) wild dance is calculated in order to "transform animalistic into mechanical energies" (AM I, 83). More critically in this respect than Adorno, Anders writes of the kind of jazz that by contrast with conventional claims evoking "'visceral memories of desert and jungle drums'; is much rather (or at least equally) 'machine-music'" (AM I, 83). This kind of dance is associated with the age of the "industrial revolution"; count the techno ways, whereby "what the dancer dances is not only the apotheosis of the machine but at the same time a festival of renunciation and coordination [*Abdankungs und Gleichschaltungsfeier*], *an enthusiastic pantomime* of its own total defeat" (AM I, 84).

Like Arendt and like Heidegger and Jonas (and so on), Anders enjoyed a classically German *classical* education, including both Athens and Jerusalem, which is why Anders speaks of *aidos*, shame. The issue is having had a father, a progenitor, rather than a fabricator, an inventor. Neither a "product" nor a god—think of Sartre's very Cartesian, existential articulation of this dream—we are just and merely creatures born with every "creaturely inadequacy" (AM I, 36).

Finite and limited, we're only human as we say. But if we were as gods, could we but be manufactured to precision standards at the consummate height of the technological engineering we are so sure is coming our way? We imagine that everything would be so much better, and the tech dreamers behind the mass vaccination schemes follow lockstep in this conviction.

Anders' figural analogy, *God = Product*, is compelling, thoughtworthy, as Heidegger would say. The product *is* God. As Anders goes on to say at this point:

> The attempt to prove his "thing piety," endeavouring an *imitatio instrumentorum*, one has no choice but to undertake a self-reformation: at the very least and in the smallest degree to undertake effort to "improve" [today apologists for transhumanism prefer to say "enhance"—BB] himself, rectifying the "sabotage" suffered owing to original sin: the legacy *nolens volens* of birth, now for once reduced to the smallest conceivable degree. (AM I, 36–7)

For Anders, we want to correct the mistakes in our make-up: the errors that cause us to become ill, to suffer, to die. An imperfect rather than a well-made product, as René Descartes had already pointed out as part of his philosopher's proof of the existence of god (the Parisian theologians did not a miss a beat with this one), a proof that just also happened to condemn God's manufacturing specs: had he, Descartes, fabricated himself, he would have done it better. We will need to return to this.

For Anders, we have already, at the time of his writing in the mid-1950s, begun to undertake this same rational and Cartesian enterprise which we call, and it is instructive for those who believe in the logarithmically accelerating evolutionary trajectory of technoscientific engineering and design that we use, rather the same terms that Anders emphasizes in 1956, and formulated in English as "Human Engineering."

As a corollary, so Anders reminds us, the human being is manifestly a "defective design," (AM I, 32) especially when regarded from the perspective of technical devices (error tends, as we know, to be "human error" rather than a result of a deficiency in the machine, whatever the machine might be and quite to the extent that to use a machine requires that we conform to it).

In this way, the first chapter of Anders' *The Antiquatedness of Humanity: On the Soul in the Age of the Second Industrial Revolution* is titled "Concerning Promethean Shame," prefiguring the evolutionary drift or longing of the human towards a literally technological rapture. Note that the word "rapture" is not an overstatement. What is at stake is the resurrection in human time, here and now, of the body, replacement, consummation, salvation, transfiguration—and like the technical problem attendant upon the theological (or Disneyesque) problem of the resurrection of the body, what to do with the old iPhone when the new one arrives? This is an already-present and growing problem for iPhone owners all over the world with the most recent and coming 5G models (no one but no one thinks of the bees), adding to the collection some owners have in a drawer somewhere.

For Anders, Descartes' musing that God had created him with deficiencies (this would be the true maker's mark, this would be the Promethean shame), can rightly be kicked up a notch. Here we see that like Arendt, Anders too is Heidegger's good student, and thus he moves from Descartes to Kant. Thus, we move, Anders argues, "into the obligatory." In "other words," as Anders explains, "the moral imperative is now transferred from the human being to the gadget" (AM I, 40).

What ought to be, and what should be, is now the tool, the device, and the gadget. We want technology; and as we ourselves become our own technology, so much the better. This is the transhumanist dream: let there be not merely the human but high technology, and let us not forget, as we reflect on this, as transhumanists are often in the business of selling technology: let there be stuff to buy.

Anders repeats for his part the maxim that Heidegger identified in his *Contributions to Philosophy* as the maxim of fascist technoscience (whatever is technically possible should be actualized as quickly as possible) which, as Heidegger had anticipated and Anders could not but corroborate, applied with fairly dispassionate equal measure to Soviet and capitalist aka American science alike:

What can be done counts now as what ought to be done. The maxim: "become the one you are" is today perceived as the maxim of the gadget. . . . Gadgets are the gifted, the "whiz-kids" [*in English in the German original*] of today. (AM I, 40)

But for all the claims that are made on his behalf (claims Anders happily echoed for his own part), to the effect that Anders opposes Heidegger, just as he similarly opposes Adorno, Anders also takes over (quite as Adorno also charged) and radicalizes Heidegger's critique.

Hence, Anders begins *The Antiquatedness of Humanity* by reflecting on the impossibility, as it were, of criticizing technology, that is to say of "refusing" or distancing oneself from technology: an impossibility that found expression for Heidegger himself in his *Gelassenheit*—and a critical impossibility that has hardly been ameliorated, let us be careful to underscore this, in the interim:

> As I articulated this thought at a cultural conference, I was met with the counterclaim, in the end one always has the freedom to turn off one's technological devices, indeed one even has the freedom to decline to buy any such, and dedicate oneself to the "real world" and just and only this world.
>
> Which I disputed. And indeed, just because the one who strikes is at much as the disposition of technology as is the consumer: whether we play along with it or not, we play along, because we are played. Whatever we do or fail to do—that we increasingly live a humanity for whom there is no longer "world" or world experience but a phantom of world and a phantom of consumption, no part of this is altered by our private strike: this humanity is today the factical with-world, which we must take into account, to strike against this is not possible. (AM I, 1)

Elsewhere I underline that to follow Heidegger always means that we find ourselves in contestation with him, just as Michael Theunissen also reminds us. But this also means that we are called to question as Heidegger questions. To this same extent, Anders thinks Heidegger's critique as Nietzsche would recommend thinking critique in his own reflections on Kant, through to its "furthest consequences."

Here we recall Heidegger's allusion to Rousseau and to Schiller at the start of *The Question Concerning Technology*, "Everywhere we remain unfree and chained to technology, whether we passionately affirm or deny it" (QCT, 4). When Anders reflects on the ultimate impossibility of denying or refusing technology, inasmuch as we are human beings in a world with others, he repeats a point Heidegger had underlined early in *Being and Time*, where Heidegger writes "Dasein's being-in the world is essentially constituted by being-with," underscoring that this remains even when Dasein is alone, "even when factically no Other is present-at-hand or perceived. Even Dasein's being-alone is being-with in the world." (SZ, 120)

But as Heidegger later articulates this problem in "The Turn," one of the original lectures he presented in 1949 in Bremen, warning in perfectly apocalyptic tones attuned to the cybernetic technology of the day and effects of which continue on the internet that is the current form of the broadcasting technology Heidegger describes: "we do not yet hear, we whose hearing and seeing are perishing through radio and film under the rule of technology" (QCT, 48).

In an age where the technical gadgets of which Anders speaks, that is, again, the *Geräte*, "technologies," as we speak of them, have become more indispensable than ever, we have cell phones and seemingly cannot live without them, facilitating government surveillance by the most complicit means imaginable, short of cyborg hacking or injection. We carry our own spying and tracking device with us, and we keep it fully charged and close at hand.

As Anders reminds us, no matter what we do, and in this he handily includes every imaginable luddite expedient, we remain constitutionally incapable of renouncing their use:

> What holds true of these devices holds, *mutatis mutandis*, for everything. . . . To maintain regarding this system of devices, of this macro-device, that it is a "means," saying that it is at our free disposal to be set to whatever purpose, would be completely senseless. The system of devices, the apparatus, is our world. And world is something otherwise than means. Something categorically otherwise. (AM I, 2)

In addition to his Heideggerian anticipation of Bruno Latour's claim that it is difficult to draw the line between ourselves as actants and "things" as actants, whereby we can scarcely distinguish between ourselves and our tools/technologies, for Anders we coincide with, we "are" the technological things of our lives, as these adumbrate our lives, and reflect ourselves to ourselves as such.

Anders highlights the already-given and determinate character of the modern consumer, determined as we are by our modern advertising. Thus, as we like to say, here making it all-too plain that we speak from the perspective of the advertisers, we live in and on the terms of and as a consumer society. This is the heart of Heidegger's analysis of *Ge-Stell* as Anders continues to analyse it, here without reference to the term per se.

> For, taken in all precision these are not just so many "preliminary decisions" but the preliminary decision instead. Yes. The. In the determinate singular. For an individual device does not exist—the entirety is what is at stake in reality. Every individual device is consequently nothing more than part of a device, merely a screw, merely one piece in a system of devices, a piece partially directed to the requirements of other devices, its existence in part exigent upon other devices which in turn compel the necessity for new equipment. (AM I, 2)

Although describing Heidegger's fate as a thinker and critic of technology and science, Anders analyses the reasons for our silence as intellectuals in the face of technology and its effects as indebted to nothing more effective and egregious or tragic than simple socialization: in order not to be supposed a "reactionary" (AM I, 3). Nor has this fear of being thought reactionary (or technologically backward) changed in the interim. Hence, his further reflection also bears repeated consideration:

> that a critique of technology has already become a question of moral courage today is consequently unsurprising. In the last analysis (so thinks the critic) I can't afford to permit anyone to say of me . . . that I was the only one to fall through the cracks of world history, the one and only obsolete [*Obsolete*] human being, and far and wide, the sole reactionary. And so he keeps his mouth shut. (AM I, 3)

For just this reason, Anders could not but be a reactionary.

As author, Anders was haunted by his lack of reception. Thus, his books, both academic and literary, would not be published and perhaps more significantly for the philosophical history of reception, his peers refused to receive his work even as he was fitfully engaged and certainly acknowledged, for example, by Adorno. In addition, and beyond academic small-mindedness—the sort that inspires quips concerning the minimality of the stakes involved—Anders also violated the pro-tech ethos ruling theoretical studies of technology before and after both world wars, accelerating post the 1950s and onwards. To this extent, as Heidegger also underlined this in his lectures to the Club of Bremen and in Freiburg, it is always troublesome, "dangerous," to question in the wake of technology.[47]

Whither?

When it comes to technology, to machines and the question of (human) mastery, I maintain the Andersesque hope that unlike Adorno, who, as some might argue, perhaps heard Anders' explicit suggestions to him as insults, we might yet find ourselves willing to take up the charge, and maybe even, as Anders suggested, to take the lead in a moment of human freedom.

Once again, we can cite Heidegger's *The Question Concerning Technology*, "Everywhere we remain powerlessly chained to technology, whether we passionately affirm or negate it."[48] The language includes the term "*unfrei*."[49] The very point echoes on the first page of Marcuse's *One Dimensional Man*,[50] and Anders takes over Heidegger's insight that it is impossible simply to renounce technology, as if this were an option and for the very early Heideggerian reason that we are human beings in a with-world, *Mit-Welt*, among other beings in the world and, above all with others: *Mit-Dasein*.

At stake is the Marxist and critical challenge of action. More than Heidegger certainly and even more than Horkheimer or Adorno, Anders was a scholar who acted on his politics, as radically conceived as they were, in the real world, the engaged life of human action. And what often goes by the title of political agency, be it reading the paper, voting in a two-party or parliamentary system, along with the everyday politics of whatever given public sphere, should be contrasted with Anders' activism as this last involved the kind of life action that would seem to have been technologically eclipsed until the events sponsored, aided by today's technology, said to have been, though this is debated, cellphones, however short-lived in the end, in the so-called Arab spring or the failed American movement, Occupy Wall Street—now a nearly forgotten venture. For the most part however, for most of us, especially we academics, we think ourselves "activists" if we sign a petition or click on an email link and hit return.

I've more than once emphasized that it cannot be said that *no one* writes about Anders.[51] As indicated by the aforementioned references in the text (and in the notes to be sure), Anders was factically "engaged" from his childhood years *qua* "known" entity, so to speak. People knew who he was. In addition, and along with Heidegger and Benjamin, Adorno and Arendt, Anders was part of a prime circle of influential thinkers. And yet, and this is the "reception effect," despite the changing regimes of academic reception history, Anders' name consistently manages to fall by the wayside. Inasmuch as I do not think that there is no reason for that constant sidelining, it may be that the easy expedient of noting his work, or "discovering" his discoveries, may not suffice to bring Anders to scholarly attention.

I have suggested, as will also be illustrated by the chapters to follow, that some part of the reason for Anders' lack of reception has to do with the scholarly range of his interests and his interdisciplinarity. Thus, scholars who variously specialize in philosophy of technology, philosophy of science, philosophy of media, political theory or philosophical anthropology overlook Anders' work. Anders' philosophical sociology of music, phenomenologically articulated as it is, only complicated matters. Music as such both charms us and confounds us when it comes to philosophy; just think of the related case of Nietzsche, where, apart from quoting his very quotable apothegm, *without music, life would be a mistake,* philosophers seem able to do little more than point to the fact that Nietzsche liked music, emphasizing his favourites by way of their own enthusiasms (*what kind?*, opera, *what kind?*, Wagner, to be sure, or Rossini, or Bizet), all the while managing to overspring Nietzsche's emphatic interest in Beethoven and in Greek music, including rhythm. Or else, if we like, more proximately, we may think of Adorno and the generations of scholars both within and outside philosophy of music to simply focus on jazz, as one imagines Adorno's views on jazz to be as one imagines these: negative and benighted and nothing more. In the case of Anders, we need for the sake of a reflection on his understanding of technology, including the antiquatedness, datedness, of the concept of humanity (as opposed to the post- and transhuman), as indeed for his radio and broadcast theory to include Anders' early reflections on music[52] as this resonates throughout his complexly stylized study of the human being in eclipse.

And we could not agree more; we are more than ready to be posthuman, transhuman,[53] waiting for an upgrade, hardware, software, wetware, the next booster shot, quite as we are primed to wait for the next-generation consumable to purchase, like an iPhone, say. If anything, here in this respect like the online complaints about the much-delayed and the much-hyped final short season of *The Game of Thrones*, badly written and even less fortunately executed, consumers turn out to seek only more of the same, intensified, more and ever more. Hints of the pitching and pacing of iPhone models seem underwhelming rather than overwhelming, just as in the case of the HBO screenwriters' hasty completion of George R.R. Martin's script inspired online expressions of anger and protests. Both the original author and HBO meant to move on to other things. But fans tend to hold—this is what is means to be a fan; this is the point insisted upon as key to the culture industry as Adorno and Horkheimer and Anders analyse this—to the patterned schema they've been "programmed" to expect. And this is how priming works.

These resonances also underscore that Anders himself hardly intends his title in the sense of today's transhumanist projections of technological perfectionism/ enhancement fantasies (how could he?) an issue when it comes to contemporary media studies. Anders is (how could he not be?) a child of his era, and yet his analysis of his era and the media of radio and television and film remain so prescient that we can benefit from an engagement, particularly where this is also echoed, and, inevitably, updated in Baudrillard's analyses.[54]

Whose Critical Theory?

There should be many different critical theories of technology; one should be able to amplify the range as circumstances change. To be sure, this is not at hand, and a good part of the reason why it is not has to do with our "enlightened" conviction that the problems that led to twentieth- and twenty-first-century wars and revolutions had a great deal to do, if not everything to do, with a failure of reason.

Thus, Habermas and co., that is, adherents of today's "Frankfurt School" professing to teach critical theory, owe their success, that is their posts as such, to their unswerving dedication to the ideal of rationality.

Anders took his stance contra such conventions, a tactic that excluded him where his inclusion ought to have been assured. But he did more than claim a counterposition. Anders wrote in such a way that the leading theorists could make no sense of him; he could not be co-opted to the standard line, the standard way of writing, standard points of view. And as more than a few commentators have observed, Anders was unrelentingly hard-necked on the position he took. Only a few commentators, like the exceptional Lüdger Lütkehaus, have had the prescience and readerly stylistics capable of matching Anders' achievement. The issue concerns Anders' style, tone, or voice as many readers find this a particular challenge.[55]

This challenge in turn underlines a point that is hermeneutically significant as one must read through one's own prejudices when it comes to reading any thinker. Student as I am of Gadamer, this is key. Even when, precisely when, we have understood another thinker, another position, a historical era, we understand otherwise. Always. What this means is that understanding is neither a Hegelian nor a Thomistic achievement: we do not attain to transparency, with the success that leaves us, this was Bernard Lonergan's modern Thomist ideal, with the sense that the inquiry has attained its goal: no further questions needed. This means that we must read *and* reread, engage *and* re-engage. Failure to do so inevitably imposes our own thinking, our own interpretation in place of the text, and then, as Nietzsche warns us, the text cannot but disappear underneath or beneath the interpretation. The only way to safeguard ourselves against this tendency is to leave the awkward aspects in place and that entails that we attend to them, hear them, without imposing our certainties in the process.

It will hardly come as a surprise to note here that Anders was not as adept in his critical interactions with others as we might imagine that he could have been, given current ideals of social interaction and engagement. Thus we earlier cited his introduction to his first book where he begins by emphasizing just this socialized ethos contra criticism, especially in terms of the precarity of any criticism of the machine ("Die prekäre Maschinenkritik" AM I, 3) and in terms of the equation of criticism with a reactionary stance ("Kritisch gleich reaktionar" AM I, 5). Anders had trouble with this anti-critical ideal as he rightly recognized this as the effective instrument of mediatic totalitarianism: we will all have (or we should have) the same tastes, otherwise the channels broadcasting the "hits" would have to be curated for as many tastes as there are listeners. And although YouTube would seem to permit such diversity, it tends to monotone virality and is, to boot, liable to censure. Thus as Adorno and Horkheimer also argued for similar reasons the tastes that conduce to "virality" were themselves already manufactured to order by exposure, conditioning the listener.

Thus, and we will have occasion in a later chapter to come back to this, Anders, citing the iconic era of American jukebox culture, argued that the very-material reality of the jukebox as such meant that tastes of one's neighbour *could not but* be imposed on any listener in the acoustic neighbourhood of the same. One's nickel or dime, or quarter (and the increasing coinage value says a great deal here), ensured the music one opted to pay for would be broadcast, imposed on all in earshot, and if a customer were to object, as Anders attempted to object (with predictably disastrous consequences), the objection simply could not hold. The alternative option given multiple table-top jukeboxes, playing a piece of one's choice, yielded the kind of cacophony we today (pre-Lockdown) could take for granted in US eating establishments, especially in New York City.[56]

These days earbuds securing a perfectly social a-sociability, as studying social media posts on our smartphones, also helps social isolation, the *Happy Days* charm of a table-top jukebox problem can seem strange to us. Today, the problem has morphed into the complete or consummate fulfilment of what Adorno named "standardized ubiquity" and which Anders described for his own part variously as imposition and complicity/compliance. The "radio," *pars pro toto*, imposes a taste and a background voice that today accompanies all one's activities in almost every respect, at home, and owing to portable radios and automobiles, this includes music streaming on demand and GPS everywhere one might travel.[57] Thus, the "compleate" audiophile is equipped with a radio circa 1950, 1960, 1970, and 1980, and if anything has changed, it will only be the narrow range of radio broadcasts/streaming/on-demand. With the new media, with smartphones and smart tablets and such, the consumer is free to take his or her cue from the very same standardized ubiquity now available "on demand."

The theme of controlling consumer consciousness is common to critical theory in its older modalities, Horkheimer, Adorno, Marcuse, and no less, Anders. In my own essays, I point to parallels with Dallas Smythe and Jean Baudrillard, but as I also argue in *The Hallelujah Effect*, social scientists like political theorists and sociologists commonly tend to walk back the efficacy of advertising. The suggestion offered by academics is thus that advertising doesn't really work. But the claim is absurd, given advertising's omnipresence and given the current pandemic orchestrated by little other than government and social media advertising.

But, say the social scientists, dominated as they are by the ideal of rational choice, advertising only serves as prompt, leaving the consumer free to act on his or her own free will, as he or she sees it: "rational choice" rules. No social engineering, no social media hacking of consciousness or perception. At the same time, a feeding frenzy on "fake news" and the so-called post-truth era would seem to put paid to this conviction. For if media influencers did not influence, be they located in the White House or fake Twitter followers, or on Facebook and otherwise, what *is* our concern?

With the exception of Baudrillard, most scholars proceed as if there were no need for critical theory: as if the thinker, the consumer, the citizen might simply read popular journals and newspapers (*nota bene*: in online versions) and watch network television or check Facebook and Twitter (surely I should provide a trigger warning here) and proceed to make up his or her mind according as he or she judges.

None of this would accord with the view offered in the middle of the 1940s by Horkheimer and Adorno (and, as I argue, also, if unexpectedly so, Heidegger), the views Anders argued in 1956 and again in 1980. Baudrillard and Kittler pick up some of the slack, but I would argue that we have all we can do to *try* to read Anders.

Media Branding: How to do Politics with Programming and iGadgets

Beyond media, that is to say, beyond the "effect" of programming social consciousness and complicity, Anders also offers a critical theory of manufacture and of mass goods, well in advance of the current empirical turn in the philosophy of technology and thing ontology more generally. Anders reflects on our way of being and having, including the having of things, which having becomes our identification with them, our competition with them and aspiration to be them: this is our Promethean shame. This last constellation is arguably the most complicated because this is his gadget-thinking, that is, his precisely *material* theory of technology and his reflection on style and allure that we associate with technology on a precisely physiognomic, embodied, material, manufactured, hyped basis. The hype in question is what I name the "Hallelujah Effect."[58]

Talking about the sound as we can hear this at once, immediately and not via some inference, rational or otherwise, which allows the listener to differentiate between different makes of automobile, Heidegger himself participates just because he replicates this same material and auratic dimension. Thus, we *hear* the mark of the kind of car that is "a" Mercedes, as Heidegger writes, quite in opposition to the brand that is an "*Adlerwagen*." Anglophone readers have little notion of the brand in question, that is to say: we don't recognize the make in question; thus, the translator (and philosopher of technology in his own right), William Lovitt, helpfully renders the term "Volkswagen," useful in spite of being inaccurate as such where it is the precision of the specific brand of automobile as we hear it that corresponds to Heidegger's point, just as it would be to the point of an advertisement for the vehicle as such, or some other market item. *Is it live, or is it Memorex?*

Marcuse, whose *One-Dimensional Man* remains to the current moment one of the best critically theoretical discussions of modern technology and culture, focuses on what he names "repressive desublimation."[59]

Marcuse rightly saw, here quite in alignment with Anders, that we identify with the products we surrounded ourselves with: these are Anders' gadgets, products Marcuse argued to have been less a matter of personal acquisitive choice than specified accoutrements, de rigeur for materialist, bourgeois life in America. Marcuse's books analysed what was thereby lost in the trade, articulating, as he wrote, the "enchantment," especially the "eros," sacrificed in the wake and working effects of what Horkheimer and Adorno named the culture industry and analysed as disenchantment in their *Dialectic of Enlightenment*.[60]

The focus here is on Anders, and Lacan for his part characterized this phenomenon as I also seek to do in terms of desire, and there are elements of that same focus, to name a current, popular intellectual (and not incidentally a Lacanian), in Slavoj Žižek. Elsewhere I point to some of the limitations of that language *qua* eros, less Norman O. Brown than Woody Allen as the "erotic" is for the most part adumbrated in and through the imaginary-for-the-male. Here, without speaking of male and female desire (as such) and the difference-that-is-no-difference (I argue that the male perspective seems to be the only one on offer precisely because the male is not offered for female enjoyment or celebration, assessment or appreciation, a point made more generically by the analytic philosopher Kate Manne),[61] there is the matter of alienation and recognition. We see ourselves in our possessions, according to Marcuse, in what we consume, and so too according to both Adorno and Anders. Anders will argue that more than identification with our technical objects, we measure ourselves by our gadgets; better said, we measure ourselves *against* our tools. And as Anders emphasizes, we come up short in our own fantastic projection of the results.

Today we use our gadgets to orient ourselves in the world, thereby changing our focus on the world to just the limited scope our gadgets can convey. Our eyes are affected, our minds are affected, our world is diminished, and it cannot be said that we notice this, much less that we mind. Here at issue is not a question of dystopian techno-analytic thinking as one may find this in the pages of magazines or websites dedicated to contemporary popular psychology. Rather for Anders, mirroring our tools, we aspire to the status of the tool and thus to make his point in contemporary language, we match ourselves to AI, even when we have not (yet, so we say) designed an AI worth worrying about (so we say) because, although sheerly imaginary, we project ourselves in the image and likeness of the instruments we make: our gadgets, ourselves.

2

Heidegger's Authenticity and Günther Anders' Neg-Anthropology

Heidegger's Authenticity and Günther Anders' "Humanism"

As Heidegger's student, Anders allied himself early with those who criticized Heidegger on a number of matters, most particularly, as Adorno would echo this, Heideggerian "authenticity" [*Eigentlichkeit*].[1] Most well-known is Anders' indictment of Heidegger's "Pseudo-Concreteness."[2] But, like the then-omnipresent terminology of "situation," as Anders also used this language, the "concrete" is complicated.

Here it can be helpful to recall Adorno's reference to Anders' essay—note, too, the spectral allusion—in a footnote to his *Negative Dialectics*:

> The word "concretion," most affectively occupied in German philosophy between the two World Wars, was drenched with the spirit of the times. Its magic used the feature of Homer's *nekyia*, when Ulysses feeds blood to the shadows to make them speak.[3]

Where both Heidegger and Husserl had sought to avoid the charge of "anthropologism," a criticism then and now indebted to a Kantian distinction, Anders, a student of the social (human) sciences, embraced "philosophical anthropology,"[4] and in this sense one may read the first volume of Anders' *The Antiquatedness of Humanity* as an explicit (if negative) "philosophical anthropology."

Adorno, as Anders' more acerbic, Frankfurt School colleague, reflects on the philosophical "situation" of our times when it comes to the social scientific discipline of anthropology, writing:

> We cannot say what the human is. Humanity today is a function, unfree, regressing behind whatever is ascribed to him as invariant—except perhaps for the defencelessness and neediness in which some anthropologies wallow. He drags along with him as his social heritage the mutilations inflicted upon him over thousands of years.[5]

We are by now in the philosophical domain well beyond the hype of the 'singularity' but still absorbed by the humanist allure of the posthuman, the transhuman,[6] typically without having first inquired into, as Adorno here asks almost as Heidegger or as Scheler might have done, "what the human is?" For however modified, trans- or post-, the human goes without saying even when undefined or indeterminate or "overcome."

The language of "overcoming" is Nietzschean enough, and yet what Nietzsche meant when he called for such an overcoming does not mean what some theorists of contemporary post-truth and transhumanism suppose it to mean. To this extent, as Anders would have recognized, it is no easier to read Nietzsche in the service of transhumanism than it is to read him in the service of theism. This does not mean that such readings cannot be or have not been offered.[7] Thus Nietzsche's parodic (an emphasis typically unnoticed) Zarathustra first proclaims his "teaching" of the "Overhuman"— "*Ich lehre Euch den Übermenschen*"[8]—with the declaration that the "human is something that shall be overcome" [*Der Mensch ist etwas das überwunden werden soll*]," a claim repeated, mantra-like, throughout Zarathustra's "Prelude" [*Vorrede*]. For Nietzsche, the great thing about the human being "is that he is a bridge and not an aim . . . a *going over* [*Übergang*] and a *going under* [*Untergang*]."[9] At the same time, a detail that causes confusion among scholars (*both* Nietzscheans *and* those who write on his work without being familiar with it)[10] this dramatic impetus—there is a tightrope dancer in the background as Zarathustra speaks, literally *going over* and literally *going under*—is not what we tend to mean when we speak of *trans*humanism for the religiously founded reason that the idea of transhumanism is itself a millenarian notion.

In the consumer culture that is late capitalism, product updates urge consumers to buy the latest thing, the newest "gadget" as Anders already wrote on this transhuman product-inspired ideal. For his part, Heidegger, who argued the untenability of "merely instrumental, merely anthropological definitions of technology" (QCT 23), included "the man at the switchboard, the engineer in the drafting room" (29), within what he named the technological constellation, that is the set-up, or *Ge-Stell*.

Triumphalist humanism is long-standing in the Western tradition which typically not only sets the human as deiform—*imago dei*—but which also drives the notion that the human can be improved upon, a point Descartes repeats among his arguments for the existence of God (once again: had he, Descartes, created himself, he, Descartes, would have done a better job).[11] In just this way, today's transhumanist ideal tends to translate to "long live the human" or in Descartes' formula: "and thus I should myself be God."[12]

To just this extent, transhumanism *is* a humanism.

Still, we have the problem of definition: *What* is human? The question, to go back to Heidegger, as Anders reminds us, is a "what"-question[13] opposed for Heidegger to the that-question and, indeed, to the who-question (SZ 45). Orienting the question of Being in *Being and Time*, Heidegger offers a traditional definition:

> *Das Dasein*, i.e., the Being of humanity is in common as in philosophical "definition" comprehended as ζῷον λόγον ἔχον, the living being whose being is essentially determined by the capacity for speech.[14]

Here, Heidegger refers to Plato and to Aristotle on language, along with Rousseau and so on, but he also, as Anders emphasizes, refers to Kant's *What is Enlightenment?* For his own part, Anders (himself a son of a famous psychologist father, "the" William Stern) emphasizes the challenge of emancipating oneself from one's minority status contra the father, in accord with Kant's language of *Mündigkeit*.[15]

To the extent that Heidegger's definition repeats an Aristotelian-cum-Platonic commonplace, one can specify ἄνθρωπον in place of ζῷον. Here, we recall that Aristotle underlines the hierarchical schematism of *plant* nutrition and growth, *animal* perception, sentient awareness, or consciousness, and *human* contemplation: thinking thinking and it is with the last that Heidegger remains.

Jemeinigkeit: On Becoming a Question to Oneself

Heideggerian hermeneutic phenomenology includes a famous turn to what he calls "mineness" speaking of *Dasein*, articulating the force of his title *Being and Time*: "*Das 'Wesen' dieses Seienden liegt in seinem Zu-sein.*" The "essence" of this entity, this *Dasein*, lies in its to-be. Dasein is at issue, in its being, for itself, distinctive in its mineness: "*ist je meines*," specifically "mine to be in one way or another [*meines wiederum je in dieser oder jener Weise zu sein*]" (SZ, 42). Here what Heidegger says is basic enough, even self-evident, drawing upon a tissue of conventional references. Yet, despite the immediacy of the being that is to be investigated (*nota bene*: for the sake of the *Seinsfrage*), the "me" in each case turns out not to be authentically mine but always already characteristically and almost incorrigibly "unowned." As Anders observes: "even in its fullest concretion Dasein can be characterized by inauthenticity."[16]

Although scholars continue to bristle at the conceptual dissonance involved when Heidegger informs his readers that the inauthenticity of which he speaks *does not* correspond to a "lower" degree of Being ["»*niedrigeren« Seinsgrad*"], Heidegger intensifies the point, as he very habitually does,[17] by explaining, note that Anders echoes him here, that it is "inauthenticity" that determines Dasein in its "fullest concretion."[18] Indeed, and contra Anders' assertion of Heidegger's merely "pseudo-concreteness," the "concretion" in question encompasses ways of being human on a day-to-day basis: "when busy, when excited, when interested, when ready for enjoyment" (SZ, 43).

To this same extent, invoking intentional exemplification, Heidegger is able to remind us of a perplexing reflex towards the conclusion of Augustine's *Confessions*, where the saint asks what, after all, could be closer to me than myself to myself: "*Quid autem propinquius meipso mihi?*" (X, 16). If Augustine's phrase is well known, it also tends to be underread, which is hardly to say that it is not discussed. To this extent, inattention remains even as the phrase is a commonplace, even as we may note the beauty of the formula Augustine finds confounding, just where the entire text is submitted before God from the outset—the entire text can read like a prayer for today's secular eyes, and still we can read the beauty and reflexive precision, a question I have become for myself: "*mihi quaestio factus sum*" (X, 33). Augustine's successive reflection concerning time and his understanding of it, which also includes a variation on this questioning after himself, echoing his own proximity to himself, compounds the problem.[19] The logical proximity of mineness, immediacy—concreteness being part of this—is key. This Heidegger seeks to unpack.[20]

Once again, somewhat contra Anders' hyperbolic rebuke of Heidegger's pseudo-concreteness, it is also patent that what becomes existentialism in France likewise attests to a certain concreteness (this is why Adorno finds it necessary to attack French existentialism and its "humanism"),[21] and other readers of Heidegger, like Husserl, fault him precisely for his "anthropologism." The same "mineness" that invites Heidegger's reader to read along with him means that we are "free" to fault Heidegger's observations (as Anders does, as many do), because we, of course and collectively, know better. Similarly, Augustine's writerly style invites the reader to follow his confessional modality which can result in a-historic solecism, at least in Augustine's case.[22] If Nietzsche's style exemplifies the same invitation, it also intriguingly, as David Allison shows,[23] sidesteps some of the same risks, although style, as such, has never managed to prevent misreadings.

For his part, Heidegger proceeds a bit as Anders does, in the mode of Husserlian phenomenology, challenging his own teacher as he challenges Husserl's project, to argue as Heidegger does that the Being question and thereby the existential analytic of Dasein, here understood in a rigorously Kantian sense as a science, that is as "*a priori basis*" has to "come before any psychology or anthropology, and certainly in advance of any biology" (SZ, 45). The emphasis contra Husserl is repeated as title for the following section: *How the Analytic of Dasein is to be Distinguished from Anthropology, Psychology, and Biology* (SZ, 45).

Anthropologism and Psychologism

Like its cognate, "psychologism,"[24] anthropologism as rebuke follows from Gottlob Frege's admonition "to sharply separate the psychological from the logical, the subjective from the objective."[25] What is sought is the ("objective") truth, not (human) psychology, not (humanistic) anthropology/ethnography.[26] In addition to Husserl (after Frege), Heidegger repeats Kant's own distinction, precluding empirical recourse to what human beings (in fact) practice. In this way, Kant distinguishes his inquiry into practical reason from "anthropology." Here we should recall that Kant emphasizes that what belongs to a science—properly said as he underlines this—requires what neither anthropology nor psychology (nor indeed chemistry, which is a topic of some contestation in philosophy of science) admit, namely mathematics, as Kant writes in the *Metaphysical Foundations of Natural Science*: "a doctrine of nature can only contain so much science proper as there is in it of applied mathematics."[27] The point of underlining that chemistry is not a science, leading the editors of a volume on the philosophy of chemistry to begin their reflections by asking a prototypically hermeneutico-phenomenological (and thus none-too-analytic) question, "But *what are all those chemists doing?*"[28] is quite that this issue can be extended to the range of other, non-physics sciences, including, as Rom Harré did not fail to note, sciences like geology. From this perspective, like chemistry, sociology and ethnography/anthropology cannot but remain "improper" sciences, as, from a Kantian point of view, the theoretically unguarded quality of the social sciences permits a fair amount of free play in its definition by its professional practitioners throughout the last century, such that, as quoted from Adorno at the outset, its topic or subject matter, humanity, turns out to be "a function, unfree, regressing behind whatever is ascribed to him as invariant."[29] The upshot "vetoes any anthropology."[30] This veto is in line with Frege.

For his own part, Heidegger emphasizes that

> in the existential analytic of Dasein we also make headway with a task which is hardly less pressing than that of the Being question itself as—the task of laying bare that *a priori* basis which must be visible before the question of "what man is" can be discussed philosophically. (SZ, 45)

For Heidegger, as already noted above, the analytic of Dasein precedes *anthropological* questions, not to mention questions of the proper/improper, for good Kantian reasons as the analysis is not to be limited to the human being per se (quite whereby Kant takes his own prescriptions to apply to extraterrestrials, explicitly so, and, supernaturally, a bit more elliptically, to the "holy one of the gospels"[31]). Today, concerned as we are with AI, we would seem to have made little progress when it comes to animal- and plant- and even rock intelligence (as Thales is attributed to have ascribed this last to the lodestone), that is, non-human intelligence, not to mention questions of value and dignity. As Adorno observes, as Adorno was more a friend to animals than Anders (or *most* philosophers to this day):

> In the experience of nature, dignity reveals itself as subjective usurpation that degrades what is not subordinate to the subject—the qualities—to mere material and expulses it from art as a totally indeterminate potential, even though art requires it according to its own concept. Human beings are not equipped positively with dignity; rather, dignity would be exclusively what they have yet to achieve.[32]

Adorno takes his own reflection in the direction of art and the aesthetic constitution of the human being after Schiller's ideal of aesthetic "education" or formation, a programmatic constitution that, as Anders could not but phenomenologically, hermeneutically observe to have fallen off, irremediably. For Adorno, unquestioning conviction had a distinct benefit, one enhanced to no small degree by failing to question:

> Under the sign of the dignity that was tacked on to human beings as they are—a dignity that was rapidly transformed into that official dignity that Schiller nevertheless mistrusted in the spirit of the eighteenth century—art became the tumbling mat of the true, the beautiful, and the good, which in aesthetic reflection forced valuable art out of the way of what the broad, polluted mainstream of spirit drew in its current.[33]

Anders has his own reflection on human dignity, and this, he argues, is abrogated by the total mobilization of humanity for the purposes of the current post-war, post-atomic, and now we can underscore as transition: today's transhuman, in an age of Lockdown, ongoing or relaxed, masked and socially distanced life. As Anders reflects, we have the ability to repress the events of the past, foregrounding one of the most conspicuous of these events and one that continues to be repressed to this day, "the events of Auschwitz and Hiroshima."[34] But the repression or forgetting is less Anders' concern, as he writes, than the fact that the repeatability of these events cannot be similarly excluded from our awareness, oblique or dark as this is. In an alarming way, so Anders argues, we have consummated what Ivan Illich regarded as the expropriation of death,[35] rending the human being as exactly outdated as Anders argues here, using not the term "Antiquierten," *antiquated*, but "*obsoleten*," an obsolete option *qua* "natural death."[36]

I have argued that the efforts of the theorists of transhumanism are ways of celebrating humanism by other means, sponsored by corporate interests to be sure. Thus, robot rights, AI values, and cyborg and transhuman configurations turn out to be human, all too human,[37] made by ourselves and fashioned in our own image as opposed to what is genuinely other than ourselves and other than what we have made, whether it be plant or animal or other "life."[38]

The agenda is clearly that of the original Frankfurt School. Adorno points out the ideological dangers already inherent in categorizing nothing more neutral, seemingly, than art movements via "isms" and suchlike,[39] and yet, and by general contrast, the term "anthropologism" (and "psychologism") foregrounds, as any "ism" can do, a deficiency or lack.

To this extent, it is useful to recall as Martin Kusch reflects that:

[a]ccording to Wundt, Husserl had exchanged psychologism for logicism. Wundt defined the two positions as mirror images of one another such that things would not seem to have progressed much since 1884: "Psychologism wants to turn logic into psychology; logicism wants to turn psychology into logic."[40]

Heidegger on Dasein and the *Seinsfrage*

It is crucial to note the risks of equivocation. This is especially true in the face of Adorno's reading of both Anders and Heidegger, characterizing the persistence of the so-called "the existentialist misunderstanding of *Being and Time*," as Reiner Schürmann (referring to neither Adorno nor Anders who happened for his part to have been influential in the French reception of Heidegger) explains that Sartre selected "some themes from *Being and Time*—being-towards-death, dread, etc.—developing them into a so-called 'ontology of human existence.'"[41]

For Schürmann, by contrast with the French existentialist tradition,

in *Being and Time*, Heidegger is preoccupied with the question of Being as such—whatever that will turn out to mean—and only therefore with the question of Dasein.[42]

Françoise Dastur exploring "The Critique of Anthropologism in Heidegger's Thought" by reading Husserl and Sartre together with the German "romantic" tradition, including Schelling and foregrounding Heidegger's fundamental ontology, goes beyond Schürmann's emphasis.[43] From a different angle, Dan Dahlstrom reminds us that "the suggestion that his thinking is alien to humanism seems *prima facie* wrongheaded and Heidegger says as much."[44]

Note again that for Heidegger, what is at stake concerns ontological, even scientific, as Heidegger always remains a Kantian, rigour. How can one raise the question of being *as a science*? How can one ask about it, *methodologically* speaking? Heidegger's first reflections begin here but it is always worth recalling Nietzsche's claim to have been the first "to raise the question of science as a question."[45] Nietzsche makes this claim in his (likewise Kantian) "Attempt at a Self-Critique," appended to his first book on tragedy concerning what Nietzsche named "aesthetic science" [*ästhetische Wissenschaft*]. In addition to his distinctive focus on questioning, Heidegger raises, as Husserl also does, the question of the human sciences,[46] noting that these depend for their data, that is, their least interpreted "facts," upon a prior or pregiven "foundational" conception of that same science itself. It is at that foundation level that Heidegger reflects

> heretofore our information about primitives has been provided by ethnology. And ethnology operates with definite preliminary conceptions and interpretations of human Dasein in general, even in first "receiving" its material, and in sifting it and working it up. Whether the everyday psychology or even the scientific psychology and sociology which the ethnologist brings with him can provide any scientific assurance that we can have proper access to the phenomena we are studying, and can interpret them and transmit them in the right way, has not yet been established. (SZ, 51)

"Anthropology" thus presupposes a defining orientation such that that anthropology stands in need of hermeneutic phenomenology properly conceived: "Ethnology itself already presupposes as its clue an inadequate analytic of Dasein" (SZ, 51). Thus, Heidegger emphasizes that "neither" the human sciences nor the so-called "positive sciences" can

> or should wait for the ontological labours of philosophy to be done, the further course of research will not take the form of an "advance" but will be accomplished by recapitulating what has already been ontically discovered, and by purifying it in a way which is ontologically more transparent. (SZ, 51)

Referring to human beings in the context of the humane sciences—*Geisteswissenschaften*—Heidegger invokes the social world by speaking of *das Man* as of a *Menschending*, a "human thing."

"Who," Then, is Dasein?

It was earlier useful to underline Anders' (fairly Schelerian) reference to the traditional distinction between *What-* and *Who-*questions. Emphasizing that Dasein is "that entity which in its Being has this very Being as an issue" (SZ, 42), at issue ultimately is the insufficiently pressed question concerning whether Dasein is to be limited to the human as such. For Schürmann, as noted earlier, the human *is* the "subject." This subjective dimension informs metaphysics, ideology, and religious sensibility. Thus, many scholars suggest that only human beings are "able" to be *Dasein*, by which is often meant, and some say as much, the literal there, the *Da* of Being.[47] But Heidegger himself is a little different, as he opens the question of the I that he foregrounds in his famous book, as the I-connect, the I-myself self of Dasein, the subject of his interrogation of the same preoccupation of Dasein with the issue of its own being, quite as this may turn out, in the conceit of the questions Heidegger didactically proposes to ask, to be other-than supposed or assumed.[48] Theology with all its advantages and all its aporia follows hard on the heels of this train of inquiry.[49]

On Schürmann's account, there is an additional reading that looks to a different conception of the subject, if all such readings "locate *Being and Time* within the tradition of the philosophy of subjectivity."[50] Thus, the great bulk of *Being and Time* seems to be "about" *Dasein*. As Schürmann counts for the reader: "of the 83 sections of *Being and Time*, 75 deal with an analysis of what Heidegger calls *Dasein*, for which there seems to be no English equivalent."[51] Indeed: once we undertake to read those seventy-five sections, our chances of recalling Heidegger's point of departure in his first eight sections "are often more or less forgotten."[52]

Rigorously, for his part, note, once again Schürmann's point that Heidegger

> is preoccupied with the question of Being as such—whatever that will turn out to mean—and only therefore with the question of Dasein.[53]

The "therefore" is scholastically key. By taking Heidegger at his word, contra William Richardson's convention of Heidegger I and II, to the extent that Heidegger himself begins by denying that "there is a break in his thought" or that he abandons "the intention of *Being and Time*," Schürmann is able to sidestep the division of Heidegger into a I and a II, arguing as the Black Notebooks would only seem to confirm in retrospect, that we will need to take the later writings into account in reading *Being and Time*.

At issue then is the question less of existentialism (and humanism) than anthropology, quite as Husserl seems to object. Accordingly, Heidegger's *Being and Time* elaborates a *hermeneutics* of phenomenology. For all the emphasis on Aristotle and given his own research on scholastic or school logicians, for all the references that we would/should be following with respect to Descartes and Kant, and hence and therefore to Husserl, Heidegger's project follows a dialectical schema, in terms of this same preliminary exposition. But just this leaves him vulnerable to Anders' critique.

On Animal *Dasein*: Anders and "Other" Others

Throughout his provocative and fairly unreceived career, Anders urged a differently minded move, not unlike Adorno's when it comes to what Adorno named a life "wrongly lived."⁵⁴ To Heidegger's hermeneutic approach to phenomenology, Anders adds an engagement with spirit, not as a reference to Hegel but rather to the religious, to deity, including a reference to Augustine and the biblical tradition along with angels, dark and light. In this angelic spirit, when Heidegger emphasizes that he means to attribute no kind of demonry to the essence of technology,⁵⁵ Anders, does not flinch at the notion of the demonic, specifically the idea of the devil, affirming the religious dimension silenced as to deity in the factic wake of history, in its complexity and essential to any philosophy meaning to consider both the dominion of the world and the damage wrought by beings such as ourselves, inasmuch as humanity, as Anders understands the human condition, has far less to do with "being" as such than with "having" which Anders takes to include reflection on what has been done:

> With these formulas—which also define our *status religioso*—a fracture in our existence (and for the first time, our current existence) has been described, a disjunction, which surpasses in importance or, more precisely, makes the fracture that once existed between flesh and spirit, or between duty and inclination, or however such differences that were once considered to be so decisive might be denominated, not appear to be so serious. What is our "capacity" for robbery or adultery, or blasphemy or murder compared with our "capacity" to commit genocide or, even worse (I must introduce this term), *globicide*? (AM II, 410)

We will need the greater part of the chapters to follow to begin to be able to consider this claim as set at the end of Anders' second volume on the outdated human.

For the moment, we note that it is our religious legacy that affords humanistic licence for global destruction; thus, Anders describes our "Promethean" "capacity"—the language is calculatedly Kantian—as *fait accompli*. To this day, we regard climate change, in the age of the Anthropocene, as having the character of revelation, a circumstance in which we somehow find ourselves and about which something might be done and today's health pandemic as it is named as such (following certain mathematical models, which means that it is more a matter of projection or anticipation such that every measure undertaken is done preventively, or, as Fuller/Lipinsky would say, in a "proactionary" mode). This is a "soft" or social media war (of all the small people against all the small people). Thus, in his introduction to his "three industrial revolutions," in a note to a section tellingly titled "Post-civilizational Cannibalism," Anders writes:

> This terrible general-license, which renders nothing taboo apart from the human and which assumes that everything has been created for the human, that is, that everything is at his disposal, has never existed apart from the monotheistic domain of the Judeo-Christian tradition (Genesis 1, 26-28): neither in the systems of the magi nor in the multifarious systems of polytheism. This is *the* defect of our "Western" ethic. Only in the framework of the anthropocentric tradition in which the world was regarded as "subordinated" to the human being, as servant, object, and means of survival; and in which the human, although still *creatura*, was not regarded as part of nature but as unlimited lord of all creation; solely

within this frame could natural science arise and with it technology and with it, finally, industrialism. That the human being should be the *goal* and the world *a means*, such anthropocentrism was the common denominator (rarely interrupted by pantheistic *intermezzos*) of the European philosophies and vulgar worldviews, whose innumerable differences are hardly significant by comparison with their commonalities. (AM II, 433)

If Anders emphasizes animals in a passage from the *Kirschenschlacht*,[56] the animal is not his concern. Rather his argument there, articulated in tribute to and in memory of Hannah Arendt, is that we have not progressed beyond the post-Renaissance heirs of Copernicus precisely in terms of biology, *qua* students of post-Darwinian evolution. We human beings are those (unfinished) "animals" who do not hold themselves to be "animals," supposing as we do that the universe itself, the globe was either specifically *designed* for us or else that it is effectively "ours" by violent default: acquisition.

Despite Anders' foregrounding of our thorough anthropocentrism, together with our fantasy of what we should, by rights be, and hence our "shame" at having been born, in our "*natum esse*" (AM I, 24), Anders himself is a humanist. To this extent, Anders remains closer to Heidegger (and Arendt, although he for his part focuses on "fathering") than he is to Adorno. Thus "incarnationist," Anders critiques Heidegger's "concreteness" as a seeming or pseudo-concreteness. For Anders, "nature" is ontologically ordered to Dasein for Heidegger, emphasizing further on the level of embodiment that Heidegger's analysis in *Being and Time* despite its explicit focus on being in the world, leaves out (this is not a matter of "bracketing") questions of hunger *and sex*.[57]

To be sure, this too we will need to unpack, but earlier I noted that what is at issue when it comes to animals is not the liberal notion of intelligence (nearly always disappointingly defined) or "moral-political agency" (a stipulated term defined in such a way that even *human beings* can be excluded, such that the whole point seems to be about privilege); the closest philosophical convention for this concern is Kantian, a bloodless and theoretical respect for the dignity of animals *an sich*, and as "other" beings.[58]

Elsewhere I draw attention to the importance of Heidegger's discussion of mechanized or industrialized agriculture given its all-too-literal force: agribusiness, the meat industry as such, is in fact, as Heidegger writes in *Das Ge-Stell*, "the manufacture of corpses."[59] Note too that is useful to follow Adorno's observation when he notes the logic of our inattention, what we manage not to notice *whenever* we dismiss what is done to animals by saying (by simply thinking) that, and after all, they're "just animals."

The locus here is Adorno's *Minima Moralia* in the title of an aphorism encapsulating Levinas' reflections on the look, the regard, the gaze. "*People are looking at you*." Derrida to be sure takes this from Adorno.[60] The context, as this is Adorno we are reading and his style escapes most readings, is elliptical or difficult to the extent that Adorno emphasizes "scotosis," blindness, as key to anti-Semitism, *qua* constitutive failure to see "Jews as human beings."[61]

> The ceaselessly recurrent expression that savages, blacks, Japanese resemble animals, or something like apes, already contains the key to the pogrom. The possibility of pogroms is decided in the moment when the gaze of a fatally-wounded animal falls on a human being. The defiance with which he repels this gaze—"after all, it's only an animal"—reappears irresistibly in cruelties done to human beings, the perpetrators having again and again to reassure themselves that it is "only an animal," because they could never fully believe this even of animals.[62]

Anders offers a dialogical reflection on "phantasmatic" realism, as what is via imagination and concomitant inattention given to be seen which is by the same token unseen, inconspicuous. The language ranges between Adorno and the same Levinas who, to be sure, translated Anders into French[63]:

> Your face. Or mine. In mine you know *me*. In yours I see *you*. With all things, more or less, as all of them have a face. Animals too. Machines too. Also a house. Even a summer's day. (AM II, 322)[64]

Unlike Heidegger who focuses on the human in order, from the positionality of the most proximate, to pose the question of Augustinian immediacy concerning the *who* that I am myself, for the sake of raising the question of being *qua* being, Anders is focused on the human *qua* human but without the usual concessions human beings tend to grant themselves: an enduring state of exception.

Anders glosses, and this is just in passing and does not quite count as a focal concern as it does for Adorno, Kant's contention in his *Lectures on Anthropology*, that with respect to "[irrational animals] one may deal and dispose at one's Discretion" (AM II, 433), pointing out in passing that today's whalers as indeed recreation or trophy hunters, including "collection" in the name of science, be it for university labs or for museums, quite as much as for the fishing industry, could invoke the same claim to justify their dealings. Thus, we saw that Anders emphasizes the crucial importance of the Judaeo-Christian tradition as assuming that "everything is created for the human being" (AM II, 433).

Nietzsche foregrounds a similarly key consonance between Western thinking in religious and moral values and Western science. Anders, as noted earlier, takes this in the same critical direction, if not towards, a "genealogy" of morals but as indicative of deficient or "negative ethics."

This Judaeo-Christian tradition of anthropocentrism Anders connects, as Nietzsche does, with Western science and technology, but Anders goes further than Nietzsche because the connection with the Western tradition turns out to be, as we well know, compatible with both theism *and* atheism:

> Today, of course, the natural sciences and technology, which would never have come into existence lacking theological anthropocentrism, have also found ground among those peoples, such as the Japanese, for example, that did not originally possess the requisite theological presuppositions for them. However, these presuppositions have also long been forgotten in Judeo-Christian cultural

circles. Moreover, the technocratic countries are no longer united by a single faith; to the contrary, what unites them is (rarely as articulated but exercised) *atheism* that (despite the occasional proclamations of faith on the part of physicists) is the basis of the natural sciences. (AM II, 433, emphasis added)[65]

Sein and *Dasein* without God

Anders faults Heidegger for excluding elements of hunger and sex, excluding the incarnate body with all its needs from his thematization of that Dasein for whom its being is at issue. Anders also shows that those human cares can tend to be concerns that are not at issue for Heidegger. If Heidegger's focus on the tragic question of humanity in his *Introduction to Metaphysics* foregrounds the uncanny, this is because his question is as as Schürmann says the "being question," hence neither the existentialist question of the human nor, indeed, as some theologians claim, "meaning."

What excites our interest in reading Heidegger goes beyond the where, the here, and the there of our being on this earth to touch the gods in flight. It is where Heidegger reads Nietzsche on the unfinished animal and it is where Heidegger reads Hölderlin, as he speaks of divinities in their passage, calling to them, where are you? *wo bist du?* To inspire readers in this fashion is no small achievement. We find it hard as Nietzsche says to see beyond our own shadow, and Heidegger seems, for some readers, to bring us a little into a certain light with his focus on the transcendental *schlechthin*. Thus, in *Being and Time*, as Dan Dahlstrom points out:

> the transcendence that makes up the very being of being-here encompasses a relation to oneself as well as a correlative relatedness to the world at large. Heidegger attempts to capture this distinctive transcendence with the metonym, "being-in-the-world."[66]

Still and at the same time, we remain lost. If the thinkers of existentialism sought in Sartre's voice at least to claim their movement as a "humanism," and if the human face became the cause of a philosophical generation, we are now in the age of Anthropocene, which is more than even Anders had imagined it, the Anthropo-obscene confronted with a different, another, and more invidious way to read the reflection Heidegger borrows from his friend, the physicist Werner Heisenberg, to make his own, to close his *Question Concerning Technology*, one of the key places he speaks decisively of the human, beginning his essay with a chained reference, unmarked, to Rousseau as also to, on the side of the promise of cultural redemption, Schiller. Few of Heidegger's readers have come to terms with Heidegger on technology but foremost among those who have, and both take a Marxist point of view, are Kostas Axelos[67] and Dominique Janicaud in his *Powers of Rationality*.[68] Perhaps this is a fault of translation or of philosophical fashions but what is at stake concerns the history of reception, meaning, once again the lack of interlocutors. Thus, if Janicaud invokes Illich, he still seeks as his title suggests a different envisioning of the future, de-technized, and his imagined interlocutors are analytic or mainstream, regrettable as mainstream scholars do not engage his work, and Janicaud was simply absorbed and co-opted into what he condemned as the "theological" turn.

Heidegger introduces his reflection with a reference to Nietzsche who names humanity a "skin disease of the earth," emphasizing the hopping and the blinking tendencies of that same terrestrial "inflammation." This is arrogance, like the tiny flying creature Nietzsche invokes at the beginning of his essay on truth and lying, arguing from a rather different point of view than does Wittgenstein when he speaks of the lion, that like ourselves, as Nietzsche writes:

> If we could but communicate with the Mosquito [*Mücke*], we would learn that it too swims through the air with this same pathos and feels within itself the flying centre of this world. (*Über Wahrheit und Lüge*)

Heidegger, speaking of what he names the "danger," "not just any danger but danger as such," explains that the human is "endangered from out of destining" (QCT, 26). Here the "fall" that threatens is a human inversion, as the human "precisely as the one so threatened, exalts himself to the posture of lord of the earth" (QCT, 27). It is no accident that Heidegger cites Heisenberg's "*Das Naturbild*," the same Heisenberg who articulates the Copenhagen interpretation of quantum mechanics as the Uncertainty Principle,[69] whereby

> the real must present itself to contemporary humanity in this way. In truth, however, precisely nowhere does the human today encounter himself, i.e., his essence. (QCT, 27)

The structure that thus unfolds, the "*Ge-Stell*," is occluded for the human, preoccupied as the human being is with himself in every case,

> he fails to see himself as the one spoken to, and hence also fails in every way to hear in what respect he ek-sists, from out of his essence, in the realm of an exhortation or address, and thus can never encounter only himself. (QCT, 27)

Heidegger challenges human exceptionalism, "For there is no such thing as a human who, solely of himself, is only human" (QCT, 31). By contrast, Heidegger thinks Hölderlin's reflexive word, *Dichterisch wohnet der Mensch*. The "questioning that is the piety of thought" (QCT, 35) is to be thought in "the presence of the gods, bringing the dialogue of divine and human destinings, to radiance" (34). In this spirit, there is no way to read the final line of Heidegger's "The Turning" other than as a kind of archaic, gnomic prayer:

> May world in its worlding be the nearest of all nearing that nears, as it brings the truth of being near to the essence of the human, and so gives the human to belong to the disclosing bringing to pass that is a bringing into its own. (35)

One can pray as Heidegger seems to do, to whatever divinity or else, absent piety, one can reflect on enlightenment without enlightenment, as Anders did in 1936, in the ambit of Heidegger's lecture *On the Origin of the Work of Art*, as of the proletariat, as Anders writes, without proletarian consciousness, the work without a worker, which is also to say, "The self-made man as mystic."⁷⁰

If Heidegger kept as clear of critical theory's disenchanted enlightenment as of the real implications of political rhetoric, Anders took a lifetime to engage the horrors of war and not less the real danger that is academic blindness to the same real, all-too-real horrors, foregrounded in Anders' 1985 reflections not so much via Heidegger but on the blindness, be it incidental or deliberate, that also characterized Max Scheler and Georg Simmel—both of whom advocated on behalf of war—cheering not only its outset but throughout (KS 54). "And that they were sufficiently clever enough to have found the means," Anders writes, "to have *remained* so very naïve? In order to allow themselves to be abused as bona fide free whores?" (AM II, 407).

The force of this language can surprise us—it is written late in life, hyperbolically, provocatively, more so perhaps than even Marcuse or Adorno, as Anders kept well to the side of all academic spheres, as it can seem (not having an appointment has this as consequence). Nevertheless, calling thinkers like Scheler *and* Simmel "whores" seems excessive (and complicated enough that we will return to it). But the language is deliberate. Thus, the reference to calculated advantage cuts to the quick of the problem, as Max Weber already underlines this in his *Wissenschaft als Beruf/Science as Vocation*, and as Lucian had already written in the second century AD in his *Philosophies for Sale*, which will always be about the practice of philosophy as profession: a paid job rather than a vocation one follows, paid or not, for the love of it, as Anders did. To the same extent, the great majority, that would be nearly all philosophers in Germany during the war, as everywhere else before and after the war, tended to take the side of what the great naïve of the 1960s and 1970s of the last century called "the establishment."

For Anders, "Scheler's dictum" was crystallized in the contrast between his own disposition to "believe" as he did "in the devil (in contrast to the theologians of his own generation who believed in the existence of god but not the devil)" (AM I, 407). Dependent on Scheler and Heidegger, as his teachers, Anders also invokes some of the most influential voices of his generation (his invocation would irritate Hannah Arendt on the same topic), citing Denis de Rougement on the idea of the devil in America society:

> The Devil's first trick, remarks André Gide, is to make us believe that he does not exist. This trick has never better succeeded than in the modern epoch. All America has fallen into the snare.⁷¹

Once upon a time, not only with Arendt but also with Maritain, and centuries of Augustinians and Thomists, not to mention the refinements of the Dominicans,⁷² the question of evil was significant. More recently, tracking the analytic history of philosophy, the question returns with Susan Neiman, upending the sense of Nietzsche's conception of evil, along with the sometimes as harmless but always by no means incidentally Hegelian question of ugliness.⁷³ Thus, the question of evil had been a theme historically and after Arendt, for the Holocaust, but otherwise, in an age of nonbelievers, it became an "outmoded" question that fit into notions of Gnosticism and magically anticipatory catastrophism—as if activism against the bomb post-Hiroshima was somehow an overstatement. Recently, Covid-19 shut down the Vatican itself, conspicuously enough for those interested in theological punctuations, on Easter Sunday 2020, and all other churches (synagogues, temples, mosques) along with it. Thus Anders' formula concerning the devil's new apartment or address has to be read in an era that takes zombie pandemics in stride, virtually speaking with, *The Walking Dead*, along with a literal televised version of the apocalypse for prime time, with the 2019 BBC series *Good Omens*, and the deaths of collective folk deities with the still-running 2017 cable television series *American Gods* and HBO's little-noted 2003 and 2005 *Carnivàle*, along with *Twilight* and *True Blood* and the magical realm of *Hogwarts*, all spells and wand-waving included. And why ever not? We have had orcs and half-orcs for years. What was once fiction for readers of Tolkien, amused provocation for C.S. Lewis with his version of Wormwood, very different from Alan Rickman's invocation, is more harmlessly, as we suppose, lodged in our collective unconscious *gaming* fantasies. I will return to this at the end, as it concerns what Wolfgang Palaver, citing Pierre Dupuy's "enlightened catastrophism," characterizes as a "prophylactic apocalypse."[74] We can read as we do about disaster capitalism as about surveillance capitalism because catastrophe capitalism, pharmaceutical and quasi-militarized enforced home-confinement, is (or can easily again be) the order of the day.

To cite Baudelaire (the source, *pace* Arendt,[75] for de Rougemont's repurposing of Gide):

My dear brothers, never forget, as you undertake to vaunt enlightenment progress, that the most beautiful of the Devil's tricks is to persuade you that he doesn't exist.[76]

Speaking of the devil, we have the same saying in English, is a German commonplace, if it also has, post Luther, a vulgar dimensionality to it. Anders takes this over as well from Goethe, who writes his Mephisto with a touch of urbanity and bumbling pathos, prerequisite for sympathy well beyond Milton's tragic Satan in *Paradise Lost*. If Anders also draws on Scheler to invoke the devil, the formula may be found in Kafka and in Buber's 1953 expression of the age of God's eclipse[77] along with Sartre's more existentialist writerly vision of nihilism, which is Hell itself (and the durably wretched claim that other people suffice for this) or the classic contrast between *Le Diable et le bon Dieu*. This is Anders' reference to the devil's "new address" (AM II, 410). For Anders: we disattend to the category of evil quite where we think we have nailed it down with Eichmann or indeed and after the Black Notebooks, with Heidegger. The point here, as both Bloch[78] and Anders remind us, is that by assigning the lion's share of evil to Hitler/Heidegger, one may be seduced into the confident illusion, we have this in the meme of the moment—*hunting* Nazis, *punching* Nazis—that one had thereby permanently categorized and enclosed it.

If everything that is to be done is in a sense, to repeat Adorno's expression as so many emphasize this, undertaken in order to ensure that "it" never happens again, one has created a pseudo primal scene: one that can be neatly cordoned off, identified, as if on an old map of ocean monsters: there is where evil lurks, but here is where we find ourselves. Do we talk about Hiroshima? Nagasaki? Fukushima? Do we talk about 'climate change' or do we mention, as few do apart to be sure from Peter Sloterdijk, the ongoing and accelerating circumstance of both weather control and weather multipliers effected by spraying accelerants on trees for years and years in the fire disasters, in California, in Greece, in Australia?

Who today, which academic, which philosopher, here and now, that is since March of 2020—and again apart from a sole exception, in this case: Giorgio Agamben[79]—raises the question of what we have done and are doing to the old and the sick, in an era of social media programmed and thus organized hysteria, pro *mandatory* masks, pro *mandatory* testing, pro mandatory, that means forced, vaccinations, all in response to a single pathogen somehow supposed singularly responsible for all death as such (this is a stunning complicity in statistics) and for overwhelming medical facilities *in potential*, unless all manner of fascist measures be taken on a global scale—including individual self-incarceration, lockdown and quarantine, the prohibition of movement and social interaction of the most everyday kind, in addition to the pursuit of one's livelihood, including art and theatre and face-to-face teaching.[80]

"Human," as we are, are we, as we prefer to see ourselves, allied with the divine? Image and likeness? As Nietzsche mused in recollecting his youthful theodistical variations on good and evil, as Anders contemplated the fact not merely of Auschwitz but also of Hiroshima and Nagasaki and the ongoing instantiation of nuclear violence via nuclear reactors in the then case of Chernobyl (which is still with us and today we also have Fukushima, but the point is not to talk about it, as he maintained in his *Gewalt, Ja oder Nein?*[81]), is there evil within us? The tradition that sets the devil as the "prince of the world"—this is not by accident the title of Jacob Taubes' edited collection on Carl Schmitt[82]—seems frighteningly fitting, again consider only the burning of Australia with wildfires abetted by months and years of chemtrail accelerants in the trees followed by the wholesale persecution of its wildlife, nothing seems to hinder our willingness, as Anders argues, to do the devil's work for him.

3

Günther Anders and Hannah Arendt

Love, Triangles, and the Political

Philosophical Triangles: Students, Scholars, Lovers

If love matches in philosophy are fairly rare—such that one can name the famous pairs pretty much on the fingers of one hand—philosophy is named for love as the *love* of wisdom. Indeed, some scholars note the physicality of love, beginning with Plato, quite as what can draw one beyond the body, a redemptive erotic that suffuses, beautifully, Plato's *Phaedrus*. Indeed, and as already noted, a 'concretized' attunement to the erotic as such inspires Anders in his reflections on Heidegger's "pseudo-concreteness," indicting Heidegger's *Dasein* as monkish, as sexless: ascetic. This conflicted point is sufficiently nuanced that I will need to come back to it in a later chapter.

Love affairs, at the same time, are by their nature, personal matters, if they also seem to make philosophy more human. Perhaps for this reason, despite the spareness of such pairs, we overstate love matches in the history of philosophy like Socrates and Diotima, as variations adumbrated by analytic readings are currently making the rounds on the internet. Apart from Socrates' famously self-ascribed friendship with his Mantinean hetaira (how, one wonders, knowing as we do how things were arranged with Thrasymachus, did he secure the wherewithal needed to recompense her time and her teaching?), literary (and phantasmatic readings include Hölderlin's *Hyperion*, and a range of unrestrained associations),[1] as well as the dramatic (and devastating), like Abelard and Heloïse,[2] or, at the level of unrequited passions, Nietzsche and Lou,[3] or else, as one dives deeper, Lou and Rainer Maria Rilke for more mutuality (if less philosophy),[4] to the sublime: Sartre and de Beauvoir.[5] To the list, as Giorgio Agamben takes care to remind us, may also be named Heidegger and Hannah Arendt,[6] for a love affair that spanned half a century, to which we may add, triangle-style, more muted but as long lasting, by way of letters and the like, fifty years, so Anders tells us: Anders and Arendt.[7]

I have been at pains to underline the considerable difficulties that cannot but complicate writing a biography of Anders by contrast with Heidegger, who taught a range of influential thinkers including both Anders and Arendt. It matters to note that Heidegger was a famous opponent of the personal *an sich*, as a follower of the Aristotelian dictates of *doxa*. For Heidegger, a thinker thinks only one thought. But the complex thinking in Anders requires as much of the personal as one can get. *In der Mitte sitzt das Dasein* [literally: in the middle sits the Dasein]—to riff a bit on Lüdger Lütkehaus' extraordinary title[8]—may capture the elementality in question, as Lütkehaus reflects on Anders and Sloterdijk, triangulated, no love lost and quite intending the pun of it, with Heidegger.

At the time, Günther Stern/Anders met Hannah Arendt in 1925 in Marburg in a seminar taught by Martin Heidegger with whom Stern had already studied in Freiburg, where he wrote his doctoral dissertation with Husserl. The coordination could not be closer. At the time, Arendt herself was already absorbed by the claims of her then-clandestine relationship with Heidegger. To this extent, the tension of the triangle adumbrated the relationship between Arendt and Stern/Anders from the outset, and, arguably, such triangulation remained between Arendt and her second husband Heinrich Blücher, by then a double triangle quite to the extent that Heidegger remained a love interest for Arendt. Hence, on her own account of it, rather than admiring Anders as a possible love option—although it matters on the lived, phenomenological level that Arendt would notice Anders' physicality—he could do handstands!—Arendt instead aspired to taking Anders' place in easy and public conversation with Heidegger as Arendt found herself jealous of his freedom to walk *and* to talk with Heidegger in the open.

It is the freedom to enjoy conversation that will always matter when it comes to affective desire, that is, the life of the heart and the mind as the erotic dimension includes both. The poet and classics scholar Anne Carson works out the lines of visual, perceptual, affective observation, and its pitfalls and its tricks, in a chapter entitled "Ruse" in her book *Eros, the Bittersweet*. Carson's book is dedicated to Sappho from whose lines the title is taken, yet most of the chapters focus on poets like Archilochos and Anacreon (and my students always notice Catullus). By contrast, this beautifully complicated chapter reads Sappho's Fragment 31: "He seems to me equal to the gods that man who opposite you sits and listens close to your sweet speaking."[9]

The comparison with Sappho, given Carson's reading as she explains "the geometrical figure formed by their perceptions of one another, and the gaps in that perception," matches quite to the level of the voice, first person, in the above account of Arendt observing Heidegger and Anders. Thus, we note Carson's analysis of Sappho's poem: "Thin lines of force coordinate the three of them. ... The figure is a triangle."[10] To this extent, the detail, self-recorded, self-reported, that Arendt herself saw, and envied, the two, Heidegger and Anders, talking in public and seen from a covert position, herself unseen, echoes the structural lines (although hardly the affect) of what is, arguably, one of the earliest expressly philosophical triangles in history, staged to be sure, as we are told from the outset that it is staged, as illustrated in sculpture and woodcuts.

I refer to the classic image (Figure 3), unfortunately but instructively entitled "*Weibermacht*," as it gives us Aristotle captivated by Alexander's beloved, and for this reason submitting to her wish, originally instigated by Alexander, to play "horsey," thus pretending to ride on the old philosopher's back as he crawled on his hands and knees to Alexander's covert observation—and hilarity—thus the triangle, not of jealousy but derision.

Figure 3 Hans Baldung Grien, "Weibermacht." Woodcut of Phyllis seated on Aristotle on all fours, with Alexander peering from the parapet. 1513. Wikicommons. Public domain.

The same triangular motif is so very influential that we are told that Nietzsche would invoke it for the sake of arranging as he did, seemingly in an echo of Alexander's instigation, the troika photograph of himself and Paul Rée, the two of them hitched to a cart in which a crouching Lou Salomé was posed with a whip entwined with flowers.[11]

It is this triangular context we would add to the further note that, as Kerstin Putz remarks, their wedding in September of 1929 "surprised" both family and friends.[12] With or without a reference to Sappho or Aristotle, triangulation as a structure nonetheless followed Arendt *and* Anders as they continued to have other liaisons and other marriages all without losing contact with one another. The rest is difficult to assess with any certainty, for many reasons but not less because such matters, to use Arendt's own language, are affairs that transpire *"unter vier Augen,"*[13] doubling the title of Thomas Hardy's *A Pair of Blue Eyes*, marked by Alexander Nehamas' formula as "two pairs of eyes," as Nehamas discusses attraction and desire — and beauty — in *Only a Promise of Happiness.*[14]

If we read their correspondence and take care to note the dates, not unlike the date stamping that increasingly marks our internet-inveigled lives, the intervals add and subtract intensity, a sign of a dead letter, a sign of unvalued presence. And then there is the matter of different styles between the two correspondents. Arendt, more successful by far than Anders, was also promoted and protected by Anders. Thus, Anders worked to get her out of Germany and then to arrange sponsorship, as she writes to him with some baffled reproach, on June 4, 1941, as if he had "prepared [her] arrival as if" she were "who knows who."¹⁵ Anders' efforts on her behalf succeeded, which also meant that she would come to have increasingly less time, a detail both she and Anders emphasize in their correspondence.

In addition, but now we tread on the completely invisible to reconstructive attempt, save liminally, there were telephone calls. As Merleau-Ponty reminds us, in a negative modality, the reductive efficacy of technology works to attenuate being by magnification, in this case via the *voice* of the other on the telephone—Joan Baez also has a song, *Diamonds and Rust*, about this immediacy and distance—that brings them to us, in their presence, a bringing that is to be sure, and this is one of the dangers of modern media programming via the internet, a summons: a call to recollection.¹⁶

We need Merleau-Ponty, who offers an ontology by contrast with an operational account of language, as Merleau-Ponty's account, concerned as it is with the mediation of telephone and the voice, turns on exchange. This exchange is key to the letters between Arendt and Anders, as intervals into which telephone calls and meetings must also be counted but which, owing to the nature of a volume of correspondence, we can read, and hence we are beguiled into thinking that phone calls of which there is, like personal meetings, one to one, which is one of the reasons social distancing was so very critical as psyop, no record, can for the same reasons of this lack of evidence, be discounted, as if they never took place. And yet, like face-to-face encounters, the phone maintains the life, the bodily presence of the other, by way of voice. And this was true for Arendt and Anders and for any twentieth-century love affair. As Merleau-Ponty writes, although he is not here foregrounding an affective context,

> A friend's speech over the telephone brings us the friend himself, as if he were wholly present in that manner of calling and saying goodbye to us, of beginning and ending his sentences, and of carrying on the conversation through things left unsaid.¹⁷

To this we add the reflections of the Italian philosopher, Giuseppina Moneta, as she articulates this in an extraordinary, if lapidary, contribution to a *Festschrift* for William J. Richardson, S.J. Moneta articulates the spacing inherent to and in dialogues *between* philosophical friends, those attuned to one another. This emphasis echoes the kind of attunedness key to Anders' phenomenological reflection on music as he underscores listening — the context is Heideggerian — for the sake of a specific *Mitdasein*, *Mitwelt*, for the listener who is a specific listener, in Anders case, one affectively, intellectually, creatively attuned to music. In this way, to parallel friends who are also philosophers, we can recall the subtext of Anders' *Kirschenschlacht*, as it offers a description of "listening to another dimension of awareness."¹⁸ For Moneta, "the soundless distances between words and sentences," calls forth and bridges the distances of time, intervals punctuating communications between lifelong friends, characterizing the correspondence between Arendt and Anders and so on, as Moneta was a friend to Bill Richardson, a Jesuit priest and expert on Heidegger and no less as she was also a friend of the Hannah Arendt expert, Jerry Kohn, and to the current author, given Pina Moneta's visits, almost a perambulation, annually, between Rome and New York City. As Moneta noted, one notices, quite as Merleau-Ponty also details in his own phenomenological account, the "voices of silence."

These for Moneta offer a staccato "interval of muteness." Here it is important that such muteness adumbrates all exchange as such, no matter whether face to face, on the phone, or correspondence via letter. One sends a postcard and the other saves this forever; one says more than the other, yet, as Moneta writes, the other, keeping silent, gives "tacit yet clear invitation to keep in check the urge of wanting to talk,"[19] and the first one is thus constrained and served by this elective muteness to listen for what is *not said*. In this way, Moneta cites Heidegger, who is for his part, himself silently—*stillschweigend*—citing Nietzsche concerning the "habit of always hearing only what we already understand."[20] For Moneta, offering a reflection that may follow all our efforts with one another in and over time: "The spoken is dissolved as soon as it is uttered and the air seems to have become a solvent to sense and meaning."[21]

The acoustically sensitive, arch-Augustinian Anders—here it should be noted that Moneta's reference is an architectural one as Anders always noted the architectural as *musical*—would have valued such observations. For my part, I think many of us might share a related sensibility when it comes to intimate exchange. For Anders, in the case of one phone call about which we do know, quite because he wrote about it, the immediacy of the contact overwhelmed him. Thus we read a painful letter to Arendt written after his marriage to his third wife, Charlotte Zelka, came to a factical close—they never did divorce—but contact-wise, as Anders recounted under the guise of an apology for his agonized crying on the telephone in a previous conversation,[22] explaining that Zelka had told him that she preferred, for reasons the music-minded Anders could never fathom, to stay in California close to her sister and to content herself with teaching piano rather than performing herself and rather than hearing others perform.[23]

Moneta is forbearingly generous in her reflection on the intervals of communication, between friends, between different and elliptical intentionalities, differing mutualities, reminding us that such intervals exceed gaps or absences. Such "emptiness," as she reminds us, "was not a blank but a stretching out of the spoken: the range and horizon of its resonance."[24] This is the *between*, Heidegger's *zwischen*, that remains or endures between friends who stay in contact. It is the *spacing* of their contact: the absences noted in correspondence, smaller intervals noted quite as much as the decades, to note Arendt's and Anders' later correspondence.

Of their phone calls, as these calls punctuated their letters beginning already in the 1950s, we have no records, although we can suppose as a matter of cost and custom that such early phone calls would perforce have been brief. It is not impossible to imagine that the FBI, as an example of one intelligence agency, might have heard more than we scholars will ever know, given suspicions concerning both Anders and Arendt. And we have little sense of the conversations that took place in their rare later meetings face to face, and what we know of these must be reconstructed, and not everything, as Goethe famously reminds us, is written down.

But there was love. And from reading the letters, as it is a characteristic of romantic and friendly love, we also know that this love affair was adumbrated by a clear sense that one loved more than the other. This is always true if only to the extent that in every love affair there is the one who is the lover and the one who is the beloved, and Anders loved Arendt as beloved, and so "more" in this fashion. At the same time, affectively, this was compatible with Anders taking other lovers for his own part, as Anders was as old fashioned as was Heidegger and as was Hannah Arendt's husband, Blücher, with his extramarital affections, not that (as the #metoo movement might underline) all that much has changed in the interval.²⁵ And although Arendt was desired by many suitors, it was Arendt who loved Heidegger more, as the constancy of her love, over intervals of temporal distance, ensured that the love between Heidegger and Arendt could endure half a century to Arendt's death. But in just the same fashion, Anders would also remain true to Arendt until her death. And in her turn, and in her fashion, Arendt remained in contact, mediated by their once upon love, one flesh, as they had married, with Anders.

Heidegger has dropped out of our reflections on Arendt and Anders and yet he remains part of what remained between them. This contested locus is one marked by negation and critical emphasis and rebuke. In this way, Anders' discussion of Heidegger's "Pseudo-Concreteness" emphasizes that Heidegger lacks any space for affect, writing that Dasein is both bodiless and sexless (PC 349). Here I've sought to underline some question of the differences between love for Arendt and love for Anders to approach the complexity of Anders' reproach (Arendt-Heidegger) as it is not simply a theme that stands on its own. In addition, as I point out in *The Hallelujah Effect*, one must also consider the "work" of desire, eros between men and women, in terms of passion and beauty.

Several scholars have noted Hans Jonas' contention that Arendt's role was seemingly subservient, intellectually, to Anders and of course by extension, though this is less remarkable, to Heidegger.²⁶ Here, too, there is a triangle, as Jonas too, so he tells us—so we also learn from Arendt herself—was one of Arendt's admirers, and jealousy (and the diffidence that can be part of jealousy) may have played a role in Jonas' assessment of the relation between Anders and Arendt. At the same time, there is the fact that the collective treatment of Arendt (the context of record was a meeting in the circle of the early Frankfurt School) is often criticized (and rectified) by scholars, but the sexism that dominates is unabated, and multifariously so.

To this extent, and this too is part of what the Hegel, and Adorno, scholar, Gillian Rose called "love's work,"²⁷ Anders' dedicated affection for Arendt, his clear admiration of her along with his own gifts as a scholar *and* as a "philosopher"—importantly, a term Arendt for her own part would always refuse—meant that Anders edited Arendt's work at the time they were together: and he assured its publication, efforts he undertook along with the support and encouragement of Karl Jaspers. The simple circumstance, the fact that Arendt regarded Anders as editor/corrector would then be part of their relationship as scholars working together on related and joint topics, one more junior in this case than the other: they were a pair, and both were scholars, but they were not equal. This aid to Arendt on Anders' part was vital, and thus Arendt's doctoral dissertation on the concept of—what else?—"love" in Saint Augustine came to light as a published book.

Triangulating the Triangle: Arendt–Heidegger–Anders

Any full reflection on the love affair—and not only the simple fact of the marriage—between Arendt and Anders must await a biography of Anders attentive to his differentiated concerns, hopefully one looking to highlight the affective dimensions in Anders' life, as such tracking is more complicated in Anders' case than it usually is (and it is always complicated as in matters of the heart we tend to lie to ourselves and to others about love, motivations, achievements, disappointments), so much so (this is Freud's great insight as Lacan only complicates it) that we remain strangers to ourselves when it comes to those we love, first loves, the love of a lifetime, a great love, and not less when it comes to wives/companions in Anders' case, given his lovers, and when he was young he was a beautiful man, with three wives to his count. It makes biographical reflections no less easy to underline Anders' personality as he was apparently strikingly difficult to be with, and that is on his own account. Anders was contrary enough, paradoxical enough to continue to act on behalf of his wives, certainly in the case of Arendt, after their divorce, and likewise for his other wives. Thus, Anders remained close as well to Elisabeth Freundlich, his second wife. All the same, he would die in the condition of loneliness he reflected upon throughout his theoretical life.

It is to be sure, as we know, but perhaps we are wrong about this originally, that it is Arendt herself who gives herself the title of a "girl from a distant land," "*Mädchen aus dem Fremde*," thereby, and this is the charm: claiming Schiller's words as her own.[28] Heidegger echoed the claim or did he originate it in conversations between the two of them, which she then took up and he repeated, words between them to express a love between them? Certainly Schiller's lines infused their love, as Heidegger who was fond of poetizing as Anders loved music, wrote a poem for her, as only Heidegger could but also as only Arendt was able to "hear." Thus, Daniel Maier-Katkin includes this as the title and key to his *Stranger from Abroad*: "Stranger from abroad, du, / may you live in the beginning."[29] Anders echoes some of this poetic thunder, underscoring his own literary gifts, as he seeks to recount, all of this only posthumously and in retrospect recounting Arendt's surprise at the same, yet another dimension of triangulation, in Anders' *Kirschenschlacht*.

No different from the majority of studies that have examined the relationship between Arendt and Heidegger, Maier-Katkin downplays Anders, whom Arendt married in 1929 to divorce eight years later in 1937 (Arendt would marry Blücher in 1940). The concern of the story, as it might be sold on the open market, is not the story of a failed or derailed or otherwise misconceived first marriage (Arendt-Stern) but the story of Hannah Arendt and Martin Heidegger, a love affair parsed as it should be, it lasted a lifetime, as *friendship*: a friendship which, given the peculiar resistances of Heidegger's character, entails that Maier-Katkin has recourse to the language of forgiveness. This dual emphasis makes Maier-Katkin's book a contribution to the debate on Arendt and Heidegger, while at the same time rendering the book difficult reading for those interested in their love affair, perhaps given the *frisson* that such a common fact of academic life can exert (think of Jacob Taubes' abusive treatment of his first wife, Susan Taubes, and his one-time lover Ingeborg Bachmann, and his eventually estranged second wife, Margherita von Brentano),[30] or else to wonder at the durability of any love (and any friendship) that stayed the course of half a century: until Arendt herself died in 1975, half a year before Heidegger himself would die in 1976.

The longer lived of either Arendt or Heidegger, Anders published the second volume of the essay reflections comprising his *The Antiquatedness of Humanity*, subtitled *On the*

Destruction of Life in the Age of the Third Industrial Revolution in 1980, four years after Heidegger's death. Similarly, Anders' research concerns, like his formation, differed from the preoccupations that would make Arendt herself a popularly contested figure, politically as well as an indispensable name in political theory following her writings on totalitarianism, on Eichmann, and what she called the "life of the mind." Given these keen theoretical differences, Anders—and note here that I am *not* speaking of Stern—and Arendt intersected via a shared concern with the *human* condition. Thus, Anders can underscore the extent to which Arendt had "stolen" from him and he from her, both of them freely. Thus, Anders can advance a lover's joke, between lovers, highlighting mutual theft of one another's ideas. And this too is part of love affairs.

The very fact that Anders and Arendt remained, arguably, quite as much as Heidegger and Arendt remained, on a "friendly" basis, is clear,[31] although it is no less evident that Heidegger, quite like Blücher,[32] maintained the upper hand in Arendt's affections. But friendship is a funny thing and can be strained between those who have been lovers, especially perhaps those who were once married. If most biographical studies of Arendt focus on her marriage to Blücher, noting only in passing or to date the relevance of her life as "Mrs. Stern,"[33] "Frau Stern" as she is reported in print among and for the adherents of the Frankfurt School (a school which itself would not be particularly friendly to Arendt, and vice versa), Arendt's relationship to Anders tends to go completely by the board: overlooked.

In addition, there is the conflicted dimension of friendship as such. Thus, despite Aristotle's emphasis on friendship as the crown of the virtues and a friend as 'another self', academics tend to worry (mostly to sexualize) friendships of even the most innocuous kind, by which I mean the fictional sort, the mythic sort: Castor and Pollux. So we wonder: Were Tom Sawyer and Huck Finn really just and only *friends*? Was there not an erotic subtext to Mark Twain's fictive friends? So too we likewise wonder about Schiller and Goethe or Nietzsche and Rohde and the ongoing enthusiasm for asserting that Nietzsche was "gay." Was he? Was he not? And how many know of these two—what was it about Rohde—named the "*Dioscuren*" by their teacher, Friedrich Ritschl, and much more commonly well known, the affection—what was that really?—between Adorno and Benjamin (it surely helps to have one of the friends die relatively "young").

Even in the age of Facebook, where the one or two or four or six friends in the course of a lifetime (and that only for *some* lucky souls) routinely morph into hundreds (more for younger users), and especially in an era where friendship and contact are officially proscribed (post-Covid), we question friendship, its basis, its truth. In this measure we may wonder if Arendt and Heidegger were really friends? What kind of love affair was it that they shared? Are there details? And again, and above all, how did it manage to last?

Even more questionable then will be the nature of the love between Arendt and Anders. A marriage that did not last, was it not undergirded by Arendt's "truer" love affair, as it seems patent that she was besmitten, with Heidegger? So, what affection would she have had for Anders? Like friendship, love always needs to beg forgiveness, even as it "keeps no record of wrongs. . . . It always protects, always trusts, always hopes, always perseveres. Love never fails" (1 Corinthians 13).

In philosophy, like Aristotle, we count as real or true only those friendships and loves that endure, or outlast utility or survive the fading of sensual fancy (and of this Elfriede Heidegger was never sure when it came to Heidegger's love affairs, even as Bazon Brock relays a certain factical *diriment impediment* on Heidegger's part).[34] At the same time, it says something about Heidegger that in spite of his character, transparent as he was to his friends, as Karl Jaspers tells Arendt ("here we are, the two best friends he has, and we see right through him"),[35] Heidegger would have friends like Jaspers and like Arendt.

I have underscored that even dedicated discussions of Arendt that focus on love and friendship say little of Anders, who, following his divorce from Arendt, married Elisabeth Freundlich, quite as already noted, in the course of their shared exile in the United States in 1945, returning with her to her native Vienna five years later in 1950 where they lived until their divorce in 1955. And, again, it matters that Anders remained in contact with Freundlich until the end of his life (Freundlich herself would die almost a decade after Anders in 2001).

But having married two intellectuals, the first a scholar of theological philosophy, as Anders regarded Arendt, the other, Freundlich, a writer herself and scholar of theatre, a dedication that ultimately collided to some degree with his own style and absorptions, Anders turned, two years later, at the age of fifty-five, so goes the affective life, to a marriage in 1957, with the much younger pianist, herself once a child prodigy, the American, Charlotte Zelka. Arguably, this third and final marriage was the worst of the three for Anders, although it would technically last until the end of Anders' life, quite because, as noted earlier, Zelka, herself unhappy, would leave him in 1972 to return to California. Their contact remained desultory.

We can add to all this, as we know from several sources and as if to counterbalance Anders' own harsh critique, that Heidegger himself hardly held Anders in high esteem. Maier-Katkin, one of the prime sources for this assessment, adds the comment that "in the end neither did Hannah."[36] Given this, and as already noted, it is to his credit that Maier-Katkin emphasizes Anders' beauty, as can also be seen in the photos of the time, quite in addition to the handstands.

Anders became a passionate theorist of the human condition as this term is associated with Arendt, thinking this in the current technological age, between Husserl and Heidegger *and* Arendt (although the last influence was surely mutual) arguing that we were, as Anders argues, "*born*" as opposed to being a "thing" (AM I, 30), of being, as it were, not a product, not fabricated to spec or code.[37] There is a certain justification for looking at Anders' analysis of "Human Engineering" (in English in the original, AM I, 35f) Here, if this chapter had been part of a study focusing on Arendt and Anders, it would be crucial to call attention to Arendt's focus on natality along with her own critical interest in technology, as this last resonates with Anders' and not less with Heidegger's philosophical interests in technology.

The challenge Arendt faced with respect to Anders, namely that of being taken adequately seriously, affected her relationship with Heidegger, although Anders read and corrected her work lending her his insights as a matter of joint research. By contrast, Arendt knew that Heidegger would not read her work. Resigned to this circumstance, she would write to Blücher:

I am, as you know, quite prepared to act with Heidegger as if I had never written a line and was never going to write one. And that is the unuttered *condition sine qua non* of the whole affair.[38]

The issue of Zionism is more than a little complicated, and it is easier to discuss with respect to Arendt than Anders, even given the complex reception of Arendt's own problematic relationship with Zionism (distinguishing Theodor Herzl and Bernard Lazare's definitions of Zionism and gently clashing with Gershom Scholem) and nationalism,[39] especially given her "calls for the active pursuit of peaceful coexistence with Palestinian Arabs"[40] and her opposition, shared, as Maier-Katkin reminds us, with both Sidney Hook *and* Albert Einstein, to "acts of terrorism by Jewish groups," as Arendt writes in a letter to Jaspers quoting Blücher: "If the Jews insist on becoming a nation like every other nation, why for God's sake do they insist on becoming like the Germans?"[41]

It is claimed the comment is intended half-jokingly. Yet, there is a serious dimensionality, to recall Jacob Taubes' account, arguably in the last book he would 'dictate' of his own accord, and in person Taubes repeated, on several occasions, the same story to the present author. Like Arendt, Taubes was a friend of Gershom Scholem and collided, as he was fond of saying this, with Scholem's disapproval of his own failure with respect to Israel—Scholem would call him a traitor, *Verräter*. This must be matched, as I believe, with the account Taubes also relates (and indeed and again more than once) of the role played by the Nazi Catholic jurist Carl Schmitt's *Constitutional Theory* in Israel as Taubes, on his own account of it, went in search for his own purposes (to give a lecture on Descartes) for which lecture he tells us he required a copy of Schmitt's *Verfassungslehre*, which dealt with the problem of *nomos/lex/Gesetz*.

Taubes relates the story, combining the force of Augustine and Plato's *Republic*, not to mention Descartes—Strauss is not mentioned for nothing:

> In 1949 I went to Jerusalem as a Research Fellow with the Warburg Prize; Gershom Scholem, kabbalist and friend of Walter Benjamin, was my patron. Not only was Jerusalem a divided city in the 1940s and 1950s, but the Hebrew University had been exiled from Mount Scopus and was located in a monastery in the city centre. The great library was locked up on Mount Scopus, where an Israeli guard changed every fortnight under the supervision of the United Nations.[42]

The English translation skips over what the verbal German report underlines as the specifically *bureaucratic* "enjoyment and sadism" of the librarian,[43] as Taubes perceives his response to his request, and in its place we are given a report of what seems an *an sich* inoffensively "neutral" response to his query:

> The chief librarian listened carefully, but explained that he was powerless to speed the book ordering process. It could take two or three months before I got hold of the book. This was little help, since in three months the semester would be over.

As Taubes tells it, repeated on several occasions in person, the telling of which always underscored the dramatic description of smuggling materials out of the library on Mount Scopus[44]:

> the minister of justice, Pinchas Rosen (formerly Rosenbluth), needed Schmitt's *Verfassungslehre* so that he could deal with some difficult problems in the drafting of a constitution for the state of Israel. The book was therefore immediately brought from Mount Scopus and had now arrived in the library on its return journey, where my urgent request had been kept against an "opportune moment."[45]

Taubes emphasizes his "bemusement" with "the idea that the constitution of the State of Israel (a constitution which fortunately still does not exist) would be drafted using Schmitt's *Verfassungslehre* as a guide."[46]

In this same context, it can be worth recalling, subject as it was to rebuke from her friend, Gershom Scholem, following her *Eichmann in Jerusalem* for her lack of love for the Jewish people, Arendt's response in reply that she does not "love" peoples or nations but only friends, "*nur meine Freunde.*"[47] It is emphasized as noteworthy, and it also important to qualify that Arendt does not always do this in her correspondence with Scholem, that in this particular case she does begin her letter: "*Lieber Gerhardt.*"

The positions here between Arendt on Israel and Taubes on the constitution of Israel are misalignments. Anders' sister, Eva Stern Michaelis, would move to Israel and, according to her son David Michaelis,[48] would receive Arendt there but precisely negatively vis-à-vis her *Eichmann in Jerusalem*—but apart from familial accommodations, there is *no mention*,[49] a deliberate silence, as if by mutual agreement, on both sides with respect to either of their books on Eichmann (Arendt) and Eichmann *fils* (Anders), such that to report these differences, Anders himself would engage an utterly different set of concerns.

In addition to such complexities, and in addition to Arendt and Adorno, there is a greater connection between Taubes and Arendt than may have been supposed, not least via Scholem and also perhaps via Löwith and Jonas, and this is worth exploring for its own sake. Here, however the concern is Anders, who remains more complex just to the extent that his own research is broader, including Hiroshima and Nagasaki as well as nuclear energy projects. It is significant that Anders' concerns go beyond the received, standard and standardizing post-war focus as this involved Europe as such and the Holocaust. We will return to these difficult themes, each of which would properly deserve a study of its own.

4

Anders and Arendt Reading Rilke

Love Songs to God

Kafka

It is common to write of Kafka—the very name is a commonplace, even *acoustically* appropriate—when it comes to thinkers like Anders, and like Adorno and like Arendt herself who writes on Kafka and about whose work on Kafka a fair amount has been written. And there is a great deal (relatively speaking) already written on Anders and Kafka, as there is to be sure on Kafka and Adorno, Benjamin, Arendt. Anders however reminds us of the need to be particularly cautious when it comes to Kafka, using the term "Kafkaesque," itself an *Inbegriff* for his times, the times of 'darkness,' and for everyone mentioned, this darkness would perforce have been their "situation." In the same way, the challenge would be less that of writing poetry "after Auschwitz" to use Adorno's famous coinage, but rather, as Kata Gellen cites Anders on the constellation and dangers of cliché:

> one could think it must have actually been "difficult" for the likes of us not to write about Kafka. However he who is forced to live a Kafkaesque life does not read Kafka and does not write about Kafka. Even K. would not, under these circumstances, have read Kafka. We had more urgent matters to attend to.[1]

Although a careful reader herself, Gellen immediately distances herself from Anders' efforts to question the often-unquestioned conviction that literature or writing on literature is a salvific or political or redemptive enterprise as such. For literary scholars, such a question can be difficult to hear let alone to take seriously, thus, although Gellen breaches the issue, she does not long consider it. This allows her to express a certain scepticism rather than to inquire into Anders' lifelong struggles with money, a point that may be compared to Kojève's explanation, although this is to be sure not Gellen's own comparison, given Kojève's abundantly influential *Introduction to the Reading of Hegel* lectures, an insult from which rank and file Hegelians have not recovered to date,[2] namely that as Gellen writes, and this would thus be the patent parallel between Kojève and Anders:

> the French academy sought a lecture on Kafka and that he [Anders] needed money: Kafka was thus "useful" as a source of income. This, Anders suggests, and not the plight of Josef K. or Gregor Samsa, is *real* existential angst.[3]

This summary of Anders' perspective is incidental in the course of Gellen's reading, which pro forma reference yields what is surely an unintentional solecism, substituting tombs for tomes:

> When standing before the greatest ethical and political crisis of the modern age, one did not waste time reading (much less writing) esoteric tombs on Kafka—or at least one should not have.

"Tombs" are esoteric by their nature.

As Anders writes on the occasion of the death of his cousin, Walter Benjamin, in a poem written in 1940, *Das Vermächtnis, in memoriam*—which should be matched with a poem written two years later by Arendt herself (their style is worth comparing)—a remarkable parenthesis: *no one steps voluntarily through the gate: everyone*—meaning even those who take their own lives—*is shoved over the threshold.* "(*Keiner / trat selbst durchs Tor. Sie warden / über die Schwelle geschoben.*)"[4]

The language Anders uses to introduce his reading of Kafka is complex, indebted to Kant but not less to the physiognomic—only Virilio emphasizes the distortion that intrudes into scientific experimentation on the vulnerable, the mad, but also of course the prisoners in concentration camps in his *Art & Fear*.[5] Indeed, in an anticipatory parallel, Anders takes the reader into a discussion of love in his 1950 book on Kafka, that is, the erotic in a literary, aesthetic context. Reminding us that it is in Wagner (Friedrich Kittler will repeat the observation, if not with reference to Anders) that "salvation and dissolution of self become identical," emphasizing further, and here the logic is Nietzschean, even if the reading is not, "it is in Wagner that the two feelings of love and pity merge."[6] Anders' summary lack of reverence for sacred cows is striking in the continuation of the sentence—reading Adorno or Nietzsche one is not likely to find a comparably cavalier account:

> so complete is this merging that it is impossible to determine from the molten sweetness of [Wagner's] music whether his heart is consumed by compassion or by sexual anguish.[7]

Kant appears on the first page with a reference that Kant himself offers early in his own first critique, and Anders comes back to the metaphor, drawing its ultimate consequences: "modern science, in order to probe the nature of reality, places the object of its investigation in artificial, i.e., experimental conditions."[8]

The Kantian reference to the "experiment," quite as trial, recurs in Anders; it is central to Heidegger and, as torture, in Anders, as has been noted in Schraube.[9] But it is also key to Anders' reading of Kafka as it allows him to read *The Trial, The Castle, Metamorphosis, Before the Law* and *The Report for an Academy* as he does:

> For even as an experimental technique produces results, we shall not have penetrated nature's secrets unless we understand exactly what our technique is and does.[10]

When Anders reminds us that interpretation, and here we might wish to draw an incidental, accidental parallel with Anders himself, the *need* to interpret, the *facility* with the same, corresponds to a lack of power—elsewhere I show that Nietzsche tracks what I analyze as an *acteurly* hermeneutic facility in the Jew as in women: a facility, as Nietzsche describes it, in the art of deception, illusion, all compelled by necessity[11]—thus, Anders explains as a hermeneutic corollary, paralleling an "unremitting search for meaning" with "a mania for interpretation."[12] As Anders continues to argue with a patent allusion to Marx:

> This mania for interpretation is thus the stigma of the individual deprived of power; of one who (to adopt an old saying) must forever interpret the world because others rule and change it.[13]

If it can be argued that Anders offers a palimpsest in his 1950 Kafka book for what becomes in the germ a certain reading of an all-too-real-world "process," *Eichmann in Jerusalem*,[14] this is not my project here.

The focus of the current chapter is not Kafka but Rilke. Nevertheless, reading Kafka, Anders not only reminds us that the two were contemporaries but points out that Rilke's "the beautiful is only the beginning of the terrible . . . for it serenely despises to destroy us,"[15] seems to serve as mirror and epitome: "these lines might have been written as a motto for Kafka's work."[16] Anders' emphasis is inherently divided: one is caught in grip of terror, and at the same time this terrifying "renunciation is" and can be "a gift of grace."[17]

Rilke

At issue is hardly the question of "choosing," as it were, Rilke over Kafka, or, as also might have been done, given Heidegger's own option for Hölderlin or a keenness for Aeschylus or attunement to Goethe, the last emphases importantly underread in Anders. Much rather it is hermeneutically decisive to note in reading Arendt and Anders on Rilke, as Anders himself notes in his own retrospective reflection, that Anders and Arendt wrote together a text in Anders' hand—we will come back to this—on Rilke. The urgency is assured by the themes of love and of voice and not less deity in Rilke.

In the second strophe of the first of Rilke's *Duino Elegies*, a second sentence begins, and we can imagine the young couple reading this together, as if addressed quite to the two of them, together: *Sterne* [stars]. The poet calls to the reader's memory, reminding the reader in intimate, direct address, that even the springtime needed "you", myriad stars wait to be sensed, while from the depths of distant memory a wave rolls in, and a violin, "as you passed by under an open window," yielded itself to be heard. All of it, the poet says, is assigned—*Das alles war Auftrag.*

Do we ever, we can read the poet as asking, find ourselves measured to, *up to*? this assignment, this task? This is "the paradoxical, ambiguous, and desperate situation" that is the first line of their joint 1930 essay, "from which standpoint the *Duino Elegies* may alone be understood has two characteristics: the absence of an echo and the knowledge of futility."[18]

We hear paradox in the first formulaic frame as we also again hear the language of "situation" as Anders would use this term more technically with reference to music—

he is not the only one, as we read the same language in F. Joseph Smith's reflections on and towards a musical phenomenology among others, including Carl Stumpf—and with specific respect to poetry we hear "situation" as the word of the time, given the political and economic circumstance of 1930, but also given the influence of Heidegger and Heidegger's later, Rilke-suffused reading of Hölderlin as a specifically "desperate" poet, complex in a needful era, we hear needfulness, "desperation."

The question echoes in the poem *Das archaische Torso Apollons/The Archaic Torso of Apollo*, and it is the reason perhaps that the poem has proven to be as significant as it has for Heidegger, Gadamer, and most recently Sloterdijk, to name the more famous of the philosophical names of those who write on Rilke. That Gadamer himself will, thirty years later, give his own answer in response to the urging of this archaic column in human form, by speaking of trembling, of shaking, and thus of nothing so articulated, as Heidegger will say of the work of art, "this painting spoke," not then of symbol, nor allegory despite Gadamer's reference to both and as Pierre Hadot has also taught us to hear the medieval resonance of the same, nor indeed in a more Platonic/Aristotelian guise of reproduction, mimesis, play, terms Gadamer invokes in short order—and seemingly invoking the beginning of Rilke's *Duino Elegies* heard through Heidegger recollecting Hegel—"Great art shakes us because we are always unprepared and defenceless when exposed to the overpowering impact of a compelling work."[19] Thus, Gadamer cites the Seventh *Duino Elegy*: "As Rilke says, 'Such a thing stood among men.'"[20] Signally, Gadamer, who thus reads between Rilke and Hölderlin (his alignment running contra Heidegger), cites the *Archaic Torso*, and it is this line in Rilke, this line in Gadamer's own essay, that stays with us: "There is no place which fails to see you. You must change your life."[21] Gadamer's closing reference is to Hölderlin, which may (or may not) suggest his loyalty when it comes to the Heideggerian question of the work of art as there are also traces more temporally proximate of Adorno,[22] "That 'something can be held in our hesitant stay'—this is what art has always been and still is today."[23]

For Anders and Arendt, writing in 1930 there is another echo as this directly resounds in the poem. The acoustic claim speaks to Anders as is evident in his own musical writing, in Rilke's reference to the violin heard in passing, as this also seems the clue to his 1930, and thus roughly simultaneously composed, two-page musical essay, "Spuk und Radio." But just as in the case noted earlier, for Gadamer, who is also alluding, in the interim, to Benjamin's later essay on the *technological reproducibility* of the work of art, if Rilke is influential here for Arendt and for (Stern)Anders, as he clearly is, and if we can show that the same poetic lineage may be traced in Adorno on music, Rilke himself already draws in the word a lyrical, musical echo of the first verse of Hölderlin's *Brot und Wein*, originally titled: *To the Night—An die Nacht*. It was this first verse that was, during the poet's lifetime, the *only* version in print, dedicated as it was to the author of the novel *Ardinghello und die glückseligen Inseln/Ardinghell and the Isles of the Blest* by Wilhelm Heinse.

Hölderlin's motifs in these few lines seem to outline questions of economy, "*Und Gewinn und Verlust wäget ein sinniges Haupt*," that is, matching Hölderlin's Suabian sensibility: weighing "profit and loss." Thus a mindful, sensible "head" attuned toward the "balance," the satisfactions of the market, thinks back on the day. Hölderlin continues, as the night goes on, before the watcher calls the hour, to trace distance, ambiguity, memory:

But the play of strings sounds distantly from the gardens: perhaps, that
    There a lover plays or a solitary man
Thinking of faraway friends and days of youth . . .[24]

Hölderlin, whose word should also be heard and for the same reasons of the musicality of the word in the first lines of Nietzsche's *Birth of Tragedy*, in Nietzsche's case: concerning the conflicts between lovers, reminds us that with music we meet an art that brings us distance, overcomes distance, in its most intimate claim. With Rilke, we have to do with angelic hierarchies (cf. Figure 4): who might there be among all of these, in all these heights, to *hear* us, were we suddenly to cry out?

We've noted Anders' reflection that in effect, and the consequentiality here is key, the religious world is "an *acoustic* world."[25] There is a clear ambiguity in relation to the divine and thus the patent or seeming lack of answer that is part and parcel of "the situation," again we note the term, in which one seeks God. As Anders writes together with Arendt, and the dialogue between them subtends the argument, in their joint reflection on Rilke's *Duino Elegies* as we have already cited this allusion to questioning in Rilke (and thereby in Heidegger), as the kind of question that must be heard in the poet's word, heard:

in the form of a question that no longer hopes for an answer. Still the question does not perish from lack of an answer; rather it survives as disquietude suddenly changing into despair at the very encounterability [*Treffbarkeit*] of God.[26]

Figure 4 Bethesda Fountain, Central Park, New York City. Photograph, Babette Babich, December 12, 2018.

The resonance seems unmistakable in connection with the beginning reflections of Anders' introduction to the text he had meant for his habilitation, *Concerning Musical Situations*. There we read the same dialectical tension, he names "Die Situation des Zugleich," the "situation of the at-once," simultaneously, at the same moment, where his specific concern is to reflect the simultaneous circumstance of being not sheerly "in-the-world [*in Welt*], but 'in Music'."[27]

The tenor of the encounter, traced in Rilke's faintingly uncanny perishing "from his stronger existence" [*von seinem stärkeren Dasein*], is thus explained by Anders/Arendt with reference to nothing other than Anders own notion of being in-the-world as in-music, a specific 'being-in-hearing.' This is unpacked:

> This "something" consists in an in-stance of hearing, being-in-hearing [*Inständigkeit des Hörens, Im-Hören-Sein*]. Today there has to be a condition and an occasion for being-in-hearing.[28]

The acoustic we have already seen but what is at stake is the difference between a world with angels, a world with creatures created, and a world with and without God:

> With the denial of the experience and existence of God, nothingness disappears as a determination of human being: the human being finds a natural home in the world. If the human being still understands himself as nothing, then it is not as nothing before God, but as nothing as such: his life no longer lives in nothingness, but in the meaninglessness of his being. When he admits this meaninglessness, he lives in nihilism.[29]

What is at stake, as we read in the last line, is not "lament over what has been lost but, rather, the expression of loss itself."[30]

Angels

Explicitly Hebraic both in Benjamin and in Anders, the angelic locus is uncanny and otherworldly, literally transcendent, these "messengers" bring in the sphere of religion—the underworld for fallen heroes for the Greeks—and the salvific seemingly by definition. Complete with an emblem, the angelic includes Paul Klee's 1920 *Angelus Novus* (Figure 6), a painting famously acquired by Walter Benjamin.[31] In this context, the tonality of Rilke's poem, as Anders emphasizes, carries the vulnerable and not less audacious intimacy of the Rilke of the *Stundenbuch*, first published in 1905.

Love letters from the soul, as one may read these, are a play on the dimensionalities between a brevier and an intimate diary, as Rilke writes to God:

> You, neighbour, God, if I thou, from time to time,
> in the long night disturb with loud knocks,
> it's only because: I seldom hear you breathe

and I know: you are alone in the hall.
And were you to need anything, there is no one
to reach to your groping, something to drink.
I hearken always. Give a small sign.
I am very near.³²

The subtitle of this first book—there are three books in total—*Vom mœnchischen Leben, From Monkish Life*, composed in 1899, captures much that would be of theological interest to Anders, informing the focus of both Anders' appreciation of the spirit and the flesh, incarnate as living intimacy with deity: "*Du, Nachbar Gott*"—You, neighbour, God."

The tentative beauty of Rilke's voice inspired Heidegger and many others: the third of the books, the "work of an April week in 1903,"³³ marks a year that would be for Heidegger of great significance as it also sees the publication of Nietzsche's *Philosophy in the Tragic Age of the Greeks* and, of course, arguably changing the world of letters in its entirety, uniting scientists and humanists, artists and poets alike, Diels' *Die Vorsokratiker* and in this same year Rilke publishes *The Book of Poverty and Death*.

Rilke's larger *Stundenbuch* appears in 1905, and the allusions to this work are significant not only for Anders but arguably these may also be heard in Rilke's own later work. Thus, I quote the next lines concerning hearing, overhearing, and what is "builded" or constructed from images and likenesses:

Only a slim wall is between us,
just by accident; for it could happen:
a call from your mouth or mine—
and it would break apart, not a whimper, not a sound.

Out of your images it is builded. [*Aus deinen Bildern ist sie aufgebaut*].³⁴

Rilke published his *Duino Elegies* in 1923, just around the time Anders was writing his dissertation. When, some seven years later, Anders came to write an essay on the *Duino Elegies* together with Hannah Arendt, it seems plain to recuperate the points already made above that reflection on hearing and assignment in addition to the theme of religious intimation and renunciation might be involved.³⁵

Here, there is a question of authorship, incidental or in passing, quite as Anders assigns authority to Arendt when he publishes the text, ascribing in a courtly reflection, the greater influence to Arendt. Questions are raised by Jerry Kohn, who points out that "it is not clear how much of it Arendt actually wrote" in his introduction to Arendt, *Understanding 1930-1954: Formation, Exile, and Totalitarianism*,³⁶ noting in the process that the original manuscript was written in Anders' handwriting.

Indeed, as pointed out earlier, the relationship between Arendt and Anders was never one of parity but that between younger and older scholar, in which the older scholar worked to promote, and that also meant to vet *and* to correct, the work of the younger scholar as we also read in their correspondence from this time.

In the case of the Rilke essay, however, such complexities of correspondence and correction, have vanished. Thus, the current translation does not report authorship as "Günther Stern *with* Hannah Arendt" but settles it as mutually coordinate: Hannah Arendt *and* Günther Stern, "Rilke's Duino Elegies," in Susannah Young-Ah Gottlieb's edition of Arendt's literary essays.[37]

One consequence of noting Anders' relative obscurity is that it permits us to see that questions of the weight, in either case, of a co-authored piece inevitably tend to privilege the better-known author. To use a famous example not referring to Rilke but the nineteenth-century so-called romantics, it can seem to be one consequence of a long (and still unresolved except in the minds of some disputants) debate concerning the authorship of the *Oldest System Programme* (written in Schelling's hand, but "anonymous"), which is attributed, very much depending on one's enthusiasms, variously to Hegel, Hölderlin, or Schelling; the most radical solution, someone other, someone else,[38] almost never comes up.

One can argue, fairly easily, that the Rilke essay follows, at least stylistically, Anders' concerns, but Anders' own remonstrations emphasize that the text was born of mutual involvement (*both Anders and Arendt*), such that we must forsake the tendency to read the work as the name placement can induce us to do so, listing Arendt first, and thus and thereby assuming it as following for the most part *Arendt's* initiative/formulation.

Elegies: Poetry and Hearing/Poetry and the *Schma*

Before continuing with Rilke, it is worth recollecting Arendt's reflection on "Kafka, Krauss, and Benjamin."[39] Arendt is talking truth (and Heidegger along with the specifically Benjaminian "secret" of truth) but it is the acoustic that it is worth noting, between accents and intonations (Arendt herself will invoke various German dialects or *Mundarten*)[40] as she proceeds to explain:

> Once this truth had come into the human world at the appropriate moment in history—be it as the Greek *a-letheia*, visually perceptible to the eyes of the mind and comprehended by us as "un-concealment" (*Unverborgenheit*—Heidegger), or as the acoustically perceptible word of God as we know it from the European religions of revelation.[41]

With reference to Anders and Arendt on Rilke—reference to religion and to Judaism, for Arendt the more important emphasis, is complicated beyond the immediate context here and must await a full, and fully integrated, biography—what is key in Arendt's distinction between the "visually perceptible" and the "acoustically perceptible" is Anders' observation that "Rilke's world is, like every religious world, an *acoustic* world."[42] And it is in this same heard or acoustic sense that we should not fail to read, or, better said, to *hear* the first lines as commentary on the prayer, "Hear, Israel," *Schma Yisrael*[43]:

The conscious renunciation of the demand to be heard, the despair at not being able to be heard, and finally the need to speak even without an answer—these are the real reasons for the darkness, asperity, and tension of the style in which the poetry indicates its own possibilities and its will to form.[44]

The kind of dialectic, the kind of ambiguity, is that in direct proximity, continuity with the divine, and Heidegger's specific kind of questioning may also be heard in the reflection on the seeming lack of answer which attends the moment—Anders would say: "the situation"—in which one prays, a petition given, as we cited it earlier, "in the form of a question that no longer hopes for an answer."[45] If all of this, including the very possibility of an encounter with deity, physical as this is, including breath, space, creaturely encounter [*Treffbarkeit*], is at least fairly esoteric, this poem also would have seemed almost obvious for the newly married Sterns.

Thus, we read a clear reflection on the audible, using Anders' language of *Zu-hören*—the text begins by emphasising the poem's "musical key"—and one hears elements of Benjamin's angel of history:

Listening is so little bound to an object that, on the contrary, it receives "*seine ununterbrochene Nachricht, die aus Stille sich bildet* [the ceaseless message that forms itself out of silence]" (1st Elegy) whenever the objects are lost and blown away: it is not a listening to a particular, articulated message; rather, it is a listening to the urgent beseeching of a heart ("*Höre, mein Hertz* [Listen, my heart]"), therefore a mode of being ("*so waren sie hörend* [such was their listening]," (1st Elegy).[46]

For both readers, and here there is the full resonance of the theological and the acousmatic-acousticological dimension that would bespeak the correspondence between the two, we read: "Indeed, as a state of being, listening is already its own fulfilment, since it pays no attention to whether its beseeching may be heard."[47] This is abandonment to deity and it is also the abandonment intrinsic to the immanence of love. We may think we can hear Anders alone as we read; it is certainly his formulation, the rescued "something" that the poet is able to wrest from the then (and still we might say) current "religiously alienated situation"; thus, we quote, again, but now more completely, the Sterns' reflection on this:

"something" [that] consists in an in-stance of hearing, being-in-hearing [*Inständigkeit des Hörens, Im-Hören-sein*]. Today, there has to be a condition and an occasion for being-in-hearing. In place of complete objectlessness, for which our heart is no longer adequate, the occasion for being-in-hearing becomes the disappearance of the object, which we pursue with our ears.[48]

This is written by the two, together, but it is also in Anders' voice: full intentionality, full musicality, a voice in dialogue with the beloved.

If we read in this way, we see that what is at stake is not handed down from

"the Angels' hierarchies," which the "Elegies" vainly attempt to woo ("*Engel, und würb ich dich auch! Du kommst nicht* [Angel, and even if I were to woo you, you would not come]," 7ᵗʰ Elegy), nor does it come from other people; rather, it comes from things.⁴⁹

Just as Anders will later recall, he first "won" Hannah Arendt with a word, a sign of his brilliance, his heart; thus, this lover's discourse between both authors. The reference to things is characteristic. For the poet, the things are of this world, and the reference to these invokes the human estrangement from, alienation in, the world:

> The fact that whatever remains of the relation to the world flees to what is relatively most distant—in any case does not turn to the other, to what is closest, but rather commits itself to this distant entity and claims closeness to it shows the extent to which human existence has here been estranged from the world⁵⁰

At stake then are echoes of Anders' acoustic focus—as he will also emphasize his own attention to tone as musically resonant in time/space and hence not as physically delimited—along with Arendt's theological concerns, and in both cases, love, here at its inception, is *seduction*, wooing, solicitude, praise. Reading Rilke, reading Arendt and Anders, Anders and Arendt, we can also hear the two together, whoever it is who writes this *text*, as they, writing on the Seventh Elegy, continue to reflect on love, particularly as inscribed in Augustinian echoes:

> Praise only grows from the futility and despair of wooing. Only in praise is there a being-heard [*Gehörtwerden*], namely the being-heard of what is told, even if it has nothing to do with being-hearkened-to [*Erhörtwerden*]. The first impulse of the call is thus a religious impulse, the failure of which gives rise to poetry, which contains a double ambiguity for this very reason: measured in accordance with its religious origin, poetry is already the falsification of that origin. As poetry, however, in other words, as expression of the interior world, it fails to live up to its own premises. "Listen, my heart, as only saints have listened"—this is the impulse that, as is shown by what follows ("*Nicht daß du Gottes erträgest die Stimme, bei weitem* [Not that you could endure God's voice, far from it]"), already contains the failure of listening.⁵¹

The reading reflects between the complexity of Arendt's own dedication to theology, Christian theology, as Anders reflects with only a hint of teasing (speaking of the subtleties of the theology "*die Du perverserweise studiert hast*" in the silence at the end of her life and towards the end of his own [KS 38]), articulating thereby what inevitably falls short, as both failure *and* redemption.

Anders begins his *Kirschenschlacht* reconstruction without the slightest hesitation referring not to Rilke but to Goethe, not with respect to himself but to Arendt, emphasizing that it was she, like Faust, who had studied theology, "*viel Theologie . . . ja christliche*, and even dissertated brilliantly and profoundly concerning Augustine" (KS, 23). Of course, to say it again, the theme was love: "Listen, my heart, as only saints have listened."⁵²

Subsequently, we find a reflection on "situation," as this is to be sure, after Nietzsche, given the scientific sophistication of the age, also a concession to a-theism:

This poetry is thus directly grounded in futility: at the point of non-differentiation in which religious intention and religious denial are sublated, a peace and balance, hence a beauty arises that has nothing to do with religion in its origin.[53]

The text goes on to refer, seriatim here, to the hero, the child, the lover. For both registers, that is, for the escteric and for the erotic at its outset (this changes), we have everything we need. Hierarchy of orders of angels, no less, music, impassioned hearts, overpowering absorption, power, breath, and beauty and terror, and the promise of devastation, rapture, destruction, all withheld:

> WHO, were I to scream, would then hear me out of the angel
> orders? And assuming one of them were to press me
> suddenly to his heart: I'd perish from his
> stronger existence [*stärkeren Dasein*]. For the beautiful is nothing
> but the beginning of the terrifying, which we can scarcely bear,
> and we admire it so as it calmly disdains
> to destroy us. Every such Angel is terrifying.[54]

And there it is with reference to angels and humans, the same wrestling once again, an allusion to what it is to be a stranger, as Arendt speaks of this, as Taubes himself, in his *Die politische Theologie des Paulus*, includes a section entitled "Fremdlinge in dieser Welt,"[55] [*strangers in this world*] alluding to Rilke's "sly animals" and the night. Thus, Arendt and Anders, cannot but write about the same lovers' night:

> O and the night, the night, when the wind full of cosmic space
> Consumes our faces—for whom *won't* the night be there, the longed for,
> gently disappointing, for which a single heart
> painfully awaits. Is it for lovers easier?
> Ah, they only cover their destiny with one another.[56]

Rilke, as said earlier, can seem to have written for them, for newly married lovers, *Sterne* (Figure 5), and this poem, new to Anders and to the world itself when he was writing his dissertation, culminates with a reference to a window, a passerby, a violin, everything overdetermined, assignment/task, and the beauty of spring, *for your eyes only*:

> Yes, the spring times needed you indeed. Some stars even suspected
> Of you, that you could feel them. Swelling
> A wave raised itself toward you from out of the past, or
> As you passed by an open window,
> a violin surrendered itself. All that was tasked.
> But did you master it? Weren't you always
> so very distracted by expectation, as if everything announced
> a beloved to you. (But where would you hide her)[57]

Their reading, speaking of the unreliability of the world, brings them once again to abandonment and to love. Here perhaps it is irrelevant to know in whose hand this text

Figure 5 Günther Stern and Hannah Arendt January 1, 1929. Photo credit: Courtesy of the Hannah Arendt Bluecher Literary Trust/Art Resource, NY.

appears (we recall that Anders would write the text), lovers as they were, writing then, here to repeat, together as lovers, cannot but be struck by love talk:

> Love, for Rilke, becomes an exemplary situation, for love is principally love of the abandoned. As a situation, love never cleaves to a single opportunity or a single beloved; these are only occasions for it. Nor is love to be understood as one feeling among others. Love overcomes and at the same time forgets the beloved, since it intends more than the accidental individual, and its horizon is obscured by the beloved's closeness ("*Ach, sie verdecken sich nur miteinander ihr Los* [Alas, they use each other to hide their own fate]," 1st Elegy). Love lies in this abandonment alone.[58]

Rilke's words are key in every love affair, especially after Freud, after the elder Stern, this must be heard; it is the oldest story of love: "they *use* each other to hide their own fate." The "unreliability" of the world as they then go on to say, as this is read back into the poet's language,

> is therefore doubly determined: things abandon us, we who are "*nicht sehr verlaßlich zu Hause sind / in der gedeutete Welt* [are not very reliably at home / in our interpreted world]" (1st Elegy), and we abandon things, "*denn Bleiben ist nirgends* [for there is no place where we can remain]" (1st Elegy). This double abandonment,

which Rilke tacitly makes into the positive quality of abandonability, acquires an independent meaning as solitude.[59]

The reading that Anders and Arendt offer takes us through the poem, almost soft-pedalling the references to love, to the night, to the girl who might be, is, was callously abandoned. Thus Rilke reminds us of the transformation, phenomenological variation of "the way the arrow, suddenly all vector, survives the string/to be more than itself. For abiding is nowhere." We need the German and we read: *Wie der Pfeil die Sehne besteht, um gesammelt in Absprung*—Rilke almost Heideggerian in tenor for all Heidegger's complicated resistance to Rilke, "*Mehr zu sein als er selbst.*"

And the word that Rilke will go on to repeat also speaks to love and the need for a shelter, already evident to Anders, who had been in France already during the First World War: *Denn Bleiben ist nirgends.*

We have already noted that Heidegger in his later reflections on poets in a needful era, *dürftiger Zeit*, is careful to ask whether Rilke counts as such:

> Is R. M. Rilke a poet in a needful time? How is his poetry related to the indigence of the time? How far does it reach into the abyss? Where does the poet arrive, granted that he goes where he can go?[60]

For Heidegger, and intriguingly his reflection only published in 1946, in *Holzwege*, offers, one can only wonder if he means the reference to be heard by Arendt (and thus by Anders) as his essay includes one of his few musical references quite in addition to his own poetically, staccato-like allusions to Nietzsche and to Hölderlin:

> Needful remains the time not only because God is dead but because even mortals barely know or have the capability of their own mortality. Still have mortals not taken possession of their own nature. Death withdraws into the mysterious. The secret of pain remains covered. Love is not learned. But mortals are. They are insofar as language is. Song yet dwells over their needy land. The word of the singer still retains the trace of the holy.[61]

The poem Heidegger does cite is an indictment of technology as also venality and finitude, mortality, telling us, in Rilke's words: "The kings of the world are old."[62]

Rilke's poem fragment reflects on the pain of things—Anders will later in his reconstructed dialogue with Arendt, as we turn to this below, *Die Kirschenschlacht/ The Cherry Slaughter*, emphasizes the bits, the pieces, *Stücke*—and here Rilke offers a reflection on the faded sovereigns, who

> . . .
> will have no heirs.
> Their sons already as boys,
> and their pale daughters yield
> to the power of the sickly crowns.
> The mob slivers them small into money,

> the current ruler of the world
> extruded in fire into machines,
> which, grumbling, serve his will
> but happiness is not with them.
> The ore is homesick. And abandoned,
> it will be trained into a little life into coins and gears.
> And out of the factories and the tills
> It would fain return to the veins
> Of ruptured mountains
> That would again close around it.[63]

The reflection on the angel for Heidegger, restores a certain balance, after alluding to Nietzsche by speaking of "calculating will"—as if this calculation had been Nietzsche's concern—and may, as it proceeds, help us to turn to *The Cherry Slaughter* below, as Heidegger quotes Rilke's "late period":

> …When from the salesman's hand
> the balance goes over
> to that Angel who, in the heavens,
> stills and gentles it with the trade of space . . .[64]

The Cherry Slaughter: Back to Love

Despite the attention to Anders' life, quite as Gadamer would highlight the critical importance of history for a text and as this matters in philosophical anthropology and the *Wirkungsgeschichte* that begins at no less world historical juncture than with Augustine's own *Confessions* as "the" literary *topos*, what we have been observing here is no biography. Nevertheless, biographical elements are involved and Anders who did not write an autobiography nevertheless found it necessary to tell himself, to tell us, the story of a life shared, *The Cherry Slaughter*, a posthumous telling that is at once a testament to love and a claim.[65]

One can read the title of Anders' text, *Die Kirschenschlacht* as the "Battle of the Cherries"—the text exists in French as *La bataille des cerises*[66]—or "Cherry War" or even, with a pop metonymic echo with the film/novel of divorce, *The War of the Roses*: "The War of the Cherries."

"Cherry Slaughter" can also convey the poetic sensibilities of Anders' remembered dialogue, reanimating the ghosts of memory, as Anders tended to remember the ghosts of past wars, Vietnam, Korea, the Second World War, including the Hiroshima so often left out, and the First World War. In addition, there is a great savaging of cherries: devouring, destoning, dismantling them for the sake of cooking, the plan is to make jam and jelly or, indeed, like as not, *rote Grütze*, a kind of jam soup served with cream, a mystical dish by all accounts.

And there are other, darker tones when it comes to the language of cherries. Indeed, Bürckhard Dücker traces the cultural history of the cherry in German letters—and the chapter is astonishingly rich as Dücker reads through a range of poets, including the Paul Celan poem that strikes Otto Pöggeler entitled "Here": *"Hier—das meint hier, wo die Kirschblüte schwarzer sein will als Dort"* [Here—that means here, where the cherry blossoms wish to be darker than there].[67]

The reference to Hiroshima makes the connection inevitable and yet Anders excludes such a resonance. How does anyone negotiate such memories at the end of a lifetime, by way of what he named, in an awkward formula, *"modish[er] Unsinn,"*[68] that is, as we may say, "fashionable nonsense" just given the mores of the 1960s themselves.[69] And we do remain limited by such a reference by the ongoing need to favour one disastrous name for an epoch above some other. Thus, Helga Raulff's reading of Rose Ausländer's unpublished manuscript, "After the World was Atombombed," raises the question of the sense in which certain terms seem imbued with one set of associations and not others, whereby it can seem to be a transgression or almost so to use one symbolic schema—"ashes and dust [*Asche und Staub*]"—for the work needed to make the more obscured horror "legible." The offence, as it continues to be an offence, is described by Anders, cited in Raulff's discussion as he writes, flatly and boldly, that "in spite of the fact that the world went to ground not through Auschwitz but through Hiroshima," it continues to be Auschwitz that strikes us "on an inequitably kipped moral level" as still more horrific.[70] Overweeningly so. As Raulff cautiously notes, the disproportion remains such an enduring taboo in the words of Jan Philip Reemtsma that most of us find it "obscene" even to mention the two names together—Auschwitz and Hiroshima—"in one breath."[71]

Anders' book would not appear in print, I noted above that it mattered that Anders had his troubles getting his writing into print, which also entails that what we know of his work we know via his editors and thus and only at their discretion. In this case the book would not be published until a quarter of a century later, now a clean decade ago, in 2011 as *Die Kirschenschlacht. Dialogue mit Hannah Arendt*, expanded with an editorial notice and an interpretive essay by Christian Dries.[72] The editorial notice is needful to mention and had to be included to make the book a book—as opposed to a pamphlet or opusculum, a term Heidegger's publishers liked to use for his shorter works, or indeed, the term favoured by Robert Hulot-Kentor with reference to Adorno, "fascicule."

Here it is important to recall the tension between members of the Frankfurt School, especially Adorno, but no less especially their handlers: Who champions Anders? Who champions Adorno or Horkheimer or Arendt and what difference does this make? It is a rare testament in Anders' sense of the same, and this for him was a matter of decency, that Anders was distinguished, a late award, painfully given as late awards tend to be, with the Adorno Prize in 1983. This late honor, the desultory quality of it, was the story of Anders' life.

The selection factor, a name prize, distinctions given, way too late for Anders in the case of the Sigmund Freud prize, and rather as if the members of the jury had run out of local names and had somehow noticed that Anders was 'still' alive in Vienna, corresponds to the force (if not the substance) of Anders' reflection on the "Irrelevance of Humanity" in *The Cherry Slaughter*.[73]

For Anders, who here silently echoes Nietzsche's cadences on the universe in the third book of *The Gay Science* ("Heaven Forfend," is the title of Nietzsche's aphorism), reflects that considered from the perspective of the universe, there is no cosmos, no coordinate, well-attuned, harmonized schema. Thus the point of departure for his reconstruction of his Drewitz dialogue with Arendt, Frau Stern, on their balcony, facing one another, stoning cherries, eating cherries, wearing cherries—in Arendt's case, over her ear—and debating windowless monads.

The fabrication of a dialogue is fair game in philosophy—we need only think of Plato, as we continue to take Plato, quite literally, at his word. Nietzsche's Zarathustra is another fabrication that could be a close second. But Plato's dialogues *are* the history of philosophy. By contrast, other fabrications must be underscored as such. Thus, there is the fabrication of Heidegger's written (but it is not such) deposition (but it is not such) said to have been composed and presented during Heidegger's supposed de-Nazification proceedings, very much as if we had in fact such a physical record (we don't). Such a reconstruction was published without flagging it and maybe thereby undermining its value as a reconstruction. Rather, presented as a "translation" of Heidegger's de-Nazification "deposition," the putative translators, the authors Valerie Allen and Ares D. Axiotis undertook to recount a series of fictive Heideggerian etymologies: "Trivium, although a singular word, already points to the multiplicity within—*tri-viaum*, three roads made into one."[74] Installed into this, *qua* "reconstruction" as we may assume, this is not said, Heidegger is reported to have instructed his questioners on the "art of teaching," detailing grammar and dialectic, including the risqué language the two authors set into Heidegger's mouth, imagining him as speaking of "putting the mare beneath the stallion."[75]

Anders' tone is different as his is a work of mourning and she was his first wife and thus he had his reasons for the telling. The book is thus meant to be, whether it succeeds at this is another question, I have already pointed to the conflict, an encomium in the wake of loss, in the wake of Arendt's death as Anders reconstitutes what he tells us is an *idealization*, admitting that he privileges his own point of view (a good hermeneutic point as both Heidegger and Gadamer would concur), where his guide is his memory. Thus, in this case, even after "fifty years of totally separate life" (KS, 11), one is, in an important sense, and still, one flesh, *basar echad*, words from the book of Genesis, the same book, especially the first and second chapters of which, Anders tells us, would always be "deep in Arendt's bones" (KS, 26).

And so Anders tells us that Arendt, while warning us that he will be the hero of the dialogue, the arguments he noted had to be, they could only be, his own, and so despite the respect and regard Anders had for Arendt, inevitably he will only be telling one side, his side of the dialogue in and through his reconstruction. What we see is invaluable, and Arendt is not, as he tells us, someone attuned to pictures or poetry or music, although she is not utterly clueless. Clearly, there is a difference of orientation, and Arendt, more than Anders himself, could have been a *child*, as he puts it, of *the people of the book*: "*Kind des Volkes des Buches*" (KS, 8). Note too that there is in this account a reflection on what Anders teaches her of music and painting, through shared concerts and conversation, museum visits and discussion, whereby the allusion is strong to Heidegger—bringing paintings "to speak," the turn of phrase, the language of speaking is key to Heidegger on art—and where, as it turns out, Anders plays the role of the translator, "*dem Dolmetsch*" (KS, 9), as indeed he seeks to do in retrieving, reconstituting their dialogues.

Derrida and Benjamin and others in a long tradition write on the role of the translator.[76] That is not quite Anders' concern in his posthumous love letter to Arendt. He remained, we know this from their letters as well as from this reconstruction, in close association with Arendt if not always with a "con-crete" record of the same for us to follow, as some of the bread crumbs have fallen to the winds, scattered here and there, lost but for memory. Thus, even as their marriage came to end, they remained in contact as we say, even as the love affair between Arendt and Heidegger ended, *they* too remained in important contact, and so, in a sense, neither relationship ended. This was, so I argue, in Arendt's case a matter of her gift for friendship and not less her devotion to Heidegger, not *in spite of* but perhaps, as the German phrase that plays between this opposition suggests, *because of* his deficiencies, his failures. But in Anders' case, Arendt was significantly not as tolerant. Here I note that this same intolerance seems to have been taken in stride by Anders, corresponding to the truth once again that in any love affair there is always one who loves more, even if the one who loves more than the other may change over time.

If Anders' eulogizing love letter to Arendt mixes praise along with qualifying reservations, he is harsher on himself than he needs to be; refusing Leibniz as he does, Anders is careful to remind her, to remind the reader, that he is no Leibniz scholar, quite in contrast with Cassirer, note here that this could function as another way to challenge Heidegger as par for the metaphysical course, in a reflection on world systems and pre-established harmony, by pointing to the utter innocence of things, ourselves included, carried by the whirl of the physical cosmos, a perspective we do not share and which, even if we believe it, we still fail to feel, emotionally, physically.

Anders' point echoes Nietzsche's proclamation of the same nihilistic abandon, with the same physical force, at the level of cosmological speed and depth—the height of the universe, the course of the stars, the death of God, both in *The Gay Science* in the mouth of a madman come to tell the tale, "too soon," as he determines faced with the incomprehension of his interlocutors, and again at the end of *On the Genealogy of Morals*, where Nietzsche speaks, as Anders does, of "atheism." As Nietzsche explains, to cite Adrian del Caro's new translation, speaking of the "ascetic ideal" typically associated with belief:

> Unconditional and honest atheism (—and *its* air alone we breathe, we more spiritual men of this age!) accordingly does not stand in contrast to that ideal as it seems to: rather it is only one of its latest phases of development, one of its final forms and inner logical consistencies—it is the awe-inspiring *catastrophe* of a two-thousand-year training in truth, that in the end forbids itself the lie of believing in God. (GM III: 27)

The Copernicus Anders invokes as counterpoint (it is important to echo this here as it permits us to note Nietzsche's influence) corresponds in Nietzsche's own text as the origin of what Anders calls "antiquatedness," outdatedness, precisely as Anders regards it as self-perceived, self-constructed, quite on the part of humanity.

"Does anyone really believe perhaps," Nietzsche writes, still speaking of the ascetic ideal as such, "that the defeat of theological astronomy meant a defeat of that ideal?" (GM III: 25). Nietzsche continues:

Have humans perhaps become *less in need* of an otherworldly solution to their riddle of existence now that this existence looks even more arbitrary, loitering and dispensable in the *visible* order of things? Has not the self-belittlement of humankind, its will to self-belittlement been on an unstoppable progression since Copernicus? (GM III: 25)

In Anders' reconstructed dialogue with Arendt, the two philosophize together over nothing less matched to the nature of love between lovers than the "problems of the universe," the world, writ large, literally. Their favourite occupation at the time would be *shared* philosophizing, "*symphilosopheĩn*" (KS, 13).

The problem with raising the question of the monads, the consciousness of mere things, or of ourselves, in the context of Copernicus, is the derangement that follows: one tends to lose thereby a "decent" (but what other kind is there?) "cosmos" [*einen anständigen Cosmos*], including, as a proper cosmos would, "a connectedness [*Verbundensein*] of all things with one another" (KS, 17). A world system in other words. In this fashion, "the Copernican roundtrip," as Anders says and as we have learned ever since we were schoolchildren, following its course "around the sun," happens to carry us along with it as "*blind passengers against our will*"(KS, 29). In this way, having set people and things, including poems and atoms, all along with the earth, under compulsion "to travel along as utter and completely irrelevant passengers, no: as baggage [*Gepäckstücke*]" (KS, 29), that is, bits on bits, mere pieces beside other pieces, thus monads, lacking cosmos, without a pre-established harmony.

As Nietzsche writes at the dramatic outset of the third book of *The Gay Science*, literary flourishes being even easier for Nietzsche than for Anders, starting with the shadows of the dead, both Buddha and the deity, to emphasize in the following section, with the suitably apotropaic title, *Hüten wir uns!*—again we quote: "Heaven forfend!"—and going on to detail the following reflection on the astral order, we meet his first articulation of the eternal return together with the blind will to power of the universe as such, regarded as a *Spielwerk*, a musical allusion to Kepler and Kircher and Copernicus, and the Milky Way:

Per contra, the total character [*Gesammtcharacter*] of the world is chaos to all eternity, chaos, not in the sense of an absence of necessity but an absence of order, arrangement, form, beauty, wisdom and whatever our aesthetic humanisations [*ästhetischen Menschlichkeiten*] may be named. (FW, §109)

Anders, likewise invoking Copernicus, and the reference to Leibniz is meant to convey this point, is all about the emotional, consciousness, or feeling of the very fact of it. Thus, in spite of all our learning in this—and to dramatize the detail, Anders tracks the effects of the sun's motion throughout the dialogue, "mere child's play," as he describes his own literary devices—quite apart from the Vatican, Anders says, we continue to refuse to give up "our Ptolemaic world sentiment [*Weltgefühl*], refusing *to coordinate ourselves emotionally with our knowing*" (KS, 29).

The relevance of Anders, for the planetary disaster we neutralize under the portmanteau as if it had nothing to do with us or with our technology or our industrialization effects or with our military activities and certainly not, heaven forfend, with the decades-old practices of geoengineering, spraying chem trails hither and yon, under the rubric of "Global Warming" or the "Climate Crisis," has everything to do with this insight.[77]

As for the rest, including obligatory references not only to Hegel and Nietzsche but also explicitly to Heidegger as to Georg Simmel and Max Scheler, and not less those girls who, like Hannah, with pronouncedly Jewish features, "even the pretty ones" (KS, 51), need its own complex set of reflections that are beyond the scope of the current study.

Thus, Anders adds his own afterword to his own opusculum. And there the monads remain, as does the reference to god, culminating in an academic reflection or afterword where Anders, giving the final word to Leibniz, writes in French: "*les monades n'ont point de fenêtres*" (KS, 60).

5

Between the Lines

Benjamin's Angels of History and Anders' Apocalypse

Having Been

As a pre-eminent critic of technology and critic of the atomic bomb, Anders regarded the bomb hermeneutically and phenomenologically in the visceral sense of being *and* time, literally more being, more time, than Heidegger's *Being and Time*. For Anders, who thought through to the furthest consequences of the dropping of the atomic bomb, first on August 6, in Hiroshima and then on August 9 in Nagasaki, respectively, in the summer of 1945 following Germany's May surrender, the sheer thatness of the bomb's *having been* used (where the Nietzschean dialectic of the "having been" reflects the essence of modern technology) coupled with the bland politics of nuclear proliferation, the "cold war," extended the wake of the war after its end. For Anders, the project of "deterrence" worked as so much programmatic aggression, advanced in the name of "defence." The tactic of sheerly technological, automatic, mechanical, aggression is carried out in good conscience. The pre-emptive strike—and with Covid-19 measures, masks and tests and vaccines, we are only living the logical extension of this same ongoing putatively pre-emptive ideal—is, as Baudrillard observed, ever a matter that may be blamed on one's opponent. The pains the enemy suffers, the pains we suffer with masks, tests, and vaccines, are simply the wages of (their) evil. Violence in good conscience characterizes the post-war, Cold War era and the present day with its mushrooming effects of neo-fascism under the titles of national security, anti-terrorism, and Coronavirus.

Karl Krauss' 1913 *bon mot* apostrophizing psychoanalysis as that same insanity for which it declares itself as cure [*Psychoanalyse ist jene Geisteskrankheit, für deren Therapie sie sich halt*] has never been more apt for political translation straight into the heart of what Lacan called the Real, which has "always been" the political register. Where Habermas and heirs have tended to disregard Anders (as they also sidestep Heidegger and Nietzsche), just as most philosophers of technology (and philosophers of science) ignored the political as well as the ethical in their eagerness to avoid any suspicion of technophobia, we continue to lack both critical theory for our times as well as a critical philosophy of technology, a conjunction incorporating Anders' complicated dialectic less of "art" in Benjamin's prescient but still innocent age of technological reproducibility, but and "concerning the devastation of life in the age of the third industrial revolution." Thus, rather than reading Anders' critique of the bomb as limited to a time we call the Atomic Age, Anders varied Samuel Beckett's 1957 *Endgame (Fin de partie)* as *Endzeit*—"Endtime," using the eschatological language of Taubes fairly as Anders does.

What is at stake is less Anders'/Arendt's Rilkean angels than Benjamin's more morose, more titanic angel of history, reflecting on war and its wreckage, detritus, a legacy after the First World War, for Anders reflecting on the bomb with his technological critique of the outdatedness of humanity as of a piece with our dedication to hurling ourselves against our own mortality. The same concern with the violence of technology, this hatred of the vulnerability of having been born and having been set on a path unto death (the mortal path that is the path of life) inspires Anders' engagement with the sons of Eichmann (heirs of those who designed and executed the Nazi death camps and extermination chambers of the Holocaust) and, banality on banality, the sons of Claude Eatherly (the heirs of those who designed the bomb and of those who as pilots and weathermen, deployed, guided the bombings that exploded the supposed stuff of the sun itself contra the Empire of the Sun in the attacks on Hiroshima and Nagasaki).

We, embroiled as we are in wartime after wartime and now suppressing public protest on a scale as never before, silencing free speech as we are, in country after country across the globe, cannot dispense with Anders today.

Angels of History

Walter Benjamin, could trace the mystical art of his possession, acquired from Scholem (who under tragic circumstances also got it back), Paul Klee's 1920 *Angelus Novus* (Figure 6). Elsewhere I have epitomized the painting as postcard, iconic, and in my own university, I have for decades passed it on a colleague's office door, Professor Anne Golomb Hoffman, a scholar of literature and a painter, and I am grateful every time I pass it. The angel of history, recollecting the word painting of the open mouth, his "eyes are staring, his mouth is open, his wings are spread,"[1] whereby we conflate Klee *and* Benjamin, one with the other. I cite Benjamin's *Theses on History*:

> A Klee painting named Angelus Novus shows an angel looking as though he is about to move away from something he is fixedly contemplating. . . . His face is turned toward the past. Where we perceive a chain of events, he sees one single catastrophe which keeps piling wreckage upon wreckage and hurls it in front of his feet. The angel would like to stay, awaken the dead, and make whole what has been smashed. But a storm is blowing from Paradise; it has got caught in his wings with such violence that the angel can no longer close them. The storm irresistibly propels him into the future to which his back is turned, while the pile of debris before him grows skyward. This storm is what we call progress.[2]

In an end time, at the *end* of time, the strobe light of horror showed the still figure, the frozen figure of the angel of every apocalypse. And of course, as we recall Rilke's *Duino Elegies*, angels were on everyone's lips: *Ein jeder Engel ist schrecklich*.

Figure 6 Paul Klee, Angelus Novus, 1920. Gift of Fania and Gershom Scholem; Courtesy of the Israel Museum, Jerusalem. Photo credit: HIP/Art Resource, NY

Of course they are.

Of course, we should not fail to to ask just where one finds oneself, to ask what has become of one's life, one's eyes, such that one *can* see angels? Anders explores the mode of such modalities, *können* and *nicht können*, *to be able to* and *not to be able to*, as opposed to Shakespeare's rag in Jack Benny's voice and the filmic icon of the same, Nazi Germany, 1942 Hollywood style: *To Be or Not To Be*, being and non-being. Non-being as a possibility, real in a different sense than it had ever been before for any time since we humans had become, in Hölderlin's words, a conversation with ourselves, for ourselves. For Anders, in 1975, all of these are old-fashioned worries, and the problem now as ever is to come to terms with what we have learned that we are able to do.

This is also the point of Anders' invocation of Goethe's "The Sorcerer's Apprentice." Note that more is involved than the simple historical detail that Hollywood had translated this figure to the film centre of a cartoon musical opera, *Fantasia*.[3] Thus, it is worth citing once again, as Anders writes: "We are incapable of not being able to do what has once been done. It is thus not can-do-ability [*Können*] that we lack, but no-can-do-ability [*Nichtkönnen*]" (AM II, 395).

Obviously, Anders notes the Goethean source of his insight which he traces at the same time, with Heideggerian precision, Nietzschean acuity—Nietzsche always claimed that one had to have many eyes—towards a distant project to be attained: the prospect of understanding the end time, just to the extent that our time is the time of ending things: everything, the world, ourselves, and every other thing on it.

Thus, for Anders, as for Nietzsche as I have argued in connection with Nietzsche's critical philosophy of science, and not less for Heidegger as one may also underline his philosophy of modern technology, what will be important is to consider the ultimate and further consequences.

To the extent that received scholarship reviews certain genocides but not others, Anders by contrast outlines the lockstep of the ability to destroy together with the inability to locate or place the blame on *this* people, *this* political constellation rather than that. And that mucks up everything for the political theorists, the political philosophers, the pundits, and the casual reader, who collectively find themselves asking: How dare he say such things?

And thus, we bring in the experts to tell us (no shortage of these) that Anders was a polemicist, a *Ketzer, Hetzer*: a pain in the neck. The bluntness coheres with the terms Anders himself would use to characterize school or university scholarship. The higher your position, the better ranked the school you find yourself at, the more you will fit the mould. No one at the top of the academic ladder does anything to challenge the hierarchy or the received view.

I earlier indicated that Anders refused appointments for his own reasons. Perhaps, one might wish to say, he suspected there would be no way to change anything from within. What is certain is that the only thing university appointments do is produce university rank and file, a lockstep as true for the most cutting-edge grad student as it is for the most distinguished professor. If few of us have bothered to read Anders, certain scholars have had recourse to him in their work from Peter Sloterdijk to the theologian Jürgen Moltmann, who cites Anders' differentiation of the ordinary thinking of end times, traditionally speaking, from the thinking of such times in a nuclear era which he thus describes, quoting Anders, as "a naked apocalypse, that is to say an apocalypse without a kingdom."[4] At the same time, what is all-decisive is the non-reading as this occludes Anders from the scholarly world view. Hence, in a scholarly world where Heidegger is read, even *qua* denigrated and discounted, or where Adorno can be read, however typically misinterpreted, and where Benjamin is even revered, there is no excuse for excluding Anders. And yet we do.

It is violence in perfect good conscience that characterizes war as it characterizes the post-war, the Cold War era but also the present day, including mask mandates and so on, with its mushrooming effects of neo-fascism under the titles of health and national security and the terrorist, from surveillance to full-body (meaning naked body) searches, matched on the health front with direct inoculation, literally in place of the metaphor of surgical strikes and individually-tailored, targeted Armageddon in the form of drones, all in the name of anti-terrorism. It is surely not for nothing that Žižek was a student of the thinking of Lacan, whatever else, perforce he also did with respect to Hegel, in Paris.

Where Habermas and his heirs disregard Anders (as they also manage to set aside or minimally to sidestep Heidegger and Nietzsche etc.), just as most philosophers of technology (and indeed philosophers of science) have ignored the political as well as the ethical in their eagerness to avoid suspicion of technophobia—a reserve that characterizes most political theory that considers technology. If the otherwise exceptional Bruno Latour is no exception in the question of such reservation, that's the academic deal;

hence, Latour says what mainstream thought insists upon, although sometimes, as in the case of his *The Pasteurization of France*, with a certain subtle excellence.[5] It will do to reread this text again given the omnipresence of vaccination by mandate or law.

Say truth to power in the academy with any accuracy, and you are out just where talking about saying truth to power is a safe bet. Anders, under his dark star, was always already "out," excluded from the academy.[6] Thus, if, as emphasized from the outset, numerous scholars have written, and continue to write, on Anders, it remains the case that Anders' work is not as yet "received." at the university level.

But this relative non-reception requires a bit of context, as we recall it again at this locus. Thus, it may help to recall that Nietzsche's work tends not to be discussed in philosophy departments unless tied to questions of freedom (the free spirit) and will, or perfectionism or nihilism, but not in terms of his writings on science *or* antiquity. This exclusion or narrow focus extends to Heidegger to be sure, whose Black Notebooks or his anti-Semitism, can be discussed perhaps, but rather less his work as such—thus, increasingly, Heidegger can be left out of discussions of the philosophy of technology. And then to be sure, there is Anders' work which is typically excluded in this case and others because of the now utter dominion of analytic philosophy even in what is today named (despite its often analytic style) "continental philosophy."

As a result of this disciplinary (analytic) turn away from reading the texts themselves, increasingly dominant, even in Germany and France, it will not be necessary to have read Adorno to claim to have insights into his work. But and this is part of the previous point concerning reception and the "mobbing" of outsider views in academia, much of my effort here may be read as a bid to count him in. The reader will, I hope, forgive me, if my style is also receptive to including other names along the way.

Time

We modern authors are used to positioning ourselves in time. And we long ago forgot Augustine's cautionary warning that we take ourselves to know what time is.[7] Even those who reflect on Nietzsche's Zarathustran reflections on time tend to skip over the literally contradictory contours of *Augenblick*, the intersecting courses, past and future, colliding in the gateway, *Moment*. Despite the warning title *Of the Vision and the Riddle*, Nietzsche scholars "solve" the problem, sure that there was never a problem in the first place. We scholars, we scientists, we "knowers," as Nietzsche says, pronounce on time: we claim that it speeds up (when we are having fun, when we are busy, when we are late) and complain that it slows down (when we are waiting for an anticipated event, when we are bored, when we are boiling water), and we descry and map the lines of time.

Time always seems to have a spatial dimensionality; thus, Anders reflects on the absurdity of defining, let alone distinguishing the two, and he reflects too on the absurdity of the project, pointing out that and just to be sure, and as the average person might answer that he has never once found himself in danger of "confusing the one with the other" (AM II, 350).

By comparison with Jacob Taubes and Hans Jonas and others of the day, including Löwith and Benjamin, all of whom wrote volumes or essays on eschatology and history, Anders offers us an anti-eschatology: reflections on the end, of the apocalypse, on annihilation, mutually assured and what not, which is to say that he writes about the "end time," to say that "the future has already ended." Where Anders differs from others is that he brings his philosophical, even his theological, reflections as we shall see, 'down to earth,' here to use Adorno's phrase. Anders, who has as little patience as Adorno with Heidegger but who, likewise not unlike Adorno, had no problem *using* Heideggerian insights wherever needed, could rebuke Heidegger for describing the human being as the "shepherd of being." And if religious and poetic associations serve the image of the shepherd well, the philosophical image of the shepherd has been problematic since Thrasymachus handily, floored Socrates: pointing out that there is no difference between shepherd and tyrant: from the view point of the ones "shepherded," that would be the sheep as it is they that are preserved for values or purposes not their own and not less because they are always brutally killed in the end.

But even if, through every bucolic register, one hears the language poetically, even if one hears the language through the tonalities of the New Testament, Heidegger's language still misses the point for Anders,

"The Shepherd of Being," that which Heidegger still yet very biblically, that is to say anthropocentrically, suggests—whereby he vastly overrates "the position of the human being in the cosmos" (which couldn't give a damn about whether we continue to exist or have already disappeared), no, we are certainly not "shepherds of being." Far rather we might consider ourselves the "shepherds of our product- and gadget-world" as a world that needs us, more strikingly than we do ourselves, as servants (e.g., as consumers or possessors). (AM II, 281)

The language of shepherd also appears contra Heidegger in the *Kirschenschlacht*. In general, this is the language of antiquatedness, outdatedness: the human being at an end, past its sell- or use-by date, as it were such that all time henceforth, is and can only be at an end, the end of days, the end time. Where traditional eschatologies take a leap into the mystical, the gnostic, the beyond, Anders stays squarely in the here and now. Because for Anders that is where the end transpires: not later, not in a world to come, but always already here.

These reflections on time are compelling for Anders above all, not for religio-theological reasons, like either Taubes or Jonas, and not even for the traditionally epistemological reasonings of a Kant, but just on moral grounds. If Anders thus begins his second volume on *The Antiquatedness of Humanity*, reflecting on the inversion of the Lords' Prayer, "Give us this day our daily bread" as this has morphed into a new mantra: "give us this day our daily eaters," (AMII, 15) what is required is the same culinary desperation Adorno also discovered at the heart of the culture industry: the world needs consumers, social followers, more than it needs products because, as Anders already noted, this is Heideggerian challenging forth replete with *Machenschaft*, as we may read this in the the *Beiträge* in addition to the lectures on technology and danger to the Club of Bremen, Anders is much punchier: we make products to make products to make products. To this extent, marketing and the production of market is our only occupation and preoccupation. The language of "climate change" covers the need for a new world order to put an end to this on the level of the everyday consumer. To this end all advertising, social media, psyop efforts in the same direction just to the extent that today's digital marketing is nothing but advertising? The same mediatic imperative holds, strangely and painfully in the era of Zoom instruction which is likewise a matter of such advertising.[8] Anders' point is that the only imperatives we know are imperatives consequent upon what can be done: if it can be done, it should be done. Heidegger says this too, of course, and to this day our sole concern is not with what one should do, what a quaintly Kantian question, but how we might do and how we might forever continue to do (this is the meaning of what we call "sustainability") what we can do: *Das Gekonnte is das Gesollte*. As a result, Anders has even less patience, if that is possible, with the idea that technology might be some neutral means (he has a field day with the language of ends and means when it comes to the atomic bomb and the point of its production) or that it might somehow be in our control or even within our purview. The epigraph Anders sets to the second volume is significant: "It is not enough to change the world. …" Writing in 1980, one is well beyond any imperative that would call for changing the world, in a good Marxian voice, just because, as Anders writes, we always do that anyway [*Das tun wir ohnehin*]. What is lacking is an interpretation of what we have done, especially in our times where, as he argues, our ability to act far exceeds our comprehension. Later in the book, although the text itself was written two years earlier than his introduction, dated 1979), his chapter on "The Antiquatedness of History" will make the same point with a trio of dated epigraphs—and, in a way, only the dates should strike us in this trifecta. I list them here and will expand later:

Politics is our destiny (1815)
The economy is our destiny (1845)
Technology is our destiny (1945)

Heidegger and Time: New Rules

The old commandments had failed; certainly, they were never observed. Anders thus took it upon himself to compose a set of "new" commandments for the new, nuclear age, not commandments issued by a creator, made for those in his image and likeness, but commandments for destroyers, made for an age when we human beings had learned to annihilate ourselves, along with other beings in the world, plant and animal, even the soil and the oceans, seemingly including the world as such. The "commandments" were originally published in 1957. Anders managed to secure English language circulation by sharing them with Major Claude Eatherly, the weather reconnaissance pilot who gave the go-ahead, or all clear, for the bombing of Hiroshima. Two points first: dropping an atomic bomb is differ from ordinary bombing missions. If, for the safety of combatants in modern warfare, using poison gas, as Sloterdijk argues in his *Terror from the Air*,[9] the bombers themselves, weathermen always played a crucial role, in this case one needed to know still more about wind and weather than ordinarily so, for precision bombing, given the nature of the bomb, had different implications. Secondly, the trajectory of flight path, immediately evasive, flying up and away after dropping the bomb, also testifies to this difference. Bombers are inevitably at a distance from the work of their actions; those who dropped the bombs over Hiroshima and Nagasaki were and had to be clear about the devastation they would bring because the backwash in this case could touch them in the sky. Eatherly was infamous not for having flown the mission, but for having had second thoughts about it; he was of course, like every successful bomber, a war hero.

In the "commandments" Anders sent to Eatherly, we can read, as if it were the highest moral imperative, and this is indeed how Anders meant it: "widen your sense of time."[10] Anders has his reasons for this as he introduces this broadened sense of time by calling for an equally broadened breadth of "moral fantasy"[11]: you must broaden your ethical sensibility "until imagination and feeling become able to comprehend and to realize the enormity of your doings."[12]

Concerned with the phenomenological effects of the end time [*Endzeit*], Anders was also concerned about what he calls the "guiltless guilty" as this ontological characteristic is now the destiny of the human, following the objective, physical, thingly circumstances of the modern technological era. Anders used the word "technicity," to the irritation of newspaper commentators: the same irritation has meant that scholars and popular authors could successfully ignore Anders just as they have ignored Jacques Ellul, and to a lesser degree Heidegger on the same topics. By instructive contrast, Marshall McLuhan has been inhaled by most media theorists, digital and otherwise, to this day, and this is part of the silent proscription of technological critique. *We think* (never mind Heidegger): technology can't be the problem: the medium is, the message is.

In his correspondence with Eatherly, which if my thesis here is correct ought to be read as Anders' way of communicating with American, Anglophone commentators (whereby his "sons of Eichmann" would thus be an address to German counterparts by the same token and logic), Anders did not make it difficult for those same commentators to dismiss him. Indeed, Anders put his key point, which was also his most difficult point, on the very first page, almost summing up the heart of the masterwork that has yet to be fully translated into English. Thus, Anders writes to Eatherly—this is a letter addressed to a former American airman, incarcerated for petty crimes in a psychological hospital or institution (where, for the most part, Eatherly would remain) and hence written out of the blue, as it were—by speaking of nothing more esoteric than "technification," speaking in a Heideggerian sense but no less in a Kantian sense of what Anders there describes as the

> "technification" of our being: the fact that to-day it is possible that unknowingly and indirectly, like screws in a machine, we can be used in actions, the effects of which are beyond the horizon of our eyes and imagination, and of which, could we imagine them, we could not approve—this fact has changed the very foundations of our moral existence. Thus, we can become "guiltlessly guilty," a condition which had not existed in the technically less advanced times of our fathers.[13]

By thus speaking of our "technification," the same technology on every social level that Ellul would for his part claim as the wager [*Enjeu*] of the century in a series of his own books,[14] or of what Heidegger far less popularly called the "essence" of modern technology, Anders could emphasize that it would be this same essence into which we ourselves would be absorbed. Thus, Ander's first letter to Eatherly patiently articulates the points Anders had developed in his *The Antiquatedness of Humanity*, points continued and especially articulated in the second volume, parts of which were written in the 1960s.

We *are* our tools, and we *are* our gadgets, our devices, our things, our objects. In this sense, Anders is far from today's object-oriented ontologists (I say this admitting the wide variability of these writers, and I say this noting that in some cases Anders is even cited—and the sighting of any citation, in the wild as it were, is rare enough). But Anders differs. He does not think that we can simply think the thing, the object, the gadget, and his reason for this reticence is the very hermeneutic and phenomenological reason that this objective is not accessible to us simply because we are already the object of technology as the subject of history, and hence we are ordered to (in this sense as we saw above we are the 'shepherds' of), we are claimed by things, by objects. The fact that we have made them is quite irrelevant, and this irrelevance is the scope, the range, the breadth, the sheer size (this is Jünger's titanism or giganticism), of modern technology. As we shall see, this same signal irrelevance of the connection between what we know and what we have made or done, *pace* Kant or Vico, likewise echoes in Anders' reflections on Goethe's *Zauberlehrling*/"Sorcerer's Apprentice."

Anders' main concern was the non-neutrality that Heidegger for his own part also emphasized at the start of *The Question Concerning Technology*. Good or bad, neutral or non-neutral, either point is committed to the same. Anders' argument is that once we have an object, we have it. Because it is the object that has us, we can—as a result—claim *neither* detachment *nor* sovereignty. Other authors reflecting on technology have made similar points in similarly uncompromising fashion, especially Heidegger and Ellul, but what bears further reflection is that Anders' point would not be directed to the ontological circumstance of doing and not doing. Thus, Anders was more concerned for very phenomenological purposes with "having." And this also meant that Anders' concern was with the inescapably moral fraughtness: of being "guitlessly guilty," and this is what it means for all of us, to accept the designation of banal evil as a descriptor for all of us, every one of us a son of Eichmann: Hiroshima everywhere.

The condemnation for Anders is the damnation of being *and* not being in the context of the things of our age. There is no way to be, simply to be, in the world in the wake of the atom bomb. Thus, Anders reflects in 1966, contra Lukács and others, that given the literally "negative religion" that was the atomic fact—and by no means only the mere threat of nuclear annihilation—everything the past century had previously considered under the rubric of nihilism, by comparison with that same "possibility of 'annihilation' turned out to be sheer culture-hall nonsense." For Anders, "Nietzsche, even the beastly serious Heidegger, come across as laughable before the madness [*Folie*] of this possibility" (AM II, 404). The "possibility" is that of a literal annihilation, the "creation," the relative *production* of nothingness, eliminating all humanity and culture and all history with it. The question of nuclear annihilation thus explicitly extends beyond the Heideggerian possibility of impossibility. This is of course the heart of what Anders, a good Heideggerian, had to mean by *The Antiquatedness of Humanity*, which is of course nothing but the "Antiquatedness of Dasein" and precisely *qua* Dasein or as such. What is at stake for us as mortal beings is no longer anything so classical as our mere mortality, that we, as beings who can die, are bound to die and bound to the loss of our ownmost possibilities for being but and much rather that today we are no longer "mortal" but have been converted into simply "'killable' entities" (AM II, 405).

Anders concludes the section by denouncing the situatedness of dying one's own death, just as Rilke had spoken of this and of course and to be sure as Heidegger had earlier made his own claim to the same. Here using a Heideggerian argument against Heidegger, Anders goes on to argue that what is singularizing about dying is that the individual's loss of his own singularity in dying is and can hardly be one's "own" (AM II, 407).

For Anders, we human beings are no longer in a position to simply regard our lifetime, even as Mallarmé might have done, as simply random, as chance tossed into the realm of possible being, or as Nietzsche wrote: "a hiatus between two nothings" (KSA 12, 473).

Anders' *Commandments in the Atomic Age* are mortal reflections as he writes to Eatherly, and as is immediately clear upon reading them, offer an array of *spiritual* exercises. Much rather than a refurbished vision of the ten Commandments, these are to be read as rules for the soul's direction, meditations of a Stoic kind, beginning, just as Marcus Aurelius begins his *Meditations*, written to himself, Book Five: *let this and not that, be your first thought upon arising.*

The point here is that there has been a reversal, a turn, a change, and things are now and forever more no longer as they were. If that sounds extreme, it is only because Anders remembers, as Benjamin does, what makes history history and that prerequisite is *always* a recording hand With an angel, we are covered even after the apocalypse. Take away the angel and you have as Nietzsche also reflects, as he writes in the parable of the mad man who comes to seek and then to announce the death of god in his *The Gay Science*, that having murdered god—"*We have killed him*—you and I. All of us are his murderers" (FW §125)—we have at the same time managed "to wipe away the entire horizon." Nietzsche continues to elaborate the very last words of the Christ as he hung on the cross, asking for forgiveness on our behalf, because we, his murderers, guiltlessly guilty, had and could have had no idea what we were doing. As Nietzsche continues to quote his madman:

"…What were we doing when we unchained this earth from its sun? Whither is it moving now? Whither are we moving? Away from all suns? Are we not plunging continually? Backward, sideward, forward, in all directions? Is there still any up or down? Are we not straying, as through an infinite nothing? Do we not feel the breath of empty space? Has it not become colder? Is not night continually closing in on us?" (FW §125)

The scene of the *Commandments* as Anders' translator put his *Meditations in the Atomic Age* is as bleak. In the wake, not of the death of God, but the explosion of the power of stars, we are, in Anders' terms, "killable": as humankind, and not only henceforth but in every other sense as well. Thus, humanity as such is not only limited to "today's mankind" or "spread over the provinces of our globe; but also mankind spread over the provinces of time."[15] The expanse is literally unimaginable—which does not mean that Anders has any trouble explaining it, and he gives Eatherly a little lesson in history as he does:

> For if the mankind of today is killed, then that which has been, dies with it; and the mankind to come too. The mankind which has been because, where there is no one who remembers, there will be nothing left to remember; and the mankind to come, because where there is no to-day, no to-morrow can become a to-day. The door in front of us bears the inscription "Nothing will have been" and from within: "Time was an episode." Not however as our ancestors had hoped, between two eternities; but one between two nothingnesses; between the nothingness of that which, remembered by no one, will have been as though it had never been, and the nothingness of that which will never be.[16]

Anders' own expression is shot through with the Nietzschean language of the door or the "gateway," the formula of the "two nothingnesses," likewise Nietzschean, but the tenor and the tone is Heideggerian hermeneutic phenomenology: a meditation on being and having been, on being and not being. This is also the Sophoclean μὴ φῦναι,[17] as Nietzsche reflects on what it would be never to have been at all, where just this is, as Nietzsche also reflects, utterly impossible for humanity, which leaves us the curiously second-best option of dying soon, as Yeats translates Sophocles into the last lines of his *A Man Young and Old*:

> Never to have lived is best, ancient writers say;
> Never to have drawn the breath of life, never to have looked into the eye of day;
> The second best's a gay goodnight and quickly turn away.[18]

Anders, who brings to his reflections *literary* considerations amid philosophical and theological considerations, makes his argument in the high spirit of the original Frankfurt School (neither Habermas nor Honneth need apply, nor, to be sure, would they wish to). Thus, Anders compares "consumer terrorism," that is compulsory consumption, to the even more significant compulsion to use. This is the compulsion of the applied. Applied terrorism is the terrorism of what happens to be on hand, what is available for use, and this applicable and therefore deployable terrorism is for Anders quite literally the reason atom bombs were detonated as they were and in the first place. One can make a similar argument concerning new pharmaceutical means now ordered for use, globally: the idea is to vaccinate the world.

In context at the time, as Anders here points this out, President Truman happened to have had two bombs available; therefore, there would have to be two targets. The only question was *where* they would be. That is the space question. The time question concerned only how soon they could be used. And given diplomacy and the ontic details of concluding the Second World War, Germany was out of the question, so the space in question, the where of the bomb, followed the question of time, the when of when the two bombs one had on hand, could be used.

But beyond consumer-terrorism and applied-terrorism, beyond having become less mortal than mere "killable" beings, Anders reflects that we are killed when we are killed by an atom bomb not by human hands, and by nothing so old-fashionedly humane as human intention or human passion. We do not die at human hands because *hands* as such do not, for Anders, enter into it at all. That's the point of obsolescence: one takes the human component out of the equation. Like Major Claude, who gives the all-clear from his plane, *The Straight Flush*,[19] and thus like the command to execute the mission, like the bombers of the Enola Gay, who dropped the ridiculously aptly named hydrogen bomb: Little Boy; such a death when it comes, would come, either shades of Eichmann (but with drone warfare the shadow falls more clearly), from agents somewhere, even thousands of kilometres distant from their target, following orders in accord with duty, or indeed through brainless and sightless machines, which have long since been emancipated from the hands and the intentions of their creators and users (cf., AM II, 406).

Far from any symbolism, the "apocalypse" for Anders could henceforth have nothing to do with any kind of second coming, any sort of new Reich, any last judgement, or anything at all that one might need to "interpret." What we no longer have is hermeneutic esotericism: there is no "meaning" in need of subtle divination.

Thus for Anders, today's end-time is of a "massive" sort, on a scale of the Lacanian Real, and beyond both the Imaginary and the Symbolic. For this possibility (and that means if it is a matter of technology: the inevitability) there are historical examples: the facts of Hiroshima and Nagasaki and today's related facts concerning the secret from-no-one calculation regarding the "overkill" capacity of today's stockpiled weapons. In our "situation," the fact that the end has yet to enter in quite as we might have expected it is no refutation of the reality of the danger, no counterdemonstration of the fact that our time is *a*, indeed *the* end time.

The "Now" of this fact of the facticity for all and for each one of us of what has been, of what has been done by human beings, lies (or better said: should lie) as a weight upon all human beings. This is for Anders the Promethean *guilt* of action, of original sin, and it has been a problem since the time of the change of the gods. For the ancient Greeks, this was the change from the age of the Titan to the Olympian gods, for Jews and for Christians, this goes back not only to Adam and Eve but above all to the time of Cain. In another way of telling the story, this guilt or acquired shame has been with us since Enkidu stopped to sleep with the woman of the city paid as she was to seduce and betray him, and who as a result lost the patience, the grace, *the time* that allowed him to run in innocence alongside the gazelle, the lion, and so on. Thereafter, Enkidu, the wild man, would not free the animals from the traps that city hunters set for them, himself caught in and by another kind of city hunter's trap, he would be lost to his forest companions, with little to do except follow the whore who had come to lure him to the city.

Sin, for Anders, Promethean shame, needs no specific confession: it is neither Jewish nor Christian nor pagan—think of the contrast Nietzsche makes between Prometheus' and his Titanic and Semitic notions of sin, as this suffuses the Judeo-Christian tradition in *Genesis*. It is the human condition that we be ashamed of having been born, as Anders writes of this, as we are conscious of our inadequacies, our frailty. By the same token, we are hell bent on becoming, at any price, more than that, more than we are. Our tools, our objects, our tanks, our planes, our bombs, our bioweapons, our vaccines—these days such things also include our digital prowess in social media psychological manipulation—seem to be just the ticket. And it all starts with a fig leaf, the flag of shame, the particular piece of technology that serves to hide our nakedness. And as Giorgio Agamben reminds us in his reflections on fashion, the fig leaf is a convention that covers over a more shameful "theological signature," which Agamben manages to connect with the *kairos* of fashion as such whereby he corrects the conventional expression of a fig leaf parenthetically:

> (To be precise, the clothes that we wear derive, not from this vegetal loincloth, but from the *tunicae pelliceae*, the clothes made from animals' skin that God, according to Genesis 3:21, gave to our progenitors as a tangible symbol of sin and death in the moment he expelled them from Paradise.)[20]

In *What is an Apparatus*, Agamben usefully explicates Anders' term: *Ge-rät*, clarifying Foucault's *dispositio*, we can overlook the constellation with technology that traces Heidegger's *Ge-Stell* as Agamben reads it, explaining that in *The Question Concerning Technology*, Heidegger

> writes that *Ge-stell* means in ordinary usage an apparatus (*Gerät*), but that he intends by this term "the gathering together of the (in)stallation [*Stellen*] that (in) stalls man, this is to say, challenges him to expose the real in the mode of ordering [*Bestellen*]," the proximity of this term to the theological *disposition*, as well as to Foucault's apparatuses, is evident.[21]

Readers have remained at the end of the array traced here, namely with the dispositive, but reading Anders we may want to read Agamben's text here from the start, beginning with the skin of an animal to flag the violence of the Fall. Thus Agamben uses the same temporal frame with respect to the Augustinian question of time as Anders also frames it but as Agamben writes "of an ungraspable threshold between a 'not yet,' and a 'no more.'"[22]

With the atomic apparatus, humanity succeeded in crystallizing the terror of laying siege to a city, wasting it, compressing it down in time and spatial act to the pressing of a button, mere minutes from start to finish. Over and out.

At least in theory—and as Anders, already writing to Eatherly took care to note (and in the interim his point has only been made all the stronger, in ways unimaginable to most of us—not that we think about it): the bomb, although hardly ever thought about (this non-thinking would be different for Major Eatherly, who knew such things far better than most) was no static achievement. Indeed, since the bomb was developed, progress consisted in further perfecting it, meaning as this was hardly lost on Anders, that that same project to develop a better bomb was all and only about increasing its deadliness, gain of function research is part of this, magnifying the destructiveness of a negative genie-in-a-bottle.

The problem with the project from the outset, following Hiroshima and Nagasaki, was only that the genie had already been out for a detonating excursion, twice over. As Anders put it:

> For the goal that we have to reach cannot be *not* to have the thing; but never to use the thing, although we cannot help having it; never to use it, although there will be no day on which we couldn't use it.[23]

It was Anders' technically attuned thinking, student as he was of Edmund Husserl—if first he wrote on logical categorization, Anders' later book on "Having" concerned epistemological ontology[24]—and of Heidegger; it was thus his techno-epistemological sensibility that led Anders to reflect on the consequences that follow simply from what we do as modern, technical human beings, living at a tempo like none before: our time is "the completely new, the *apocalyptic* kind of temporality, *our* temporality."[25] This temporality is the end time: all time henceforth must be counted from here and accordingly, and because we are at the end, we affect the future, any *possible* future, like no other epoch in the history of humanity.

Anders offers one of the first articulations of a point we now so take for granted that we simply refer to the concept by a number, counting generations—we count, biblically of course, seven generations, and then because it is now a cliché, we stop thinking about it. As Anders explains:

> the people of the Western world, since they, although not planning it, are already affecting the remotest future. Thus deciding about the health or degeneration, perhaps the "to be or not to be" of their sons and grandsons. Whether they, or rather we, do this intentionally or not is of no significance, for what morally counts is only the fact.[26]

For his own part, already in the *Gelassenheit* lecture and afterwards, Heidegger too calls attention to the manufacture, as it were, the technical production of, the human.

The point here is that the only thing that matters is our objects, that is, what we have, what we possess, and what we have done. Therefore, there is no question of intention, there is no question of rightly or wrongly deploying such objects. Atom bombs, napalm, or, quite to make it real in the current day, drone strikes, fracking, nuclear power plants, GMO crops, but not less immediate or real: chemtrails *and* 5G *and* HAARP and medical incursions contra personal freedom, including intrusions on the body itself (Foucault's biopower cannot compare) and thus the elimination of personal privacy, intruding indeed on even the possibility of love, überveillance, the new viral vector/GMO style 'vaccines' and so on. These things, these measures, *cannot* be used well.

Time/Space

Time, as we have seen that Anders also reflects upon it, is always found to have a kind of topology, a spatial dimensionality, complete with the topographic features of a particular landscape—think of Dali's *The Persistence of Memory* or think of *The Twilight Zone*'s milder television metaphors: we are time-travellers of an antique adept's variety, less the high future of a *Star Trek* cruising the edge of a singularity in space–time or even *Doctor Who* than the late nineteenth-century future of a Jules Verne. Invoking Schlegel's description of the historian as a backwards-turned prophet, an image doubtless precisely relevant for Benjamin's description of the facing orientation (Figure 6) of Klee's "Angel of history," Anders suggests that we need to demand the same of today's prognosticator or futurologist.[27] We "hitchhike" in our fantasies equipped with an imaginary scientific vision, a mere hundred years old, of time-travel via rocket ships and jet-powered speed, to take us, thank you Albert Einstein, back in time without noticing it. Thus, suitably steampunk, we prefer nineteenth-century cabinets, and our *Dr. Who* needs no spacesuit, and Bill and Ted make do, American Style, with an aluminum, midwestern phonebooth. Nor is it an accident that the latest language to describe the (imaginary) transforms of the digital are borrowed—hat tip to Evgeny Morozov and Jussi Parikka—from Harry Potter's creator, J.K. Rowling: the *horcrux* is the perfectly articulated image for our multitasking minds.

Rowling, creatrix, as it were, of the "horcrux," had her own borrowed rabbit (or lion) up her sleeve or tucked into her hat, even if she did not name the master of Wonderland and its topographical transforms, morphological shifts of size and form, down the rabbit hole and all. The mathematician author, Lewis Carroll is thus the poster boy, the ideal author of the digital era because even with no acquaintance with Alice, and no acquaintance with any of her adventures (who was the rabbit? who was the walrus? who needs any of them?) we have Angelina Jolie forever—in her gaming avatar *avant la lettre* in *Lara Croft: Tomb Raider*, we have the very idea. Mentioning, the mere mention of the wondrous is all we get *and* all we need: we know everything we need to know about the mathematico-logical transform of our new projected selves.

We are, aren't we, by now? Transhuman, posthuman, humanity 2.0? Surely we're due for an upgrade to humanity 3.0 or even 4.0.

And then, just for the locus of the boggart in the wardrobe as such, Rowling also had her C.S. Lewis. I have mentioned boggarts and wardrobes (it is a wonder that I do not invoke Alan Rickman in the same breath), cabinets and time-travel, because when we shift levels (and note that we are still talking topologies), one should be struck by the persistence of our representation of time as time *in* history, always a picture, an image, iconic. As if we might be surprised that anything with two dimensions might be other than a picture.

Adorno would give Anders' competition on the question of time and technology, as he too was also struck by iconic, canonic time, as Berthold Hoeckner rightly notes.[28] And this is always a claim with particular insistence in Adorno where music is, of course, the *art of time*. With music we are also always, and even if Hoeckner is, like most musicologists, most philosophers, most academics, inattentive to Anders (Stern), speaking about Anders, who also (as Stern) offered his own reflections on time, musical time,[29] as phenomenologically, as hermeneutically as Hoeckner himself.[30] Hoeckner, like Anders, like Adorno (if also, although Hoeckner does not note this, like Nietzsche), attends to the time of the now—*Jetzt-Zeit*—in his discussion of the "star" in Beethoven, echoes of constellations important for Adorno as for Benjamin, Anders, and, indeed, Schoenberg.[31] Quoting Adorno's "aesthetics of appearance" (under the important presumption of an allergy to Heidegger that spares any engagement with the notion as it also appears early in Heidegger's *Being and Time*), Hoeckner characterizes Adorno's "aesthetics of *Augenblick* as an aesthetics of apparition: 'the artwork as appearance approaches most clearly the apparition, the celestial vision.'"[32] Just these lines of thinking are also to be found, traced, and elaborated in this context in Anders. For Hoeckner—and here one misses a discussion of both Heidegger and Nietzsche—what will be needed is a "hermeneutics of the moment."[33] With this desideratum, the author must disentangle himself from Adorno, who exemplified perhaps more than any other author the lived anxieties of influence (not only Heidegger and Gadamer and also Anders and the same Habermas Adorno had intellectually discounted but also, and certainly, whether we like it or not—and we do not like it—Hannah Arendt as well). In addition, there are other authors who also write on dialectics and time in conjunction with Benjamin, making very close arguments for Hoeckner regarding Adorno's supposed lacks, as Günter Figal has analysed these. Focusing, as Hoeckner does, on Adorno's attention to the standstill, Hoeckner disagrees with Figal. There are less lacunae in Adorno than an abundance of eyes, as it were—the image of the Argus-eyed is significant as it should be for Hoeckner's reading—than a veritable constellation of insights into that same dialectic. Thus, we read that "what intrigued Adorno was Benjamin's objectification of the historical process in the image."[34] The key passage everyone cites from Benjamin's *Passagen-Werk* is thus worth citing here:

> What has been coalesces in lightning like fashion with the Now. In other words, the image is the dialectic at a standstill. For while the relationship of the present to the past is a purely temporal one, the relationship of what has been to the Now is dialectical, of a pictorial rather than a temporal character.[35]

The point made here overlooks a key point in Nietzsche (and it is instructive that authors, for all their enthusiasm, are at pains to keep Nietzsche's points at a distance from their own). In addition, there is the eschatological as such, in this case the very picture of it, which is the picture book Dante, in the images inextricably associated with him since the 1850s, not only for us today but for Anders, and Adorno, and Benjamin ever since Paul Gustave Doré's illustrations came to stand in Dante's name and place, an achievement arguably to match that of any other illustration in any other book.

Doré's pen drawing of the Empyrean in Dante's *Paradiso*, Canto 31, published mid-nineteenth century (Figure 7), combines as rebus both the ideal of heaven after death and the power of the sun. The same figure, the same combination can seem to have been articulated by J. Robert Oppenheimer using the language of the Vedic tradition, "Now I am become death, the destroyer of worlds."[36] As Peter Sloterdijk takes up this same association, as we may compare this with the iconic image of the Trinity explosion on July 16, 1945 (Figure 8), the "Bomb is really the only Buddha that Western reason could understand. Its calm and its irony are infinite. . . . As with Buddha, everything that could be said is said through its existence."[37]

Figure 7 Paul Gustave Doré, Dante, *Paradiso*, Canto 31. The Saintly Throng in the Form of a Rose. Wikicommons. Public domain.

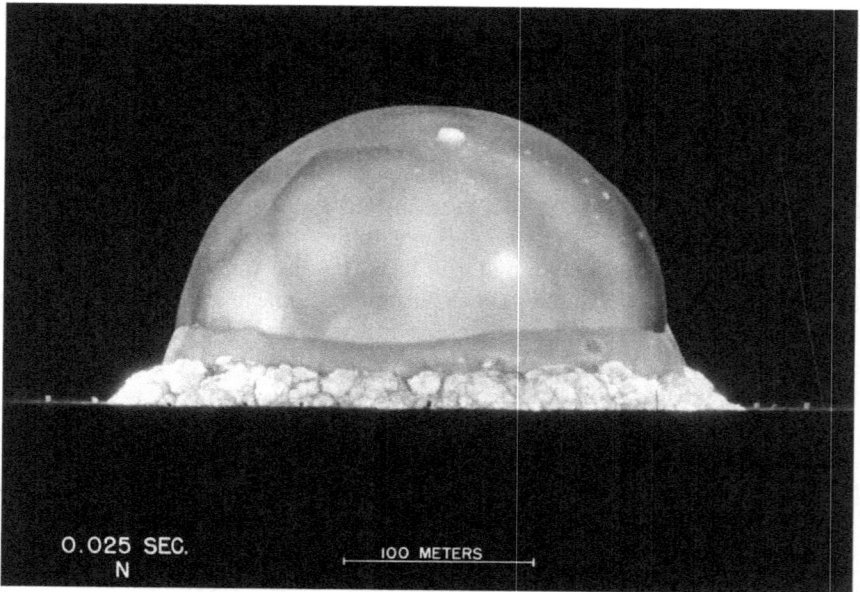

Figure 8 Trinity atomic bomb explosion, Rapatronic image. July 16, 1945, 05:29:45, Mountain Wartime. Alamogordo Test Range, Jornada del Muerto Desert. Courtesy of US government Defense Threat Reduction Agency. Wikicommons.

Anders' *Endzeit und Zeitenende*[38] requires reference to Nietzsche's eternity as this is the moment, the *now*. Again, and as already intimated at the start of this chapter, this is mapped out in space, a space of infinite dimension, fore and aft, as Nietzsche depicts it and without which dimensionality it is impossible to think the *Augenblick*, the winking, the blinking of an eye, as Nietzsche also names the moment. Time stands still and in what Nietzsche could describe as two roads, mapping infinities past and future, the crossover, the junction is the moment, again: *Augenblick*, the same word Adorno uses.

And why not the moment, the blink of an eye, an image which already closes off the seen, relegating it to a lost glimpse? Why not in Anders' time, in Adorno's time, Benjamin's time: a time when the apocalypse seemed sure just because as Anders emphasized with respect to Hiroshima, and although we scarcely like to talk of this at all, in Vietnam, or in Iraq as Baudrillard did not fail to try to tell us, or closer to home for the German Anders, already in Dresden, as Winfried Sebald has reminded us, it had *already* taken place? For Anders, starting with his own experience of it, the First World War had already done that and the Second World War, as that came and ended, not once, but twice, with two bombs, could not but repeat the same message, once more with feeling, and a reprise, *da capo*. The encore at the end of the Second World War, and the constellation, the order of events would matter for Anders, changed everything beyond rectification, beyond redemption or correction. Anders thus reflects on shame or denial: on what has been and what we have done.

This is, for Anders, in his retrospective reflections on the "Antiquatedness of Space and Time" part of the problem, emphasizing, as Gadamer would also always do in his lectures when I was a student, the importance of consummation, *satisfaction*, fulfilment, what Anders simply called "having." It is instructive that Anders begins his 1959 reflections with the illustration of "*Schlaraffenland*" (AM II, 335) but it is even more significant that we can barely translate this term into English, although we Americans have perfected its realization on earth arguably more than other people, at least in the Disney version. *Schlaraffenland* is a world where sausages leap perfectly broiled, perfectly willingly, into our always-hungry mouths, no effort at all, guiltlessly, automatically, and in this child's fantasy (not really for children because there is beer that has the same eager proclivities to satisfy any thirst we might have); the only name we have is Candyland, or the media obsession with the heaven of a certain confessional persuasion: complete with a given number of promised virgins springing, not unlike the sausages, unbidden, uncoaxed, compliant into the martyr's arms.

Our age crosses space and time, obliterating, as Anders also emphasizes all distances, spatial and temporal. We are effectively, as he argues, rendered by technological means into spaceless, timeless beings, not in the sense of transcendence but imperviousness, blindness. This is *apocalyptic blindness*, and thus we no longer have any sense of history or memory. But the problem of the modern time-less (lacking time as we do), space-less (lacking a sense of the world around as we do) way of being is precisely that it transcends nothing at all. We are, as Anders goes on to argue, mediated in all of this by our technology, which is always to be found just where we put it: precisely, exactly "in the 'middle' of the fulfilment of needs or 'facilitating' ['*Vermitteln*'] the manufacture of products" (AM II, 336).

Whose Holocaust? Which Genocide(s)?

Who shall count? As Anders asks with respect to everything from music to political sensibilities, *who* gets to be counted?

If some have followed the apotheosis, as it were, of the cattle car as this was borrowed, along with the entire factory slaughterhouse project, technique, assembly line-layout, and so on, from Chicago's stockyards and thence to Auschwitz, Dachau, Buchenwald,[39] we can also trace the lines, the tracks of the trains that ran throughout a war of destroyed transports. These tracks that could have been bombed were never destroyed, and Hitler not only got the trains to run on time, but the trains that fed the final solution ran without fail. A transport always arrives at its destination, to vary Jacques Lacan while keeping the same spirit. In the same spirit, these are the ashes of which Derrida also speaks, Anders talked about things not even a Klee could illustrate.

No paintings are possible; one is immediately moved to cite the film *Hiroshima, Mon Amour*, and even that shudders. Meshes of non-representation. Hiroshima, Nagasaki, and we have no idea what we are talking about. And then students of Adorno prattle about a *Bilderverbot*. God forbid that we care to speak of this, of these people, foreign to us, in foreign places, alien beings; who are they?

We continue to require both critical theory and a critical philosophy of technology, a conjunction incorporating Anders' complicated dialectic less of art in Benjamin's prescient but still innocent age of technological reproduction but and much rather Anders' reflections "on the devastation of life in the age of the third industrial revolution" (AM II, subtitle). Thus, Anders would talk not about enemy fascism (which was an easy sell as many authors know to their advantage) but and much rather the American, the good-guys, the non-fascist, non-(supposedly)-totalitarian, but very *democratic* (despite its complete secrecy) contraversion of just-war ideology, transforming it into just and only a war after the war had ended. For all by themselves, in the midst of the Japanese effort to surrender—surrenders are diplomatic things, that take diplomatic intervals of time, negotiation, the business of sovereignty and legitimacy—the bomber's planes would fly as for weeks, indeed for all the years of the Manhattan Project, it had been planned to fly just those planes, to send them somewhere appropriate just in order (that would be the end in question) to drop the winged death, the apocalypse itself. The end fruit of that same project was two bombs completed just prior to the end of a war (but when does anything end?) that was finished just a touch too soon before the planes (these would be the means) were nonetheless launched to destroy cities full of people.[40]

If scholars dispute whether one can claim that ordinary Germans knew or did not know about the Holocaust, Holger Nehring reminds us that in this case there is nothing to dispute. For more than sixty years, German authors have been at pains to argue, like Nehring, that no one can make that statement about Hiroshima, about Nagasaki.[41] And yet even this point can miss the point. We are, we remain still in the dark about the atomic attacks on Japan. Thus, if the aforementioned description of the timing or the necessity for the bombs dropped on Japan sounds like an overstatement, that is because, as Americans, we continue to be in denial; we, as Anders offered Eatherly a diagnosis for his mental distress at a distance, are traumatized. And this trauma today is the result of, as trauma always perpetuates itself as trauma, by means of suppression.

The development of the atom bomb was a secret during the Second World War (not only the project as such was a secret but three different locations were created, likewise in secret, in Oak Ridge, Tennessee (uranium), Hanford, Washington (plutonium), and today the best known of these: Los Alamos, New Mexico). As one cultural scholar, David S. Bertolotti, has observed, the development of the bomb, which involved building the aforenamed cities *from scratch*, was arguably the best-kept secret of the war. I have cited Bertolotti's analysis because he does not adumbrate a theory to reconstruct in retrospect, one way or the other. Bertolotti argues by citation and object illustration, via actual newspaper publications from the day, the very old-fashioned kind of media archaeology, by way, with object documentation, of a study of print media as the means of both suppression and controlled dissemination (translation: that is propaganda, translation, to borrow the language of the masthead of the *New York Times*, that is "all the news fit to print") during the Second World War.[42]

The closest we have ever come to this was Dresden, also an aerial destruction, angels again, firebombed by the British bomber Harris, who, it is said, knew what he was doing. Winfried Sebald, in the English version of his book *The Natural History of Destruction*,[43] used the nihilistic language of Lord Solly Zuckerman, the architect of the Dresden firebombing, to title his book, and it is an uncanny title. Sebald quotes the Swedish journalist Stig Degerman's 1946 report of nothing so much as a landscape of destruction, a landscape at which no one of the inhabitants turned their heads to look:

> writing from Hamburg, that on a train going at normal speed it took him a quarter of an hour to travel the lunar landscape between Hasselbrook and Landwehr, and in all that vast wilderness, perhaps the most horrifying expanse of ruins in the whole of Europe, he did not see a single living soul[44]

The whole point, the whole purpose, the sole, the one and only end, of waging war is terror. The reference is to Löwith, to Jaspers, and, indeed, to Herman Kahn.[45]

The point can and must continue to be made, and thus I will return later in the text to Alan Rickman, not with the popular reference to one Severus, named as Marcus Aurelius thanks his own tutor for teaching him the principles of Stoic Republicanism, but for his last film deployment in *Eye in the Sky*, serving propaganda quite as the culture industry always does.[46]

Once More, With Feeling

A student of Husserl (again it remains important to say this first when it comes to the prospect of articulating a phenomenological socio-anthropology), Anders was also a student of Heidegger, although that requires a separate study, and a student of Max Scheler (again, another study is essential, as Anders owes some part of his own ethics to Scheler's influence in addition to his own spiritual and practical sensibility). Here, Heidegger is central to the reading I have offered. If Anders' scholars tend to eschew Heidegger (and if Heidegger scholars return the favour by ignoring Anders), Heidegger's reflections on technology remain decisive for Anders. I have argued that one needs to keep Heidegger's criticisms in mind to read Anders as these are crucial for Ander's critiques of Heidegger. To do this, it is necessary to go beyond the limits of Heidegger scholarship as even Heidegger scholars show little patience for the sustained and thoroughgoing character of Heidegger's interest in technology as indeed in modern science, both of which Heidegger thought in terms closer to Anders' preoccupation with the same. Heidegger scholars can be the least valuable resource owing to their concern to excavate their personal favourite theme, which means they tend to cut references to Nietzsche, to either overfocus on or else to exclude Hölderlin (because who understands him?) mixing and matching Rilke (why ever not?), to nail that same personal interest to the wall as the whole of Heidegger: be it being, be it meaning, be it objects, be it god or God, or anti-Semitism or what have you.[47]

A full elaboration of the relationship between Heidegger and Anders surely remains to be articulated but elements of such a reading are offered here, and other aspects glimmer in Sloterdijk's recent work. This is the sheer thatness of its *having been* (where, once again, the Nietzschean dialectic of the "having been" reflects the essence of modern technology) as deployed, as put to use, as this also functions as programmatic aggression advanced in the name of defence and deterrence. Thus the tactic of automatic aggression is carried out in good conscience. The very notion of the "pre-emptive strike," as Baudrillard observed again and again towards the end of his life, absolves the perpetrator (the English would not be blamed for Dresden, and the oddity of Eatherly's conscience was not that it, in Anders' expression, burned but its contrast with most Americans who have or feel they no blood on their hands for Hiroshima). The blame, again, is assigned to the opponent: such are the wages of evil.

The claim of innocence did not sit well with Eatherly,[48] and similar discounting claims, as if nothing had happened, were hard on soldiers who had fought in Vietnam, especially after their return to everyday life in the United States, and the same dissonance of oblivion continues—today we name it post-traumatic stress, classified as a "disorder"—for today's fighters in the Gulf, Afghanistan, and so on.

Sloterdijk analyses this "shock" at the end of his book, *The Critique of Cynical Reason*, even going so far, and the present author is grateful for this, as to invoke Anders. But Anglophone readers looking for the next new thing have never read Anders (who was never the next new thing, perforce not, having never been translated into English) or Sloterdijk (who was), and those looking for *today's* next new thing (and it is only today's new thing that matters) cannot go back and read what they did not read in the first place. Thus, we trust young scholars who, as Nietzsche once expressed it, "have thoroughly unlearned the art of reading." And by the time anyone notices a lack, those same scholars will have moved on to where they wished to be, in time, as already noted above, to be replaced by the next set in turn.

As Žižek has observed, as Sloterdijk observes both of them following Virilio on the same theme, the war on terrorism is infinitely fightable and wildly adaptable, transformable. Indeed, our enemies are beautifully invisible: a powerfully convenient antagonist, and the invisible and therefore omnipresent enemy serves as today's transformation, the perfection of the sheer automatism of war. Today of course in the era of Coronavirus the logic of the same war on terror is turned against ourselves and our neighbors. We are all of us the target, assuming that everything is the same for everyone, an easy assumption if you believe what you see on the internet.

The invisible enemy all around us is the equivalent of the acephalic and therefore perfect soldier of past war fantasies as Sloterdijk invokes these to conclude his *Critique of Cynical Reason*.[49] Of course there is more, as the NSA has undone the old joke—we have met the enemy and he is us—by making it come true, literally so. Add to that the new laws hastily instituted everywhere, criminalizing protest and "outing" anonymity. Only today's virus pandemic outdoes as it also fulfils the "war" on terror. In any case, the war on terrorism, like the Coronavirus "war," is one like most wars fought in good conscience. The perfect war for the "guiltlessly guilty" who fight infinitely, without remorse.

But the full, as we cinematically say, technological metal jacket entails that we today use other means, geological, meteorological means for waging war, and we pretend that we have no choice, we pretend that we need energy (although Anders pointed out that our perpetuation of our supposed "need" for energy was a calculated choice of a particular and particularly non-convivial way to live, to use Ivan Illich's language as he makes a similar argument, a result of a politico-economic option to ignore the abundance of energy for the economic sake and advantage of the industry strictures, the restrictions of monetizable specific, limited resources that would then justify the utter destruction of the earth, water, air, everything). Obviously, I am speaking of fracking but also deep-sea drilling to go with the heedless destruction of the seas by industrializing fishing to reach proportions of the same apocalyptic force that is the theme of the current chapter and to which, with a different emphasis, I will return in the concluding chapter.

"What is decisive," as Adorno wrote, "is the absorption of biological destruction by conscious social will. Only a humanity to whom death has become as indifferent as its members, that has itself died, can inflict it administratively on innumerable people."[50] And I would extend this, as Adorno would, to animals as I would also extend this, as Nietzsche would, to the earth itself. Our trouble, and hence our continued interest exactly in Eichmann—and not as Anders would say in "Eichmann's sons," for we are, although commentators dispute just this, all of us his children—where Eichmann is only *pars pro toto*, a signifier for the story we tell ourselves that all our troubles in war, past and present, are always and only about the enemy: the Nazi, the Russian, the phantom Al Qaeda operative—like a Teenage Mutant Ninja Turtle, an invisible, omnipresent opponent so convenient that we could hardly resist inventing him—and so we did.

If civilian death and the destruction of human, individual habitations and the conditions of maintaining a life was always both deliberate and regretted and thus a problem in war and so a necessary evil in the case of Dresden, Hiroshima, Nagasaki, such deaths need no longer be regretted because they are no longer collateral. We send drones to kill civilians, we attack supposed "terrorist" sites and cells and incidentally, having to search them out at night, kill and rape women, children, and so on. We listen to Žižek because we no longer have Baudrillard to make these points, not that scholars ever listened to Baudrillard in his lifetime. And as we have seen, and for the same reasons, university scholars paid no attention to Anders in his living years.

6

Anders and Adorno

Genocide

On What We Have Done

It is significant that publishers gave Anders as much grief as they did. Thus, as already noted, his Munich publisher, C.H. Beck, refused until the very year of his death to bring out his novel, *Die molussische Katakombe*. I have already suggested that rhythmic forms matter for this but here we note a proverb from that long-standing project as he sets it into the closing lines of the antepenultimate paragraph of his 1980 volume of *The Antiquatedness of Humanity*:

> In Molussia there was a proverb, which in German went as follows: "*Die Dinge foltern, bis sie ihr Geständnis ablegen.*" [Torture things, until they confess.] (AM II, 429)[1]

The present chapter raises questions that are *awful* in every sense with regard to what has been done to human beings—done *by* human beings *to* one another, *to* animals, *to* plants, that is trees and grasses, and every flowering thing, *to* the world around them. Here it matters to note following the previous chapter, that typically, despite the dedicated rapacity often involved in such doings, these are for the most part acts undertaken in/with a good conscience.

At issue is a question of who we are as human beings and what we, as human beings, have done. Here, Anders insists upon the difference, as did Gabriel Marcel who, as has been noted, himself takes over the notion from Anders. Now Marcel is explicit about this borrowing, and he cites "Dr. Stern" advanced as a free option, to suit own thesis,[2] between *being* and *having*. The *tolle lege* moment, as Marcel describes it, refers to the focus on corporeality that also struck other French philosophers of the day, here however not a matter of the flesh as such, or the facticity of apricot cocktails as de Beauvoir quotes these for her part, but as Marcel quotes Anders as reflecting on what it is reflexively, subject and object at once, to be bodily as such, not as being but precisely *qua* having:

> We have a body. We have.... In ordinary talk we are perfectly clear about what we mean by this. And yet nobody has thought of turning his attention upon what, in

common parlance, is intended by the word "have"; no one has attended to it as a complex of relations, and asked himself in what having consists, simply as having.³

Before Marcel proceeds to emphasize that what follows is deeply steeped in Husserlian phenomenology and that "the language of German phenomenologists is so often untranslatable,"⁴ a reflection which authorizes Anders' insight for Marcel's purposes, Marcel observes that what is at stake for Anders involves a crucial reflection on a certain relationality beyond intentionality, beyond immanence:

> when I say, "I have a body," I do not only mean "I am conscious of my body": but neither do I mean "something exists which can be called my body."⁵

More than Husserl to be sure, Anders' point with respect to bodily having is closely aligned with Heidegger in the wake of Heidegger's emphasis on Being (and the circumstantial horizon of having as what belongs to embodied being-in-the-world), in the phenomenological and hermeneutic context that was *Being and Time*. If Anders does not always pick up on the Heideggerian hermeneutic, it will be because Anders holds the question as having to do with the (perfected) facticity of having done what has been done.

Here once again: we have a reference to the *Sorcerer's Apprentice* (the parallel Anders makes), or indeed as we may extend the reference to Goethe's theodistically confounded Mephisto (although this last will not be Anders' concern as he focuses, as we shall see later, on the apprenticeship to evil, witting or not, and maintains his own reference to the devil himself): "*Teil von jener Kraft Die stets das Böse will und stets das Gute schafft*"/"Part of that Power which would/The Evil ever do, and ever does the Good"].⁶

Given what *has been* done, how are we to face the fact of the same? This is what Nietzsche called the *es war* [it was] of the past, as this also inspires Freud. This is the force of asking, as Adorno famously asks, whether poetry could ever be possible again? Whether poetry or art or music or philosophy or even ratiocination itself makes any sense at all, in Adorno's words, "after Auschwitz?" Or else, as Anders would remind us, *after Hiroshima and Nagasaki*?

Adorno, who began in philosophy by concerning himself, quite as Heidegger had concerned himself, with the theory of knowledge,⁷ articulated what would become the ground plan of critical theory itself—even if adherents of the Frankfurt School after Habermas and Honneth, have not been critical for years— or at least not in Anders' or Adorno's sense of the word. To this end, who maintains this question as a question: How do we know what we suppose ourselves to know? In other words, and as Nietzsche also reflected: How are we as sure as we are that what we take to be true *is* true? In truth? How so?

There have been many answers offered in response to such epistemological reflections—the bulk of which fall into the flat thoughtlessness of an uncritical realism that (unintentionally and in this measure or to this extent:) ironically calls itself "critical thinking" in Anglo-American philosophical parlance (for the sake of what is surely "barefoot" expediency, having the benefit as it does of easing the burden of thinking "positivity" or "positivism," to use Adorno's generic terminology as it was also Arendt's and Anders' and Heidegger's word for what can be just as generically designated "analytic" philosophy: the same style of thinking that continues to pre-empt other modes and styles of thought with such efficacy that it dominates the philosophical mainstream to this day—even so-called "continental" philosophy, including critical theory, just to the extent that today's critical theory is itself articulated in terms of this restricted schema). The bits that remain are vapours gathered under the weasel name of metaphysics.

Adding the question of genocide to Adorno's critical epistemology, including the "crisis of causality," Adorno's enigmatic: "the answer is false" raises the question of genocide as a question. But it will do to excavate certain genealogies of the concept, the word, the practice, the act of genocide. Yet if it is right to raise the question, as it is, it is just as certain that the answer cannot but be as Adorno fairly affirms: "false." And quite in addition to promulgating a false answer, false from as many perspectives as there are eyes, there are also many feet to be stepped on and there will also be the indignation of egos seeking to have their own and only their own distinctions heard above the fray.

"Genocide studies" bristles with canons, various ones to be sure, complete with canonic disputes, duly settled on the terms of its practitioners. Thus, in *The Meaning of Genocide*, Mark Levene offers a definition, set off in an epigraph to his eponymous central chapter:

> Genocide occurs when a state, perceiving the integrity of its agenda to be threatened by an aggregate population—defined by the state as an organic collectivity, or series of collectivities—seeks to remedy the situation by the systematic, *en masse* physical elimination of that aggregate, *in toto*, or until it is no longer perceived to represent a threat.[8]

This definition could be named with the word *Auschwitz*; or what would, for Levene, be better termed "the" Holocaust as if it had been or could be or better said, should be, the only one. By contrast, the 1948 United Nations Convention on the Prevention and Punishment of Genocide defines genocide as "acts committed with intent to destroy in whole or in part, a national, ethnical, racial or religious group."[9] For Levene, this definition's breadth is *too* limiting. And this limitation remains, no matter how much we might add to its breadth by including Syria in 2015 or—and this remains the limit case, especially since July 2014 and ongoing—Israel contra Palestinian Gaza (and Jerusalem and elsewhere May 2021). And now let us have what can only be an incomplete listing: after Cambodia and Rwanda, after Chechnya and Bosnia, after Iraq, Darfur, Afghanistan, Libya, and now, Semite on Semite, one must once again add Gaza's catastrophic loss of life, which includes—as almost all these instances include—a majority of civilians, as Gazans and Syrians spill over their borders and create a refugee flight that cannot be restricted to a single locus. Now that what is visually and mediatically disseminated (though this was never not the case) includes children, a people and a world—from nations and cities to neighbourhoods and hospitals, schools, universities, museums, and so on—we might speak of being subject to what—despite all this dissemination—remains strikingly *nondisseminated*—just to speak in journalistic terms, given all the news we do not think fit to print—a disaster *du jour*. And part of the problem, as Adorno would have asked that we note, is that these (invisible or denied) events themselves correspond to instances of numerically and thematically explicit *genocidal* conflicts.

Anders uses the term "conspicuously" with reference to US military deployment in Vietnam when he publishes the second volume of *The Antiquatedness of Humanity* to exemplify the thesis of his 1968 book, *Visit Beautiful Vietnam*[10]:

> However, even after having expressed what I believed could be said about the nuclear threat, my immediate "return home" to philosophy was still blocked. For the second time I "was distracted" (if it can be said that the call of duty is a distraction), as in the sixties I met another emergency, one, which also had to do with the main concerns of *The Antiquatedness of Humanity*, namely, the demand to participate in the struggle contra genocide [*Genozid*] in Vietnam, which, indeed being implemented by means of machines, was a grievous exemplification of my machine-philosophical theses. (AM II, 12)

The focus for Anders ties the observation of genocide together with his overtly critical philosophy of technology, and his consciousness of overstepping the bounds of what intellectuals prefer to focus on, is also in evidence:

> To be sure, likewise, here one can draw no exact line of demarcation between theory and praxis; on the basis of this activity a book also emerged, a partial representation of a "Critique of Technology": concerning the language developed by the homicidal technocracy of the United States and deployed in part to disguise and in part to justify its acts of devastation and genocide [*Völkermord*]. (AL II, 13)

Wide ranging as the scholarly absorption with Holocaust studies attuned to Adorno's own question of Auschwitz is, it must also follow contemporary reflections on colonialism (including the Atlantic slave trade, which some scholars still cannot tolerate being named a genocide, just as in the United States we aver that all the autochthonous peoples we call Indians, who once lived where Americans now live and where ranchers mean to ranch their free range cattle and miners mean to mine, and frackers mean to frack, and so on and so forth—and not merely somewhere in some so-named "public land" or "reservation" but everywhere and anywhere we now happen to live in the United States and Canada—just to keep to North America, in every locus, from sea, that is, to shining sea, and up and down the continent—simply moved on, just as we claim with respect to animals, such that by this shrinking away the "West" was "won," that "they" altered what we suppose to have been their "range" as we plot it, or else we are told that they simply died of smallpox, say, or of other diseases, rather than as the result of a deliberate and sustained genocide).[11]

I will have a bit more to say on the Atlantic slave trade later in the text, yet for all that, what I say articulates what is, in Adorno's sense, "false," for all the vanished peoples—and I include here as well the vanished animals of the land in which I myself was born, on the island of Manhattan, the borough of Brooklyn, on their ashes, their bones, their unmarked graves.

Although Adorno declared impossible any ultimate comprehension of Auschwitz, this, Rolf Tiedemann writes, does not licence the surrender of attempts "at understanding, if not the *fait social* of Auschwitz itself, then at least how it could have come about. Adorno set as one of the tasks of philosophy after Auschwitz 'to comprehend the incomprehensible, the march of humanity to inhumanity' it is a task that, since the genocides in Cambodia and Rwanda, in Iraq and Darfur, since the uncountable victims of terrorism in Israel and Chechnya, since the attacks on September 11, 2001, and in the face of the hundreds of murdered in Madrid's and London's subways, is farther from a solution than during the middle of the 1960s of the, in the meantime, past century, when Adorno first formulated it."[12]

Singulare Tantum: Whose Genocide? Which Genocide?

The thing about genocides is that they can be denumerated. Perhaps this is part of the appeal of legislating that only one can or should matter. The Second World War itself was punctuated, woven through, with an unthinkable number of genocidal ventures, against Jews but indeed also against Slavs (of course), against Gypsies (of course again), and against Catholics (of course). The last matter-of-fact, parenthetical "of course" is owing to the fact that the locus of such genocides was Germany: land of the one true and enlightened Protestant faith that had its beginning and its success in the suppression not only of the church, figuratively speaking, but of Catholics—Adorno could take his mother's name, but it was his father's confession and not his mother's that, after the war and for the rest of his life, exacted upon him that guilt through which he came to see himself as a *ghost-thinker*, competing not only with Anders but so many others, *posthumous* for two more decades; and therein resides the complex of the superior said to be inferior, the chosen, the elect: the Jew in Adorno will always outrank the Catholic in intellectual, moral esteem.

It is still only one genocide that matters in this or any other discussion. This singularity may even be the sole reason we discuss it philosophically. Emmanuel Levinas, when asked by some visiting students if, perhaps, the Palestinians might count as the "other," demurred: "No," it's said, he replied. And it is this reply, this "No," that remains to be thought.

Derrida would not, I think, differ. Nevertheless, when it comes to the massive scale of animals slaughtered on every level, for every reason, consistently, repeatedly, on and on, Derrida did reflect precisely on the issue and question of industrial agriculture what is at stake, not merely slaughter but life, including an excess of overproduction, bred for death (note that this can also include canned safaris or 'snuff' or 'petting' zoos):

> No one can deny seriously any more, or for very long, that human beings do all they can in order to dissimulate this cruelty or to hide it from themselves; in order to organize on a worldwide scale the forgetting or non-recognition of this violence, which some would compare to the worse cases of genocide (there are also animal

genocides: the number of species endangered because of humanity takes one's breath away). One should neither abuse the figure of genocide nor too quickly consider it explained away. It gets more complicated: the annihilation of certain species is indeed in process, but it is occurring through the organization and exploitation of an artificial, infernal, virtually interminable survival, in conditions that previous generations would have judged monstrous, outside of every presumed norm of a life proper to animals that are thus exterminated by means of their own continued existence or even their overpopulation. . . . Everybody knows what the production, breeding, transport, and slaughter of these animals has become.[13]

Thinking the Holocaust

"The" Holocaust should occupy this specific and singular place within our conception of genocide. Rightly so. Otherwise, the numbers to be counted, the incessant siege against peoples, would shock our sensibilities. We human beings who have been at it since we began to be human, communicating with each other (again, in Hölderlin's words, *Seit ein Gespräch wir sind*) at the same time as we kill one another, might do well to recall one of the first biblical justifications for the razing of a town: for killing every last man, for raping and then either killing the women or enslaving them so as to continue the process over the generations, through genocidal rape.[14] But in this case the variety specified proved too great to encompass down to the last detail.

I refer to the biblical account of record, Samuel 15. The letter of the law required killing everything, down to the children and babies, including—just as the civilized world always ends up including—all the animals. Thus, the Lord, speaking through Samuel, commands Saul to take utter retribution, to exact total annihilation as just vengeance[15]:

> Samuel said to Saul, "I am the one the Lord sent to anoint you king over his people Israel; so listen now to the message from the Lord. This is what the Lord Almighty says: 'I will punish the Amalekites for what they did to Israel when they waylaid them as they came up from Egypt. Now go, attack the Amalekites and totally destroy all that belongs to them. Do not spare them; put to death men and women, children and infants, cattle and sheep, camels and donkeys.'" (1-3 Sam. 15)[16]

Saul does not follow this command to the letter but does take the Amelek king, a common form of tribute in war, as well as another plunder of war: the living animals, set aside and reserved for his own people. This departure from the letter of the command will prove to be Saul's ultimate error (and thinking about this logic may yield insights into the current and ongoing state of hostilities in Israel today):

> He took Agag king of the Amalekites alive, and all his people he totally destroyed with the sword. But Saul and the army spared Agag and the best of the sheep and cattle, the fat calves and lambs everything that was good. These they were

unwilling to destroy completely, but everything that was despised and weak they totally destroyed. (7-9 Sam. 15)

Even after Saul defends keeping the cattle and sheep, claiming he intended to kill them ultimately, reserved for sacrifice on another day, even after the subsequent execution of the captive king, it is still too late for strictures of the law. Thus Samuel 16 sees the change of kingship to the house of David, he who was brought into Saul's house as a musical antidote to the "evil spirit" of God.

Something similar can be read in the limpid clarity of Homer, the epic of epics, the ultimate end that is the story of Troy, the end that is the story of the beginning of Rome, a tale of origins that launched what may very well remain our sole image of empire and remains, arguably, our founding narrative to this day.

For Anders, for us today, genocide is part of the logic of techno-efficiency. Thus we mean a global reset, thus we mean to combat 'climate change.' To this same extent, in his 1978 chapter from his second volume on *The Antiquatedness of Humanity*, on "The Antiquatedness of History," and beginning with a timeline of dated epigraphs, as already cited above—"Politics is Our Destiny (1815)/Economy is Our Destiny (1845)/ Technology is Our Destiny (1945)" (AM II, 271), dates which can be aligned with Napoleon, as with Marx, and with Hiroshima/Nagasaki—and where his reflections range thetically from "Truman to Kissinger to Carter" (AM II, 280), he argues in a section dedicated to the "un-modernity" of the neutron bomb, given the tactical, so it is politically argued, unusability of new bomb products, that these engender a new array of product needs and product consumers:

> Thus, the annihilation of people is thus certainly not the main objective of contemporary production—naturally, something that I do not assert in order to make excuses for the latter, inasmuch as *genocides*, like those in Vietnam or Cambodia, are *unhesitatingly accepted as collateral damage*. (AM II, 282)

In the midst of this argument, beginning from a reflection on technocrats as leaders of nations and not less as leaders of industry, Anders' term is the "Shepherds of Products," which terms inevitably includes reference to the Heidegger, who had died just a few years earlier, and his "Shepherd of Being," as Anders reflects that we fall exactly short of this. Where Heidegger installed his conception of the "shepherd of Being," as Anders observed, "in good biblical fashion, namely anthropologically," and in the process exaggerating—this is a Nietzschean echo as we may recall the beginning of Nietzsche's essay on *Truth and Lie in an Extramoral Sense*, the significance, from a cosmic perspective, of the "place of the human in the cosmos" (AM II, 281), we are the shepherds not of being but of our things, dedicated to tending technological product lines, where it is significant that we refer to modern technology not only generically, cell phones, computers, airplanes, than by the names of their makers: Samsung, Apple, Lenovo, but also Boeing and Raytheon, and so on.

Here, we can see the sense of Anders' subtitle: *On the Destruction of Life in the Age of the Third Industrial Revolution*. Anders adverts to an inherent contradiction in the political sphere, where the policy of mutually assured destruction as was the rhetoric of atomic escalation, linked, as war has been throughout the twentieth century with "safety" and "survival." Dissonant, as Anders argues, given that the neutron bomb was designed to destroy people while preserving structures:

> The most irrefutable proof of this triumph is the manufacture of the *neutron bomb* (already [as of 1978–BB] invented about fifteen years ago), *which treats our technological facilities* as *taboo* but us, to the contrary, as *"expendable"* [in English in the original]: formulated theologically: it treats our *manufactured products* [*Gemachtes*], the *opera creata*, as survival-worthy, as *ontologically more important than we makers*, the *creatores*. In the same way, via this invention and the unqualified express readiness to put it to use, the word "inhuman" has taken on a meaning it did not possess even during the classical years of annihilation, between 1941 and 1945. If a crowning witness to "the antiquatedness of humanity" were needed, here it is. (AM II, 282)

Anders' footnote here is shocking (only Virilio's reflections on the contemporary manufacture of swine-human chimeras for the sake of "harvesting" for xenotransplantation, in direct historical alignment with Auschwitz, is comparable):

> Auschwitz, where one classified hundreds of thousands as mere containers of hair and gold teeth, containers which were annihilated in order to obtain those materials, can be regarded as a testing ground for this "pure human annihilation" (the neutron bomb, in fact, is recommended as the "cleanest weapon" [*"saubere Waffe"*]). (AM II, 452)

For Anders, the issue of scale and distance, the conviction that we are dissociated, at least in principle as consumers, as citizens, from the effects of consumption, the actions of our governments, as of international corporations—as in the case of the explorations for oil and gas as well as the destruction of the rain forest for the sake of palm oil and beef and lumber—implicates us still more powerfully in the logic of the gadget coupled with the homeworker and the intimate labour of the same taken to its more complete extension, as he writes in 1972, towards the end of a long chapter concerning the antiquatedness of "meaning," in a section entitled "The Iteration," and emphasizing his repeated recourse to this same illustration:

> Again take my previous example: the meaning of my work on the mechanical part of a machine, whose meaning consists in manufacturing another part of a machine and, moreover, for a machine necessary for a weapon of annihilation, whose meaning in turn consists in the liquidation of millions of human beings—I say, the ultimate meaning of a job so seemingly innocuous to the extent that its product is a miniscule component, consists in the final effect of the final product, despite the fact that the latter is temporally and spatially distant from me, and despite the fact that during my work I concentrated exclusively on the flawless production of that first element. *The ultimate meaning of a modest manual gesture* [Handgriffes] can be called "genocide." (AM II, 389–90)

Colonialism and the Exploitation of the Globe

The discovery of America as this was named to claim the continent as a found new land, like a new Jerusalem, by discounting the original inhabitants already there for whom the lands were not 'discovered' but lived, inhabited, was also the discovery of myriad opportunities for different varieties of genocide—rampant, multifarious: so many different peoples, so many languages and names, so many trees and plants, so many animals, including birds, fishes, and variously furred beasts,[17] enough to create a profession for those who lived by excavating the flesh of wild beings, all the professions attendant on all those hunters, trappers, fishermen, whalers, and sealers. Thus the technology for the circumnavigation of the globe would wipe out entirely different vistas, distant parts of theretofore new worlds and hemispheres, landscapes and populations; and on into the modern era, for the benefit of that industrial revolution, which, of course, as we read in our history books, *depended* on slaves and was indeed founded upon slavery (this foundation continues in wage slavery and in the unnoted, uncounted, thankless, utterly invisible work, or let us just, as Arendt does and as de Beauvoir documents, call it the "labour" of women), where to be "dependent on slaves" is to depend upon enslavement—this, then, is the benefit, the profit of genocidal practice.

The history of humanity is the history of extermination, of extinction, yet we blanch before naming *it genocide*; historically, however, *merely historically speaking*, there are and cannot but be parallels. For this reason it is worth invoking again the animals to whom we tend to refuse this parallel most violently (because the term *holocaust* derives from ancient Greek *animal* sacrifice, which involves the burning, in its entirety, of the whole animal (ὁλόκαυστος), and because the contrast is so abysmal). While Adorno drew this parallel on several different levels, the historian George Patterson has expertly traced the technique of the Holocaust itself back to the Chicago stockyard's development of industrial agricultural efficiency—its mass transports, sorting lots, and techniques for efficient killing en masse, detailed in *Eternal Treblinka: Our Treatment of Animals and the Holocaust*.[18] To be sure, there is certainly a relation of continuity: in the one, the unrestricted killing or hunting of animals to the point of their utter destruction—it is the *entirety* of the destruction that makes the difference, the fact that none survive, that none remain on this earth (and increasingly we have reached this point in terms of "extinction," a term we use to mask the fact of extermination, persecution, hunting, poisoning, which we call "control"); in the other, the devastating crossover to human beings.

This is one part of the historical argument advanced by Briton Cooper Busch's *The War Against the Seals*.[19] If it is common to suppose that the problem concerning the assault on the seals was born of the sensibilities of those who founded Earth Day in 1970 or those who still publish images of baby seals brutally clubbed to death—a systematic slaughter funded entirely by the Government of Canada for the sake of a fishing industry stressed by its own overwhelming successes (this is what *overfishing* means, and its result is the depletion of herring stocks that have yet to rebound)—then the question is more a problem for ecology, for the "management" of animal "resources" and for environmental studies than it is a concern related to the question of genocide, no matter how numerous are the variety of kinds that die—or shall we insist as many do, that they simply perish, and are in any case no more. We see no animal as suited for any other purpose than to serve our needs and an animal that competes with us for what we regard as ours (insects and other 'pests'), we destroy with no further thought, poison, traps, etc. fully meaning to eliminate the threat of any competition, the merest presence at all. Fishers fish fish, seals fish fish, and so there should be no seals. And philosophers, of course, make distinctions to keep their own minds (and consciences, one supposes) clear. But historians and social scientists muddy things.

Or if they do not, as is alas often the case, they should.

For my own part, I cannot but think that if all members of a kind are killed, then it is only human hubris that speaks of this extinction as though something other than a *holocaust*, a *genocide*, were involved in that slaughter. Like whales and wolves, pigeons and bison, Busch's seal wars thereby constitute a kind of genocide. Genocide is a matter of totalities, and these totalities are estimated by numbers and achieved by a determined and consummate execution: it is an irrationality quite rationally calculated to consummation, as Anders reminds us and as Adorno says.

For his part, Busch estimates—though no count can be correct since too many factors make every estimate at once too low *and* too high (all the while noting, as everyone does, that some species, whole kinds, including, as we can be sure, kinds never known, were driven to extinction, and for them there is no measure)—and counts up 5.2 million seals. But surely, we say, these do not count. Even if all of them (*all* of their kind) died, we do not count them. Indeed, in the recent and unprecedented 2019 bushfires in Australia, ecologists *doubled* the millions who perished from 500 million to more than a billion in the space of a week by an estimate noted as conservative.

In a recent study following Busch, Greg Grandin draws out the political context and consequences of this devastation, which Grandin reads for its parallels with the contemporary American political context. Quoting Busch, Grandin highlights his own point:

> "On island after island, coast after coast . . . the seals had been destroyed to the last available pup, on the supposition that if sealer Tom did not kill every seal in sight, sealer Dick or sealer Harry would not be so squeamish." By 1804, on the very island where Amasa [Delano] estimated that there had been millions of seals, there were more sailors than prey. Two years later, there were no seals at all.[20]

To turn from history and its seals to the literary mirror of that era, Melville's *Benito Cereno*, is to turn to the circumstances of a related "genocide," namely that of the transport, abuse, and devastation of whole peoples that was the slave trade.[21] American Indians ran away from enslavement when they could and because they could (which did not mean that they survived as a people), while African slaves were imported to meet the needs of a growing nation. If we do not care to count the seals (and it is hard to find those who do—or can), then perhaps we might, perhaps we should, count the slave trade. At least. But to say this is hardly to settle things since, as we began by noting, Levene, like other scholars, is aware that some hold the Atlantic slave trade to be an act of genocide—but he rejects such claims.

If some (not all) are disinclined to name the Atlantic slave trade genocide and if at the other extreme it seems certain, despite koala deaths along with other wildlife perishing, that the bushfires in Australia will not be so accounted, it is difficult to claim as genocide the obliteration of other kinds of human beings in different historical contexts, different loci, never mind—as my reference to the loss of Australian wildlife indicates that we do not mind, no more than we minded in the United States and Canada—species different from our own.

For Anders, the slaver's logic, like the sealer's and, as he extends it, quite contra Melville, the whaler's logic, has an impeccable philosophical pedigree or justification, even if it is one Anders refuses, writing with respect to Kant's moral prohibition against using anyone, meaning any person, as "a means only,"[22] that is,

> It is true that we could never subscribe to Kant's claim that, as he says elsewhere, "[with irrational animals] one may deal and dispose at one's discretion"["Lectures on Anthropology", 7:11] (a claim that could be invoked by the exterminators of whales and seals). (AM II, 432)[23]

A contribution to studies of territory and geography: the slaver's trade and thus the whole enterprise of slavery, as well as all those nations that grew fat—that would be an entire nation, a continent, the West as such we might say, beginning, as Robert Bernasconi has argued with Locke's own justifications,[24] nor is Locke's the only name—on the obliteration of humanity and the rights of so many, beginning with Egypt and Assyria (if we do not count other continents and peoples or indeed those in the same locus of Western antiquity), like those of ancient Greece and Rome, and like every feudal society up to and including the tremendous powers of our own world—the sealer's trade only came to an end once sealing had utterly demolished (and that means, again, to the very last seal) the possibility of its own practice. Fishing boats and whaling boats faced similar problems of redundancy and were redeployed for other purposes. Thus, it was neither enlightenment qualms nor any new fashion, new tastes, or new sensibilities that brought an end to whaling certain whales as whaling continues to date. In fact, at the time, and much rather, the target, the so-named right whales, were hunted to their death. We are, more broadly, as there remain some other whales, not yet done with that consummate destruction, nor are we finished with deep-ocean fishing in general as we continue to dredge and drag and sonic blast the seas, at this point at such a volume and so extensively that whales and dolphins are dying at unheard of and certainly untold numbers.

For Anders, we are insulated by our valuation of the practical and the useful. To this extent we are well advanced; think of our new vaccines for a hint of our progress with DNA and mRNA. To this deployment of viral genetic modification must be matched, the technique in question depends on it, the related project of creating human–animal chimeras for the sake of "harvesting" their organs; the more human, the more useful it is. The scenario is not merely akin to one taken from a horror show, and Virilio writes about just this connection quite specifically in connection with concentration camps in his *Art & Fear*, reminding us, and we can remember that Walter Benjamin makes a similar reference, if only to the Italian Futurists, in his essay on art in the age of "technological reproducibility,"²⁵ paralleling Anders' emphasis on the "cleanliness" of the neutron bomb:

> The slogan of the First Futurist Manifesto of 1909 "War is the World's only Hygiene"—led directly, though thirty years later this time to the shower-block of Auschwitz-Birkenau.²⁶

For Virilio, as we will need to return to this point later in the text, experimentation on animal chimeras is part of the Nazi project and has continued since then to be part of modernity.²⁷

Anders, using the complement *homo creator, homo materia*, follows the logic he identifies—instructively using without remark Heidegger's own language of the "manufacture of corpses" along with his distinction from "killing" as such—as following from the modern, industrial repurposing of humanity:

> The *transformation of the human being into raw material* (if we overlook the age of cannibals), began in Auschwitz. That from the corpses of the inmates of the concentration camps (which themselves were already products, because human beings were not killed, but corpses manufactured [*Leichname hergestellt*]), certainly hair and gold teeth were definitely removed and extracted, probably also the fat rendered to be used as a raw material. (AM II, 22)

Anders, if anything, offends our conventions and expectations with his next sentence as he informs us that "everyone knows" what everyone in the United States, even those engaged in Holocaust studies do not "know," as quoted already at the start of this book, reminding us of

> the American soldiers who returned from the Pacific with the gold teeth of Japanese soldiers: with my own eyes I have seen bags full of teeth, the GIs showed them to me, as unbelievable as this sounds, blamelessly. Blamelessly, because to them it was obvious to view the world as raw material and they also took it for granted that their Japanese counterparts (who, of course, had previously been demoted to "monkeys" via a systematic campaign of defamation) should be regarded to be part of that same world. (AM II, 22)

Sealers and slaves, dolphins and whales, Auschwitz and human–animal chimeras—who makes distinctions?—and yet race studies does not focus on the colours of humanity, black or brown, red or yellow or pink (or beige or pasty mauve), but "white." And as the now-effectively *exterminated* American Indian once observed—attending with good cause to white practices, and thus exceeding the insights of any ethnographer then as now (as Latour says, "The investigatee always knows a great deal more than the investigator"—an observation that has not endeared him to his fellow anthropologists)²⁸: the whites, we whites, and note that this includes any and all nations with industrial fishing vessels of any size, only leave off killing when and only when they have, in the words of the poet who, *qua* mathematician, was also good at counting, "eaten every one."²⁹

While we—noting the conjunction in the title of Lewis Carroll's poem, "The Walrus and the Carpenter"—pretend that there is convergence and competition. Like the seal (prey, like Carroll's oysters, to our harvest), the walrus is supposed (by us) to vie with us for prey we regard as "ours" by some deranged right. We thereby pretend that animal predators are our competitors. And we will not stop (and, again, in this, whatever our ethnicity, we are all white) until every last one is gone.

7

From Anders' Sexless Capuchin to Virilio's Chimeras

One conjunction that remains problematic is that between the aesthetic and the political as a category. This is a theme for Anders as for Benjamin and Adorno and not less a theme that to a certain extent may be read in and through Heidegger's own (albeit oblique) reflections on the "origin" of the artwork *qua* "overcoming" of the aesthetic, and thus it is connected quite as Gilles Deleuze argues to Kant on the one hand and Nietzsche on the other. In addition to Heidegger's reading of Sophocles' *Antigone*, there is also the question of the body as this is differently understood in different traditions, some of which entail others (physiology, physiognomy, social psychology, and the economics of affect and eros, the once vital and now oddly faded tradition of *Lebensphilosophie* as this mattered for Anders especially with reference to Scheler and others). Here, too, though this cannot here be completely explored given that it proceeds in very different directions (although one of the reasons for the complex array of footnotes is to offer some small indication of this), connections may be forged, from another perspective, with Jean Baudrillard, who also writes on sex (in a masculinist rather than feminist voice), and, as we have already seen, Paul Virilio, as well as, and we have touched on this earlier, the question of love via Heidegger (in connection with Arendt) and Agamben (who writes on both Heidegger and Arendt).

Economy, Power and Possibility, Impotence and Sexuation

I have underlined the importance of noting the influence of Heidegger's thinking on Anders' thought, and I would argue that Heidegger's claim that much of Anders' thinking can be read as derivative is not groundless. Perhaps for the same reason, Anders, who was early on a very enthusiastic follower of Heidegger, could not but encounter a lack of reciprocity on Heidegger's part. All of this plays into Anders' sustained and uncompromising critique of Heidegger's thinking overall. In this chapter, it is worth noting that in addition to his strikingly *severe* criticisms regarding sex and love, Anders foregrounded the limits of Heidegger's philosophic reflections on human cares and woes, noting Heidegger's inattention to the worldly, the economic order quite where, occasioning a certain conceptual dissonance, Heidegger invoked explicitly economic language [*Sorge*] in *Being and Time*. This complex thematic is also part of the problem of Anders' never-published novelistic work, that the text includes an allusion to Anders' own "Molussians," which he describes with respect to his analysis of "consumption" (AM II, 432). Quite where Adorno, for his part, speaks of the "culinary" with respect to mediatic culture, Anders foregrounds the economic dimensionality of ontology:

> As a matter of fact, Heidegger's trick consists in re-coining every <u>possibilitas</u> into <u>potestas</u>, every <u>Möglichkeit</u> into <u>Macht</u>. . . . It is very characteristic, indeed, that the words "*Eigentum*" (property) and "*Eigentlichsein*" (being proper, authentic being) stem from the same root. The "*Dasein*" that, according to Heidegger, first finds itself as stranded good ("cast into the world") becomes authentic by making itself its own proprietor. (PC 352)

One can use the language of the "sexuate" (here I cautiously borrow Luce Irigaray's word, although Irigaray herself does not invoke Anders) as Anders' foregrounds the bodily lived dimension contra Heidegger and to the extend the quote we have already cited: "All want is wanting; thus sex, too" (PC," 346). Accordingly, so Anders argues, Heidegger's Being-in-the-world manages to give us an utterly disembodied Dasein, bereft of any account of body, let alone of the body as lived by beings who actually *have* bodies, once again as beings subject to both "*concupiscientia*" and "toothaches." Anders refers to sexuality to refer to everything enfleshed, "One is tempted to vary the famous French word '*ni homme ni femme, c'est un capucin*' into: '*ni homme, ni capucin, c'est un Dasein*.'"[1] Anders repeats the point in a note to his introduction to the second volume of *Die Antiquiertheit des Menschen*, where Anders characterizes "Being and Time as lacking reference to the economy as to hunger and sexuality" (AM II, 432).

There is here a certain insinuation of sexuate ablation, and it is patently relevant that Anders had good reasons for his sustained conflicts with Heidegger, given the triangle between himself and his former teacher with Arendt. Intriguingly, Anders' own metaphor reads Dasein as quasi-Augustinian "angel"—that is, non-sexuate, which is exactly, and by definition, *sexless*:

> Nowhere is it mentioned that Dasein has (or is) a body; nowhere, that it has, as it was called in more than two thousand years of philosophy, a twofold nature, rather than being as the human being had been for more than two thousand years of thinking. (PC, 348–9)

In addition to abolishing, quite in addition to the ontological difference, the difference between body and spirit, Anders counts off the roster of Heideggerian omission: of "*caritas*, or friendliness, or duties, or the state" (PC, 349).

Anders began his reading of Heidegger varying Husserlian intentionality using the example of hunger and its object, thus he wrote on the erotic and its complexities, not unlike Georges Bataille. But writing on love, closer now to Arendt but also intriguingly to both Nietzsche and Derrida (and Agamben), Anders reminds us that the word of God, to which Israel and the entire Judeao-Christian world have been listening for millennia is not speech but a telephone voice. And as he muses, desire itself is difficult to maintain—a point Jacques Lacan notoriously underlined in his seminars in Paris as an imperative. For Anders this difficulty yielded the strangely necessary prayer: "*Give us this day our daily hunger*" (AM II, 15). Today's economy is a consumer's economy of end products "which are no longer means of production, but means of consumption, that is: means as such that are *consumed by being used up*, like *breadstuffs or grenades*" (AM II, 15). Beyond the routine contradictions of capital as might be noted by Marxists of the day (then and now), what was evident especially in the United States as in post-war Western Europe, also an insight animating Marcuse's notion of "repressive desublimation," whereby a concerted "effort to maintain production via consumption, at least in capitalism is today's 'concern' ['*Sorge*']," is attributed to Heidegger "as an existential melancholy" (AM II, 432).

From Consumption to Biotech

The anti-Heideggerian note is needed to sharpen the political point and the turn to the social focus of the Frankfurt School regarding industrial technology as a machine dedicated to the reproduction of itself by all means possible, consummate in every way, and that is to say, *consumed* at every level, inevitably involves us as "eaters," in Anders' terms, that is, as consumers/fabricators of ourselves (and today—and this does make it sound vastly better—we say *curators* as if a museum display or a schedule of exhibition were somehow involved), a *self-curation* which, given the current conditions of social media, also extends to our consumption/fabrication/curation of one another, in virtual and thus effectively literal terms.

Note that the complex constellation is also one that Adorno notes at the start of his own posthumously published *Aesthetic Theory*, where Adorno begins by reminding the reader, just as he would remind his students in his lectures on expressly *philosophical* aesthetics, of the relevance of Kant's legacy and the notion of "disinterested delight or pleasure" that is the beautiful:

> If according to its own concept art has become what it is, this is no less the case with its classification as a source of pleasure; indeed, as components of ritual praxis the magical and animistic predecessors of art were not autonomous; yet precisely because they were sacred they were not objects of enjoyment. The spiritualization of art incited the rancor of the excluded and spawned consumer art as a genre, while conversely antipathy toward consumer art compelled artists to ever more reckless spiritualization. No naked Greek sculpture was a pin-up.[2]

Anders writes post Adorno as we read him, and the implications for media aesthetics together with political reflection and ethics overall are significant. Anders sets his analysis in the context of a society oriented to mass destruction on every level, ecological but also explicitly military-industrial, unflinchingly describing the logic of what journalists can write about without blinking an eye: "mutual destruction" as political détente. For Anders, this is not an accident of a particular party in power nor is it limited to a given nation or a specific regime: it is built into the logic of technology. Like Heidegger, Anders is unimpressed by the West's sense of a keen difference between the United States, say, and Russia or China, say, as all regimes today believe in the same salvific powers of technology. This is the logic he thematizes from the start of the second volume of *The Antiquatedness of Humanity* in 1980, echoing concerns from the first volume and his writings in the 1960s and 1970s as well. What's doable is what should be done [*Das Gekonnte is das Gesollte*] (AM II, 17) The can-do- (to catch the Americanism of Anders' era) *is what gotta be, oughtta be*.

We tend to omit the logic of this claim in our own thinking on technology, even as we debate the virtues of displacing ourselves to another planet (should the technology to do so become available) or clone ourselves (ditto) or upgrade our brains (ditto) all as if there were a consideration of ethics somehow to be brought to bear on the question; thus, for example, with reference to cell phone technology, is the coming shift (already in place) to 5G safe? As it seems, it has been deployed, as all cell phone-related technologies have been, without any such review, no more than the effects of widespread herbicide in addition to other industrial poisons and pollution on the world at large, or the less and less relevant question of cloning and the rather more relevant question of xenotransplantation (human-pig to human) for the sake of medical-grade organ technology?[3]

That "the possible is generally accepted as the compulsory and what can be done as what ought to be done" (AM II, 17) has its most obvious extension in biotechnology. But biotech concerns are messy while at the same time confirming the language of "obsolescence," as many scholars prefer this term to render Anders' analysis of our sense of ourselves as outdated, antiquated, as Anders uses this to underscore our desire to be "corrected," transhumanized. Here it is essential to remember that Anders uses the term "obsolete" to characterize the vaporization, as he explicitly names them, of "religious and philosophical ethics" "exploded with Hiroshima and gassed with Auschwitz."[4] To this same extent, a focus on nuclear armaments may offer a clearer understanding of the dynamic of technology as perpetual motion machine. Again:

> in order to satisfy the needs of technology, that is, to fabricate what can be manufactured [*das Machbare*], weapons are produced that make it possible to annihilate humanity many times over—that is, a situation after which there will not only be no demand, but in which all demand *can* no longer exist, or, more accurately, a situation excluding any survival of industry (and not only that). And yet not only must *everything be produced that can be feasibly be made, but everything that has been fabricated ought in fact be used for the purpose for which it was designed*; not only is it the case that no weapon has been invented that has not also been effectively produced but every weapon that has been produced has also been effectively utilized. (AM II, 17)

Commentators prefer to demur, thus securing a cautious stance for themselves, if only to avoid being branded "negative." Surely the threat of atomic war remains an empty one? However, today it can seem, to the contrary, that Anders retains his relevance as the current political dynamic plays itself out between the United States and same players as ever before, North Korea and including discussions with Russia and China.

Tests and nuclear proliferation from nation to nation, arms races, are exactly nothing if they are not all about deployment and manufacture. There are clear parallels with today's mRNA and viral vector vaccines, but for Anders,

> demand for such products is often produced *post productionem*; and this is done owing to the desire for profits on the part of the producers, who deceptively present their own interests as a national necessity. Thus, the lobby for heavy industry in the United States created—and not just during the Cold War (which was also one of its products)—a need for security and protection in the "free world" by way of the production of false statistics regarding Soviet arms production and thus the most savage production of the most monstrous weapons and their acquisition by the Armed Forces was justified and set in motion.[5]

The ongoing war on terror, which remains under various transformations, with or without being named as such, increasingly without media notice of such, as of 2019 (although all such worries seemed to vanish with Covid-19), *nota bene*: pre- *and* post- 9/11 as Jean Baudrillard (among not too many others) has analysed this,[6] offers us yet another version of just this eternally recycled demand for the eternal production of the products of mass destruction. The logic is the one Anders takes to its furthest expression in his own writing on Auschwitz, Hiroshima, but also the ongoing or the third world war, as he called it, which Anders regarded as underway throughout the 1950s, 1960s, 1970s with the Cold War and Korea/Vietnam, and so on—and, today, with Iraq, Afghanistan, Syria/Gaza, we can say, without Žižekian irony, "and so on, and so on."

Where positive readings of technology highlight human intervention and power, Anders emphasizes technological determinism in an even graver fashion than Heidegger's "questioning" or Ellul's systematicity, where Anders offers his own version of his reflection on the likely possibilities or "futures" of *Gelassenheit*. Again:

> *Not only is it the case that whatever can be done ought be done, but what should be done is also inevitable*. And this is not just a rule, but a postulate, that says: *Leave nothing unused that can be used!* (AM II, 17)

For his own part, Heidegger emphasized that in addition to our disposition to the earth which we regard "as a gigantic gasoline station," which may be a generic metaphor for our treatment of the earth as repository of "natural resources," to be exploited or conserved (meaning, to be saved for later use) at our politico-economic pleasure (AM II, 32), the greatest danger may be our tendency to see ourselves and even more generally to see *other* beings (or bits of them), as 'resource,' better said: spare parts. Today and for a long time, this has included, as Paul Virilio likewise underscores in addition to Anders, what Heidegger pointed to as the human being as biological resource, not in terms of some ontological idea but factically, ontically, fetal cells and serum (no Jesuit ever made a finer distinction), and such like extractions essential to vaccines and assisted fertility and so on: none of these interventions is without the need for a great deal in the way of animal and human, cloned and otherwise *harvested* "resources." This economic and ethico-political perspective takes us back to the trans-human-swine chimeras bred for human medical exploitation or the human fetal tissue or stem cells to be similarly deployed: to quote Anders, who varies this point, it is a seeming techno-imperative, or commandment: "*You must not refrain from utilizing that which can be utilized!*" (AM II, 32).

To this day, a version of this imperative rules the so-called "war on terror" and its logic (an infinite undertaking, waged as it is against an invisible and thus omnipresent enemy—exceeding even George Orwell's imagination), key to the ecological anxieties concerning climate change, anxieties which for the most part steer clear of any suggestion that governmental geoengineering might not be well in place, with devastation done to habitat and wildlife and human populations long established, is the notion that everything remains a threat: meaning *avoidable*, if (and only if) one undertakes to put this or that counter-mechanism in place, purchases or sets funds aside for this or that presumptively salvific strategy. The technology, so it is assumed/claimed, is there (or can/could/might be developed) to save us, and the disaster thus is one that can be avoided. It is the illogic of this thinking that Anders challenges. His negativity consists in daring to think the thought of a time after or following an end time, and his phenomenological perspicuity consists in laying out the consequences of apocalyptic time backwards and forwards. He writes:

> The door in front of us bears the inscription "Nothing will have been" and from within: "Time was an episode." Not however as our ancestors had hoped, between two eternities; but one between two nothingnesses; between the nothingness of that which, remembered by no one, will have been as though it had never been, and the nothingness of that which will never be.[7]

The allusion to an eternity between "two nothingnesses" is an Augustinian echo of Nietzsche's coiled reflection on lifetime in his *Dionysus Dithyrambs*, the penultimate closing verse of the poem, "Between Birds of Prey" in his last published work:

Jetzt —
zwischen zwei Nichtse
eingekrümmt,
ein Fragezeichen,
ein müdes Rätsel—. [Now—/between two nothingnesses,/crumpled,/a
    questionmark, a tired puzzle—.][8]

Already cited above, Nietzsche's *Nachlass* notes gives us an interval, an emptiness between nothingnesses [*ein Hiatus zwischen zwei Nichtsen*].[9] Anders' reflections on the "third" industrial revolution highlights some of the more patent contradictions of the same: the need to produce weapons of mass destruction and the need to constitute opportunities to deploy said weapons (including the absurdity of "needing" to replace said items on a regular schedule with yet better weapons of yet more comprehensive mass destruction [AM II, 20]), pointing then to an even greater challenge to the social task of creating the consumer in the image and likeness of the machine.

At issue, as Anders had noted in his first volume, is the transformation of the human being, now switching out the position of maker and made and so defining the human as *Homo faber*: "the essence that fabricates its fabrications" (AM II, 25). Anders reprises this insight in his second volume as the "immense fact that the human being has been transformed into a '*homo creator*'; and in the fact, no less unheard of, that he has transformed himself into *raw material*, that is, into *homo materia* . . ."(AM II, 21) In spite of his criticisms of Heidegger, Anders follows Heidegger's lead as we ourselves become raw material for our own consumption. This is biotech, novel vaccines and nanotech, GMO and cloning, stem cells and the ultimate hybridization of humanity on the level of and beyond the machine.

Manufacture and Art: *Homo Materia*

In the previous chapter and in passing earlier, I noted parallels between Anders and the late French sociologist and philosopher, Paul Virilio, with respect to the theme of human fabrication,[10] upgrading the antiquated human being, emphasizing; this is the point of speaking of "antiquated" by contrast with the Francophone tendency to speak as the French translation of Anders' *Antiquiertheit* speaks of "obsolescence." Thus, the technology needed to engender humanity as "transhuman" is not a recent invention or possibility or new dream consequent to the Human Genome project but begins as Virilio argues (and documents) even prior to Auschwitz. In the section, *A Pitiless Art*, Virilio means to emphasize the silencing of compassion, and a parallel seems patent just where Virilio askes:

> Hasn't the universality of the extermination of bodies as well as of the environment, from AUSCHWITZ to CHERNOBYL, succeeded in dehumanizing us from without by shattering our ethic and aesthetic bearings, our very perception of our surroundings?[11]

The aesthetic reference takes us back to the beginning of the last century, as Virilio underscores Marinetti, who writes as Benjamin cites him: "War is beautiful because it establishes man's dominion over the subjugated machinery by means of gas masks."[12]

Not unlike Anders, insisting here on points we would rather not see, points even theorists are inclined to discount,[13] Virilio reminds us that Auschwitz-Birkenau was a leading research laboratory for both pain *and* genetics.[14] Like Anders' largely ignored reflections on the nuclear violence inherent in the "peaceful" uses of atomic power in *Gewalt, Ja oder Nein?*, Virilio underlines the calculated construction of

> the balance of terror along with the opening of the laboratories of a science that was gearing up to programme the end of the world notably with the invention, in 1951, of thermonuclear weapons.[15]

The point needs repetition inasmuch as we can overlook "Nazi science," which is thus less about what would have been called French science or, as Pierre Duhem emphasized per contra: "German science,"[16] or what the Nazis later named Jewish science, but a still ongoing enterprise dedicated to developing the raw materials to be had in the human being, *qua homo materia*, to use Anders' language. Virilio stresses the use of chimeras, mosaics, hybrids, and what is done to create systematically monstrous combinations of life that only our insensitivity to animals allows us to ignore.

Like the historian, William Patterson, who reminds us in his *Eternal Treblinka*, of the direct commonality—mechanism and industrial technique—between the Chicago Stockyards and the death chambers—[17] Virilio's analysis of the aesthetic ideal of silencing in art seems to echo our disinterest into the conditions of scientific research, reminding us that as we ignore what we do to the animals manufactured as so many corpses for consumption, our very dispassion—silence—as spectators of art confirms Kant's definition of the judgement of the beautiful as disinterested interest better than Kant's own argument. As aesthetic spectators, we watch as artists harm themselves. Without thinking, without a word of protest, we bracket the thousands upon thousands of animals used for artistic work (Damien Hirst, Joseph Beuys, and unnamed others in uncountable ways), not including the inks and paints and brushes themselves, the gessoed canvases, the glues and lacquers, and so on. In addition to a performance video-artist's self-castration (and accidental death on camera) and other instances of self-harm, Virilio recounts a museum exhibit that consisted, that was the whole of it, in the artist's chaining of a stray dog to a museum wall until it perished of hunger, without anyone, curator, visitors, janitors, lifting a finger to protest or to help the dog: disinterested indeed. It's *art*, is it not?

There are all kinds of reasons we would prefer not to talk of such things when it comes to writing philosophically on technology. When it comes to science, the prohibition is, if anything, even more pronounced. It's *science*, after all. Scientific research to be *scientific* research must "use" animals and, to follow Anders' argument as outlined in the last chapter, and by the same token, humans as well. Thus, the human being after two world wars had already long become for itself—this is the ultimate significance of our "antiquatedness"—a repository of so many replacement parts, that is, potentially as source material, potentially in need/hope of the same. Anders would thus update his earlier mid-1950s reference to our Promethean "shame" at lacking such replacement bits, by the time he begins the second volume of his *Antiquatedness of Humanity*, highlighting our ambition to exceed anything merely human—all too human—with reference to the logic of a productive economy and thereby a productivity dedicated to its own perpetuation. Thus, we consume and create the need we then fulfil and consume, to the production of ourselves, which already now entails the mass production, already in progress, of spare body parts. The point is that it is nothing new. Already underway for some time, as Virilio writing in 2000 points out, silenced as this has been, the spare part industry is not some future possibility of science to come but extant biotech. And recent years have seen advances in the mass production of an animal–human "chimbrid," raised for industrial slaughter, pigs, once again, 80 percent or more human, to provide medical organs (hence the need for/value of such a high human to pig component).[18]

I've said this before, relegated there to the context of a footnote, that the onetime Sirius executive and now biotech industry leader, Martine Rothblatt can, on the one hand, smoothly inform both shareholders and journalists alike of a minor obstacle she calls the "yuck factor" in order to reassure us that once consumers see the potential ideal use (real results in real practice or real life, as always, will vary), users/consumers will not be bothered by such visceral details.[19] Here we recognize Anders' observation, "what can be done has already been done," now transformed in keeping with Anders' matching industrial imperative, "what is doable ought be done." If Anders kept the point on the level of theory and economic and practical feasibility and moral righteousness, what *can* and *should* and *ought* to be done, that might have been quite enough. But Anders provokes, there's that negativity again: reminding those of us in the United States, to cite this once again, of "American soldiers who returned from the Pacific with Japanese gold teeth [*japanischen Goldzähnen*]." Here, we're not speaking of artefacts on display in a Holocaust Museum; these are not Nazi acquisitions but war booty, "harvested"—this is the force of Anders' point here—by American GIs who collected them from the jaws of Japanese soldiers. The "yuck factor," so to speak, could not be more in view, and we ourselves are implicated as Anders reminds us that the logic of war and despoiling is not limited to others but something we have done/are doing/will do.

Virilio updates this parallel, writing about art as he segues in the process of reflecting on the artist's self-exploitation in the service of his art, to the war's legacy via Nazi science for human xenotransplants. Human stem cells that promise, among other things, the next biotech revolution, exemplify recycling products of abortion and other organ harvesting—but generally speaking, where do cells come from in cytology research, and how are they obtained, preserved, secured; the answer involves fetal tissue serum or fetal broth, which involve the slaughter of animals, in increasingly hideous ways, as we ought to reflect as we consider the prospect of "meatless" meat, extracted from the calf still in the body of the pregnant mother[20]—can serve to instantiate yet another industrial, biotech revolution were this not, according to Anders, only the third such industrial revolution, well underway, already in research production for decades, with pig-human chimeras only another sort of manufacture of corpses of an "animal" kind.[21]

Anders goes on to remind us, not that we speak of this to this day in the United States, exactly not in the United States, where looting on another level has accompanied every intervention in the name of the war on terror since the escalation of the standard Gulf War, which, as Baudrillard told us, again: not that we listened, "did not take place," now transformed into the new stakes of the new Gulf Wars, from Iraq to Syria and now to Gaza and so on. In his reflections on his GIs' display of their looted gold teeth, Anders emphasized their lack of guile, naiveté, they showed their souvenirs to Anders, who promptly proceeded to write about it.

Thus, it is worth noting, given the dates of Anders' writing, that supposedly "new" transhuman experiments to create improved human beings, were already underway in his day and not merely on the drawing board, already less about the creation of the superhuman transhuman than the creation of medical products for sale and routine surgical deployment: part of today's industry of biotechnology, vaccines and all.

As Anders writes in his 1980 book of this imperative: *what can be done ought to be done*:

> I am speaking of so-called "*cloning*," of *genetic manipulation*; that is, to the possibility of producing new, "unprecedented" and unforeseen genera or species, or even duplicates of extant individuals. Whether human beings have already been cloned is not known to me. But inasmuch as we know that today's imperative recites "What one can do, one ought to do", or, "What is doable is compulsory", what was until now merely possible now looms as a breathless Omen over the present-day. (AM II, 24)

The point made here has to do with an inherent politics inbuilt into technology and an inherent trajectory. To have a tool, to have a technology, means that certain possibilities latent or inherent in the technology become realities. We imagine that human cloning is a sheer idea, empty, and yet reports of cloning humans crest from time to time on our news horizons, and already in Anders' day it was not unheard of; indeed, Heidegger points to nothing but this in his own *Gelassenheit* address on the ten-year anniversary of Hiroshima and Nagasaki, that is, the end of the Second World War. Thus, Anders can observe, in the full sense of Kant's actuality entailing possibility and the very Heideggerian imperative that is what calls for thinking through that same possibility as possibility, in this case: "the possibility of producing new, 'unprecedented' and unforeseen genera or species, or even duplicates of extant individuals" (AM II, 24).

Anders' parallel here reflects some of the experiments undertaken by Nazi medical science for the sake of fully developing or exploiting the so-called *Lebensraum*, new researches perforce explorations of *terra incognita* and yet, as he emphasizes, not for the sake of traditional colonization but towards the end of the new schema of mass destruction and as a new means of mass annihilation. Thus, Anders highlights the contrast:

> Whereas nuclear war means the annihilation of living creatures, including humanity, cloning signifies the annihilation of species *qua* species, and conditionally, the annihilation of the species of the human by way of the production of new types. (AM II, 24)

And to the same extent, all previous concerns, be it the Sartrean anti-essentialist spirit of humanist existentialism, or contra the Schelerianism of Anders' own day, would henceforth be transformed as the radically phenomeno-hermeneutic formula, "The human essence consists in not having an essence," vaporized "on the day that the human is used as a raw material *ad libitum*" (AM II, 25). These past debates now appear as harmlessly naïve phantasms, that is, "that the theory of evolution be opposed to the biblical idea of man in the image and likeness of God!" whereas Anders compares the proto-humanism of Darwinism to the

> genetic manipulation which might *produce the inhuman* and indeed by manufacturing beings that would be "images and likenesses" or copies *of types to be desired on political, economic, or technical grounds*! (AM II, 25)

Anders opposes the transhuman imperative because the aspiration to the superhuman—*nota bene*, Goethe and Milton had already trod this same ground—is as opposed to the human as it is to the degradation to the subhuman.

Allergic to the sin of thinking negatively regarding technology—thank you Don Ihde—we remain aversive to critical thought, avoiding the idea that we might well challenge the "yuckier" dimensions of biotechnology, quite as Virilio underlines the parallel between the art of cloning (and Auschwitz) and animal–human hybridization for medical purposes, whereby reference to health, as the Coronavirus crisis has dramatized beyond all doubt, should any have remained, is a trump card for effectively everything, from lockdown to easing our qualms regarding mass surveillance/übervei llance[22] as well as mass vaccination, compulsory to be sure, in the case of 'health,' and much as the museum ambiance can induce ethical somnolence in the case of art.

Anders transgresses on all sides and has not been heard. It is as if philosophers, especially philosophers of technology but also critical theorists and political philosophers, have simply opted to look away.

This option is convenient. Nothing is easier than to ignore a scholar: just overlook their work and it is as if they never drew breath. But as Anders points out, foreshadowing an imaginable, if contrafactual, debate with theorists like Steve Fuller but also (and likewise potentially) in accord with Nick Bostrom and others who reflect on the desirability of this or that singularization of a new anthropic ideal, a new androidism, a new "digital" humanism, remembering that all that is needed for this is a cell phone with an internet connection:

> And even then, if the products that might result from such attempts did not turn out to be subhuman beings but "superhumans" ["*übermenschliche*"] (what the technicians suppose to be "superhuman", as Superman-like [*superman-haft*]), if one thus for example, were to synthesize "creative beings" ["*creative beings*"] (the putative ideal in the land of the conventionalised), music- or math geniuses—the sacrilege against humanity would still not be any less than were they to attempt to create the ideal of a semi-simian machine operator. (AM II, 25)

To this same extent, it makes no difference whether the post- or transhumanization promised by a cloned replacement biotech body or a hybrid fusion of human and machine is a Frankenstein-style disaster or the seamless promise assumed by most discussions of the human-robotic mind-meld. There is an ambiguity inherent in the notion of such "improvement," better for whom, one might ask, as the recent Covid-19 health crisis can raise the question of the sense of better: do we mean better to track or trace the human being,[23] or functionally better from the viewpoint of the health of the subject? What about gain of function research, what about side-effects?

Anders' third revolution ought to be the last revolution inasmuch as it changes the way we see the world, which was previously designated in terms of Heidegger's ready to hand [*Vorhandenem*]: "If one poses the question today [concerning the essence of the world], the answer can only be '*raw material* [*Rohstoff*]'" (AM II, 32). As a result, the world becomes not a world in-itself [*an sich*] but, increasingly so, simply a world for-us. Thus, we focus on "development" as we also focus on climate change. Anders' supposition, as we shall see, may itself be too optimistic, and we may be caught in the gyre of supposed revolution after supposed revolution.

The trouble is the one Anders took over from Heidegger, as he might also have borrowed it from the other members of the Frankfurt School, including the human, now very literally as we see in the new focus on transhuman mice and swine, as just one more "resource" among other resources:

> Over the course of the history of the mechanistic natural sciences what has transpired is that the human, too, is regarded as a machine ("*homme machine*"), as the exception that would have contradicted the principle; and this is repeated today on another level: given that the world is principally regarded as raw material, so, too, must that piece of the world, "humanity," in order not to injure this principle, likewise be treated as raw material. (AM II, 26)

Part Two

Anders, Media, Music

8

Radio Ghosts

Ghosts

Ghosts are spectral affairs, if we believe in them, if we do not believe in them. And there are instances of radio ghosts, "real" radio ghosts as we might say: *qua* phenomena manifestly accompanying radio broadcasts as such "ghosts" are perceivable and replicable as phenomena. Thus, Anders might have known of the Norwegian physicist Carl Størmer's account of shortwave radio echoes which the aptly named Størmer analysed in connection with his first-time measurement of the atmospheric height of the aurora borealis.[1]

Conversations and memories, past ghosts, are named as they return to us. Thus, Nietzsche speaks of a leaf floating down to us writing on history and life in his *Untimely Meditations*. With the mystical allusion to a full moon, to cite this one more time, Joan Baez sings, "Here comes your ghost again," in a song about Bob Dylan and memories of casual male insensitivity in *Diamonds and Rust*. And to talk of the *Psychology of Everyday Life*, Freud uses the same language Anders uses to speak of his radio ghosts, citing Goethe's *Faust II*: "Nun ist die Luft von solchem Spuk so voll, Daß niemand weiß, wie er ihn meiden soll."[2] In *Politics Without Vision*, Tracy Strong cites Sigmund Freud citing Goethe, using the English translation: "Now is the air so filled with spooks/That no one knows how he might keep clear of them."[3]

The tradition of radio "echoes," still unexplained to this day, evokes theremin-like spookiness, but the language of "radio ghost" points to an echo protracted over space and in time, the same time, as it were, key for an acoustic phenomenology: both Husserlian retention and protention. Indeed, the sensibility conveys a material notionality, related to the *epoché*, and this is likely why Kittler also speaks of the same ghostly echoes, specifically auditory, specifically with reference to radio, in the first chapter of *Gramophone, Film, Typewriter*, reminding us that the 'spooky' phenomenon, in all its ghostly resonance, can be summoned to this day at will: "If you replay a tape that has been recorded off the radio, you will hear all kinds of ghost voices that do not originate from any known radio station, but that, like all official newscasters, indulge in radio self-advertisement."[4] In *The Hallelujah Effect*, I have sought to explore related issues with respect to Adorno specifically regarding the radio 'voice'; here I am concerned with the essay on radio ghosts written by Anders as this also, as we shall see, influenced Adorno.

Radio Transforms: Politics and Music

Anders' (Stern's) brief essay "*Spuk und Radio*"[5] appeared in February of 1930 in a monthly journal dedicated to what was then the cutting-edge of new or modern music, *Anbruch*. The essay appeared as the closing essay in an issue dedicated to the "so-called opera crisis." The same journal featured an essay by Theodor Adorno (on Wagner), and there is an important parallel given Adorno's focal attention in his own writing on music as he would proceed to develop a political, sociocultural "phenomenology of radio." Note that both Adorno and Anders use the language of phenomenology.[6] Elsewhere I take up this very question of a musical and perforce hermeneutic phenomenology in Adorno focusing on his *Current of Music*, which he composed while living in the United States from 1938 to 1941 under the auspices of the Princeton Radio Project.[7]

Anders' title is a short one, and if the two loci of the title include variant parsings "*im*" and "*und*" (*on* and *and*), the challenge is to translate "Spuk" in Anders' case just where 'spook' seems perfectly apt in the case of Freud citing Goethe. Here "ghost" seems more accurate to the extent that what Anders conveys in his expressly phenomenological reflections on broadcast radio sound is less the effect of theremin-like spookiness than an acoustic double, quite in the sense in which he will later indict our all-too-human "second inferiority" (AM I, 50), or "industrial Platonism," a word for what Benjamin named the age of technological reproducibility, *qua* "immortality" and "'industrial reincarnation,' the serial existence of products" (AM I, 51). At issue in Anders' 1930 essay is an echoing trace, more rather than less material, haunting, a "ghost." And Kittler in a related reflection writes about the phenomenon of "radio spectres" in repurposed military equipment and "state security."[8]

Here, in addition to the radio "voice," it is worth recalling the radio "face" as Adorno emphasizes what he calls "radio physiognomy," as one may also highlight the concentric dimensionalities of the phenomenology of a Merleau-Ponty or the hermeneutics of a Ricoeur.[9] Like Heidegger,[10] Anders and Adorno follow Husserl and thus, only negatively, Heidegger. On the level of "reception," socially, politically culturally, Adorno explores the cultural and theoretic dimension of what Anders invokes under the double terms of listening-to (including a specific or dedicated phenomenological analysis published in 1927, by no means an accidental year) as well as what Anders formulated as the compound being-in-music in his own philosophical reflections on music and its "situation" as well as via the human sciences of psychology and anthropology in his more specifically outlined reflections on "Musical Sociology."[11]

It matters that the bulk of Adorno's reflections on radio, *Current of Music*, would be published in English from the start.[12] Anders himself likewise published several texts in English, and questions of translation make up no small part of the history of post-war critical theory. To this extent, the difficulties Anders repeatedly encountered, blocking his efforts to get his texts into print in both German and Anglophone publishing contexts—and one may argue that this is related to his post-war efforts through to the end of his life to secure support for/acknowledgement of his becalmed *Habilitationschrift* quite specifically concerning the "Musical Situation"[13]—are hard to overstate: dominating his correspondence with Arendt as he also sought, unsuccessfully, compensation from the German government retroactively, quite as others could (and did) lay claim to official programmatic efforts at "*Wiedergutmachen*."

Anders' brief essay on radio describes a phenomenology of radio in the lifeworld including elements of active and passive variation in addition to a Heideggerian hermeneutic of "listening to" the radio *qua* musically attentive listener, articulated and conceived with reference to a then-new technology (the first commercial radio broadcast began with a single sender in Berlin in 1923). Using the psychological framework of acoustic perception together with a musically reflective hermeneutic phenomenology, Anders relates the change(s) in musical space–time experience, that is, the lifeworld transformation of listening adumbrated by modern technology. On its own terms, radio technology would revolutionize everything on nearly every level: from communications and wartime tactics/techniques to propaganda and the political. In addition, as Anders underscored this, it was also to transform the human relationship—both phenomenologically and sociologically—to music.[14]

It is perhaps a sign of that same accomplished transformation of life and especially of musical experience that we today may find it difficult to imagine the difference radio broadcast would have to make for the experience of listening. Yet many who lived through this—Anders, Adorno, Rudolf Arnheim, as well as Bertolt Brecht, and not excluding Heidegger himself— seem to have written about radio.[15] In addition to what would soon be used to transmit sociopolitical, that is, Nazi, and other propaganda, the very fact of radio transmission permitted the same musical performance to be broadcast to a range of tuned radio receivers in a given community, at a given time. Thus the ghosting, echoic effect.

Anders analysed the "spooky" play of experiencing the same music playing on the same station as one might move from place to place in the world. This is the acoustic ghost effect Anders describes moving in his own apartment, listening to the radio from room to room, and in its phenomenological, retentional/protentional, extension as he, after leaving his apartment, still hearing the fainter sound of his own radio playing in the distance, encountering, as he passed his neighbours, the stronger sound of other radios playing the very same music in other apartments.

To compound this phenomenological analysis of what might be assessed as an otherwise insignificant artefact of acoustic experience, it is important to underline that loudspeakers were specifically engineered and developed for mass placement at party rallies in the political sphere.[16] Writing his infamous Black Notebooks a few years later, Heidegger will decry Nazi party gatherings for what he describes as a "gigantic 'prostitution' into noise,"[17] gatherings he had earlier characterized in terms of their "organized screaming,"[18] a phrase echoing, as it does, some of Theodor Lessing's language on the social role and efficacy of noise,[19] which might also be applied to today's riled-up crowds, the far-right raging to boiling, the collectivity of the "they." That loudspeakers are the dedicated means to collimate this screaming is the point Albert Speer argues, characterizing Hitler's dictatorship as the

> first dictatorship in the present period of technical development, a dictatorship which made complete use of all technical means for the domination of its own country. Through technical means like the radio and the loud-speaker, eighty

million people were deprived of independent thought. It was thereby possible to subject them to the will of one man.[20]

Loudspeakers, large and multiply positioned or located in the case of party gatherings, more minimal in the case of the single radio in Anders' living room, entail an acoustically transformed world. As Anders observes (and Adorno and, later, Kittler will only continue this), they also change, precisely beneath our notice of the fact, the way one hears. We listen not merely with our ears or even with our minds, but also—and this is something Heidegger highlights in *Being and Time*—with our bodies, to and with the world around us and as we live in and find ourselves bodily in this world.

In the political sphere, as this also becomes an intimate correspondent of and to the private sphere in the twentieth century, loudspeakers allow us, or, better perhaps, compel us, to hear more resonantly than heretofore, with sonic sub- and super-liminality. As Anders emphasizes in his essay, as Adorno will later write about this: space-time, world-space, and world-time are altered by sound. The way we hear loudspeakers, the replication of and placement of loudspeakers and, *ceteris paribus*, earphones, shifts everything.

The space and the time of the everyday world, Husserl's lifeworld, is charged by sound as sound specifically mediates (and is mediated by) both space and time, and both space and time are transformed by technology. To this same extent, the crowds gathered in Nuremburg would not have (because they could not have) listened to these speeches in the same way, in the space, at the same distance, and in the same lived temporal relation as one would have heard the speeches of Pericles in Athens or Lincoln in Gettysburg. Thus, in antiquity, the reference to Pericles being more than rhetorical, and it is also for this reason that I continually note my earlier study of media and influence, *The Hallelujah Effect*, the Greeks themselves were aware of and designed magnifying sonic effects as these could be induced via stone itself and the resonance of certain enclosed spaces—think tholos tombs—and, hence, the design of the theatres themselves, open spaces, and also focused via theatre masks.[21] Radio broadcast speeches make the point still more clearly, if the specific sound analysis is yet more complicated quite as Anders writes and as Adorno writes in the same modality, using in his case his own terminology not of a *situationally* attuned listener but highlighting "radio physiognomy" instead.[22]

Seemingly anticipating aspects of today's debates on "Post-Truth," although perhaps missing the force of Trump's America and Boris Johnson's Brexit, Anders writes:

> no one will deny that to produce the kind of mass man that is desired today, the formation of actual mass gatherings is no longer required. Le Bon's observations on the psychology of crowds have become obsolete, for each person's individuality can be erased and his rationality levelled down in his own home. The stage-managing of masses [*Massenregie*] in the Hitler style has become superfluous: to transform a man into a nobody (and one who is proud of being a nobody) it is no longer necessary to drown him in the mass or to enlist him as an actual member of a mass organization. (AM I, 104)

As Anders goes on to say, all you need is radio. In the case of listening to music, what had been a phenomenological analysis of a lived-world experience of (a live) broadcast radio performance, untethered to its actual locus and thus able to haunt or follow Anders as he walked down the hallway in his apartment building, as a trailing echo (this was the "spooky" experience of the radio broadcast), became quite another thing, ubiquitous (as both Anders and Adorno argue) as a means for the crystallization of political opinion, before and after the Second World War, and not less for advertising. Thus, with a reference to Viktor Frankl, Anders reflects in 1979:

> Inasmuch as in the age of electronic media there is no longer any place where one cannot be informed/dis-informed [*informiert bzw. desinformiert*], or, more accurately, where one may escape the obligation to be informed/dis-informed, therefore no provinces—there are also no places where one's ears are not filled with idle chatter concerning the "loss of meaning" by vulgar philosophers, psychoanalysts, and radio preachers. (AM II, 30)

Phenomenology's Ghosts: Anders' Phenomenology of Radio Listening

If Anders goes on to offer a critical take on the ubiquity of information/disinformation (what we call "fake news") of electronic or digital media, Anders' original 1930 essay on radio is remarkable as a performative phenomenology of radio listening. In this way, Anders is a witness to the acoustic transformation of the lifeworld effected by radio broadcasting for experiencing sound as perceived in the lived world. This acoustic lifeworld is the technological condition, existentially and politically, for Aldous Huxley's *Brave New World* (1932).

Patently, with the ubiquity of media all around us today, broadcast as it is, "streamed" as it is, directly into our ears, we take all of this for granted. From phenomenology's ghosts to Anders' "schizo-topia," Anders undertakes his discussion of radio ghosts via the phenomenological modalities he learned as a student of both Husserl [23] and Heidegger. Thus, Anders offers not only a picture book phenomenology of music and spatiality, positionally differentiated in space, but takes care to highlight the curious circumstance whereby it is specifically technology that "accidentally brings ghosts along with it."[24] Radio ghosts, like the more familiar projection artefact of television ghosts, are part and parcel of broadcast phenomenology—a point Kittler also foregrounds.[25]

It is the phenomenological "variation" that adumbrates Anders' reflection on radio. Phenomenologically, Anders observes, music can be acoustically located in space, from the perspective of the subject, the listener, who can thus characterize what is heard directionally and in terms of distance, just as one might, sitting in a street café, hear a busker play on the sidewalk or, for Anders' example, the sound of a street organ moving past. Anders thus illustrates the method of Husserlian free variation, but Anders notes that listening to radio broadcast made possible an acoustic experience that had never been heretofore experienced, and was thus a definitive artefact of modern technology. Beyond listening to a street organ rumbling by, or else, as in Rilke's *Duino Elegies*, hearing a violin as one passes under an open window, one can also vary one's own disposition or position in the world by walking into and out of range of broadcast radio music, now complete with—and thus the difference between hearing a single source, Husserl's *postillion* horn or Rilke's violin or street organ in the distance—the novel opportunity to encounter a musical twin or triplet [*Doppelgänger oder dreifachgänger*],[26] undergirded with the distant echoes of the self-same sound.[27]

Where the visual object is specifically to be seen here and not there, or not seen at all, as in the case of Plato's cave (as Østergaard emphasizes) or as in the case of the picture behind the viewer, Heidegger refers to this in *Being and Time*, and thus is fixed or localizable, music tends rather to permeate space, producing in accord with the design of the composer (and the dimensions/materiality of the room in question) a more or less "voluminous" space, "massive" or widely dispersed or more or less narrow, but largely spatially encompassing, and thus neutral in the sense of being not as such here or there but which "real presence" is annihilated by radio. Note, again, that this is a doubling in ghostly fashion which may, shades of quantum theory already in the air, may be here and there, adding overlay patters as Anders emphasizes these. Anders, as already described, offers the reader an account of the then remarkable experience of hearing the radio still playing behind him as he leaves his apartment. With the door closed, the sound is muffled but able to be heard as Anders walks down the hallway. Note, to be sure, this is the relevance of distance and directionality, that the sound he hears—think of the experimental psychologist's (his father, William Stern's) tone variator—as he steps out of his apartment is the (now) distant sound of the radio heard in his living room. This Anders names the "schizoid-topic" moment, the trail of sound in the distance that accompanies his hearing, repeated as moves to hear once again, still with overtones of his own radio, as he meets and passes the same radio music playing in the next apartment.

Husserl uses parallel examples in his 1905 and 1911 reflections on internal time consciousness. In *Being and Time*, Heidegger adverts to the direct apprehension via hearing of a "creaking wagon or a motorcycle" or "the column on the march" (SZ 163). Heidegger emphasizes that we never hear "tones" as such, but always lifeworld indices, the voice of the lark, a certain airplane overhead; later he will refer to the sound made by a specific make of automobile. Thus, via Heidegger, Anders argues for an acoustic phenomenology yielding radio ghosts and "virtualities."

It is sound, as both Husserl and Heidegger attend to this, but with Anders' attention to being-in-music, noting the spatiality of tuning along with the active listener's attunement ensures the kind of sensibility, for better or for ill, that allows or permits the kind of claims that Adorno makes, because the jazz he analyses is very specifically not a "live" performance but broadcast on a television or radio programme. In his "schizo-topia," offering an echo of his analysis of simultaneity—*Zugleich*—Anders reflects on the acoustic and affective phenomenon heard on the radio, in whatever event (say a sports broadcast) or musical experience or an exceptional moment, *qua* transmitted.

Later, writing on the "antiquatedness" of the idea of the masses as radio has transformed the "masses" by 1961, describing the changed ingredients needed to make a world-class event: listening-together, *en masse*:

> Even while vacuuming her rug, the housewife is not only at home between her own four walls but is simultaneously under the vast cupola of St. Peter's, since the organ music accompanying the Coronation of the Pope is playing in her house. (AM II, 85)

Via the music, one is both where one is and at the same time one is elsewhere. To this extent, listening to the music one first heard as a young person can, in a locational transposition, take one back, as we say. The language is spatial, further specified: in time—again the term is "schizo-topia," as Anders explains, and a great deal about modern social media's experiential phenomenology is thereby illuminated:

> Because the radio can be installed anywhere, at home, in the car, or in any public place, in a way it represents a common denominator device which is capable of neutralizing the differences between these places where we spend our time. And since with their help we are always somewhere else, that is, we are never home, we are also always and everywhere at home by virtue of their help. (AM II, 86)

Observing, "If the radio is the embodiment of his de-privatization, the car is the embodiment of his always being-at-home" (AM II, 86), Anders invokes the complete coordination of these two instruments, radio and car, just so radio and airplane, as this describes our current digitally adumbrated, mediated human condition. Heidegger too, will also indict the prevalence of radio sets seemingly everywhere, in all the rooms of the house, as an indication of "planetary idiotism."[28]

One cannot, as psychologists who study the phenomenon of multitasking tell us, do two things at once, not really. This applies to listening to music and the phenomenon of ASMR, if studied phenomenologically, might illuminate this but even more saliently, Anders' schizo-topic phenomenon may account for at least some of the uncanny effects of ASMR. Anders' "Prometheus effect" both domesticates and disenchants the world, to use language Adorno and Horkheimer and Marcuse use to explicate related phenomena. Baudrillard explored this condition as endemic to modern digital media with a seemingly paradoxical description as foreclosing responsibility. The paradox is only apparent given that we seem (to ourselves) to be more directly engaged with media than ever before in our socially mediatized lives, is its one-way directionality, which Baudrillard names, I discuss this elsewhere and it matters immensely for an understanding of media, "speech without response."[29]

Baudrillard writes:

> The mass media are anti-mediatory and intransitive. They fabricate non-communication—this is what characterizes them, if one agrees to define communication as an exchange, as a reciprocal space of a speech and a response, and thus of a responsibility (not a psychological or moral responsibility, but a personal, mutual correlation in exchange).[30]

Media archaeology for Baudrillard "always," by design, "prevents response."³¹ The media thus make

> all processes of exchange impossible (except in the various forms of response simulation, themselves integrated in the transmission process, thus leaving the unilateral nature of the communication intact). This is the real abstraction of the media. And the system of social control and power is rooted in it.³²

We are inclined to overlook the one-way glass that is the mediatic screen—which is surely part of the way it works—whenever we tweet or post on Facebook, if we can hardly limit this scopic directionality to television (cable or network), YouTube, and streaming music feeds, not to mention other social media, where we post and wait, usually bootlessly, for a response to our responses which social mediatic forms³³ may thus bootless be counted as so many varieties of Baudrillardian "speech without response."

The transformation is hard to see because it has succeeded—as Nietzsche says of the ascetic ideal in his *On the Genealogy of Morals*. And for Anders, in consequence, we are everywhere and nowhere at home:

> Partly, today, we are nomadic, because, even when we are home, we are, at every moment, residing somewhere else; and partly sedentary, because, even then when we are actually driving through a foreign country, we can consume the comforts of being at home; and this means, paradoxically, that we also have the possibility of finding ourselves in another, that is, transmitted place. (AM II, 87)

In the same way, Anders analyses "background music," that is, the ineliminable music of the elevator variety we have no choice but to "enjoy." In this measure, and so far from offering us an antipode to Adorno, Anders allows us to understand what Adorno names the "culinary" with respect to the culture industry and to music in particular.³⁴ In just this measure, explicating both the culture industry and the culinary, Anders invokes Adorno's reference to breath (and suffocation):

> We have been plunged into an industrial oral phase, in which the cultural pap [*Kulturbrei*] slides smoothly down our throats. In this phase, what is supplied does not even have to be perceived, but merely absorbed. (AM II, 254)

No sooner do we note that association than we also may observe that Anders here similarly echoes Adorno's charged reflections in *Minima Moralia*, as so many variations on "air":

> The supplied commodity is, for the listener, "air"; and indeed in a twofold fashion: 1) it is something taken for granted; but 2) one cannot breathe without it. This kind of destruction, the annihilation of the object, via dissolution, liquidification, is not some kind of special feature of radio and television, but is characteristic of all production today. (AM II, 254)

Anders' analysis includes multitasking and displacement. If the later Anders reprises Adorno, he had earlier drawn on Heideggerian motifs for his 1959 reflection on the "antiquatedness" of space and time, nevertheless combined with a Frankfurt School-style critique of the culture industry as this industry commandeers the space of our lives and our minds, as we know, post-pandemic:

> Everyone knows that life today often proceeds down two roads simultaneously. Thus, e.g. while we travel along the track of our main occupation, we listen to music on the parallel track. And do not believe that we only love this double road of existence merely because we want to cover over the necessity of work with the sweetness of *energeia* (music enjoyment). (AM II, 346)

This is like the schizotopic moment analysed earlier, but this doubling can also be applied to the self as we seek to distract ourselves from ourselves.

In addition to the increasingly consummate deployment of the human being as the creator of himself/herself as "consumer," curating himself/herself for others, that is, "broadcast" for similar consumption (this is the point of social media) and as himself/herself reproducing himself/herself as a homeworker on himself/herself as consumer (in the scheme of industrial consolidation), we distract ourselves from our surroundings as from one another.

The world ignored becomes a spectre, estranged and distant as Anders observes in "The World as Phantom and Matrix," a phantom quality that is also a direct consequence of technological determinism. Thus, Anders argues—and note the claim as he means it literally and as it illuminates Adorno's related arguments on the same point concerning music as practice and as lived culture, in everyday life, at home— "the phonograph and radio have robbed us of live music performed in our homes" (AM I, 18).

It is not Anders' claim that we lament the loss of this kind of "live music"—what's to miss? Even Anders himself in his 1949 essay on the "acoustic stereoscope" argues that there is effectively, given the right kind of stereophony, no difference at all. We have never had more "music" in our lives as we can arrange via a variety of streaming services to have music with us anywhere we like, screwed into our ears, synchronizing our thoughts on nearly any occasion, all on demand, better said: on auto play.

For Anders, "wired" for sound in the open world, we have changed our relation to ourselves and to one another and to our listening, specifically to our musically attuned, oriented, or "active" capacity for listening-to-music. To this extent we hardly need the loudspeakers of Huxley's 1932 *Brave New World* or Orwell's *1984* (inverting the year of its publication in 1948) for the sake of social conformity.

Anders' concern is with what thereby becomes obsolete or outdated: autonomy, *Mündigkeit*, as Kant speaks of it: human self-determination. The eclipse of self-determination could not be more radical. As Kittler reflects:

> The literally unheard-of is the site where information technology and brain physiology coincide. To make no sound, to pick your feet up off the ground, and to listen to the sound of voice when night is falling—we all do it when we put on a

record that commands such magic. . . . As if the music were originating in the brain itself, rather than emanating from stereo speakers or headphones.[35]

Kittler emphasizes the role of wartime innovation in such technologies, reflecting on the precise triangulation of wartime bombing missions, sound and the pilot's brain, all one synaesthetic event, and it is important to cite this at length because "the hypnotic command" of which Kittler speaks was as automatic as a pressing a button:

> Long before the headphone adventures of rock'n'roll or original radio plays, Heinkel and Messerschmitt pilots entered the new age of sound space. . . . Radio beams emitted from the coast facing Britain, for example from Amsterdam and Cherbourg, formed the sides of an ethereal triangle the apex of which was located precisely above the targeted city. The right transmitter beamed a continuous series of Morse dashes into the pilot's right headphone, while the left transmitter beamed an equally continuous series of Morse dots—always exactly in between the dashes—into the left headphone. As a result, any deviation from the assigned course resulted in the most beautiful ping-pong stereophony (of the type that appeared on the first pop records but has since been discarded). And once the Heinkels were exactly above London or Coventry, then and only then did the two signal streams emanating from either side of the headphone, dashes from the right and dots from the left, merge into one continuous note, which the perception apparatus could not but locate within the very centre of the brain. A hypnotic command that had the pilot—or rather, the centre of his brain—dispose of his payload.[36]

As Kittler remarks, "Ever since EMI introduced stereo records in 1957, people caught between speakers or headphones have been as controllable as bomber pilots."[37]

For Kittler and for other philosophical analysts of music and war, "The world-war audiotape inaugurated the musical-acoustic present. Beyond storage and transmission, gramophone and radio, it created empires of simulation."[38] Like Anders, Kittler means this literally:

> *Funkspiel*, VHF tank radio, vocoders, Magnetophones, submarine location technologies, air war radio beams, etc. have released an abuse of army equipment that adapts ears and reaction speeds to World War I. Radio, the first abuse, lead from World War I to II, rock music, the next one, from II to III. Following a very practical piece of advice from Burroughs's Electronic Revolution, Laurie Anderson's voice, distorted as usual on *Big Science* by a vocoder, simulates the voice of a 747 pilot who uses the plane's speaker system to suddenly interrupt the ongoing entertainment program and inform passengers of an imminent crash landing or some other calamity.[39]

Kittler's methods of counting historical intervals match Anders' own reflections,[40] especially where Anders insists that we explicitly, albeit unwittingly, "program" ourselves:

> The pairs of lovers sauntering along the shores of the Hudson, the Thames or the Danube with a portable radio do not talk to each other but listen to a third

person—the public, usually anonymous, voice of the program which they walk like a dog, or, more accurately, which walks them like a pair of dogs. Since they are an audience in miniature which follows the voice of the broadcast, they take their walk in company of a third person, not alone. (AM I, 18)

To do the work of this programmed constant accompaniment, Anders invokes the quintessentially innocuous transistor radio, no *Brave New World*-style loudspeakers needed. Today, we programme via subscription or playlist, whatever curated soundtrack we prefer, of our own design: more homework, as Anders describes the undertaking involved to do this, from music to meditation apps. Thus, we use the fetishized distraction of our devices to undo the distraction of our devices, and today, ensuring that we are never alone, and thus that we never fail the task of being an "audience in miniature"—this is FOMO, what we call *fear of missing out*—today's wandering lovers, if they are allowed out during lockdown, might tune in to a GPS system so that they do not lose their way along the Hudson/Thames/Danube. For Anders, the point of this constant attending, audience being, guarantees that "[i]ntimate conversation is eliminated in advance" (AM I, 18).[41] In general, when people today drive, often to destinations they happen to know, they tend to choose to listen to GPS guidance, as opposed to speaking with one another. For Anders, the radio voice/GPS lady voice, is an invited third, as we transform ourselves of ourselves into passive—as opposed to active—listeners. This is being-online, being-connected, getting-a-signal. And as Anders helps us understand with being-an-audience, that is, being connected, this is an intentional status to be worked at.

Broadcast radio changes in its effects, in its dynamic working, by drawing on, transforming what Anders described as "being-in," the matrix that is absorbedness in music. Listening to music on television, or as Anders' example detailed the vacuuming housewife as listening to a mass (the papal coronation) or listening to the news effects a similarly transformed experience of being in the world. Here the first few stages of what Anders—who was more than perhaps ordinarily fond of lists—counts off as a collective move to a virtual or phantom existence are worth noting. Beginning from a crucially Heideggerian *In-Sein*, Anders explains:

> 1. When the world comes to us, instead of our going to it, we are no longer "in the world" but only listless, passive consumers of the world.

To this extent, the world becomes a phantom as he writes in his title, that is, a ghost, an illusion. As in-being, *qua* being-in, the modality or tone of our own existence is transformed.

> 2. Since the world comes to us only as an image, it is half-present and half-absent, in other words, phantom-like; and we too are like phantoms.

Anders, drawing upon the early years of loudspeaker and radio technology, continues:

> 3. When the world speaks to us, without our being able to speak to it, we are deprived of speech, and hence condemned to be unfree. (AM II, 20)

The result Nietzsche had already diagnosed as the reaction condition of modern impotence, and for Anders this impotence changes our way of relating to the world:

> 4. When the world is perceivable, but no more than that, i.e. not subject to our action, we are transformed into eavesdroppers and Peeping Toms. (AM II, 20)

Again: we are such "Peeping Toms," as we find ourselves, often without intent, "eavesdropping" constantly. We do this supposing our own invisibility/silence whenever we watch whatever video catches our attention when we "check" Twitter, Facebook, Instagram, what have you. What Anders writes about the radio and television receiver applies to our cell phones and tablets, any internet connection. Anders argues and Baudrillard echoes the point that just in the measure that we are "connected," such connections "gradually deprive us of the power of speech" (PM, 17). This passivity, making us voyeurs of the world (*qua* phantom), is what it is to be an "audience in miniature" minus the community building efficacy that it would be to be present one to another, listening together, as Heidegger describes this, as does Anders and as Adorno similarly writes of the communal power that is listening, specifically to music, together with one another. This cannot be done virtually or at a distance (literally, or else it is not a way to build a community, to use Paul Becker's language when he speaks of the symphony's "*gemeinschaftsbildende Kraft*," just as Adorno cites this in his *Current of Music*).[42]

Elsewhere I highlight that the trivial thing needed to convert the user into a voyeur—Illich will use the same macabre language in his *Medical Nemesis*, speaking of a patient dying in hospital—is the slightest chime or signal, "notification," that permits us to suppose ourselves engaged.[43] All these means of Baudrillardian "speech-without-response" replicate a scopic transformation. As audiences, we continue to relate to the world in the same incorrigibly unfree fashion Anders analyses, "spectacle" as Guy Debord contended in the 1960s, or via what Illich in the 1970s and 1980s named the "Age of the Show."[44] And like Adorno, Anders emphasized the transformed experience and the space–time of the same for the lifeworld of the human being.

In this way, in a section entitled *The Return of the Soloists*, developing his 1956 essay into a 1961 reflection on the general outdatedness that it is to invoke the "masses," Anders offers a phenomenological sociological analysis whereby what is abolished is the external world, namely the Heideggerian *Mitwelt*: the public world. If Arendt herself focuses on the political public sphere in this exchange, Anders helps us see, as Nietzsche in his own century responding to a differently totalized experience of entertainment and culture, that in the wake of the loss of the public sphere we are increasingly given over to what can seem an utterly self-focused absorption, in which one also loses the private sphere and thereby the private self as well.

Media-Induced, Collective, "Autism"

Anders refers in the 1960s to American jukebox culture, the diner culture continuing throughout the 1970s, only to end with the 1979 development of the Walkman, "The customers behave in public as at home" (AM II, 84). The focus is a shift from what can be regarded as the private and the public sphere whereby self-absorption—and the exclusion of social focus on others—acquires a technological articulation and as technologically induced, a normalized justification. In an interesting way, the 2020/2021 lockdowns, sporadically ongoing in 2021, worked as seamlessly as they did because the phenomenon of lockdown had, in effect, already been a quasi-virtual rule, think only of the title of Sherry Turkle's *Alone Together*, although, and to be sure, this is also the point of many of Ellul's as of Illich's reflections, along with Baudrillard and Kittler as indeed of Adorno's *Current of Music*. Anders' reflections contribute to this same tradition in the philosophy of technology. By using earphones, the consumer focuses on himself or herself, alone, or together with his or her companions. To this extent, the "privacy or isolation of the receiver disguises the mass character of the commodity and the mass character of its transmission" (AM II, 80). We permit ourselves the same isolated self-absorption when we focus on our cell phones/iPads at home, with our families, as in office meetings, on public transportation, and so on. The games we play, the music we listen to, the social media we "follow," permit us to participate, even before Lockdown: in our own homes, in this very same mass character:

> each of us is supplied personally not only with our de-individualization and our form of mass existence, but also, at the same time, with the illusion of privacy (insofar as it is generally a question of a dual conditioning). (AM II, 81–2)

Horkheimer and Adorno were concerned with the same questions of inductive conditioning in their analysis of the culture industry and its industrial product, namely music, entertainment of all kinds, newspapers, and so on. Anders argued, as we have seen, that we create/invent/form ourselves in the image of the culture industry, which today extends in our digital era beyond what had been a certainly burgherly limit to hours of the day and the night, such that today we think nothing of interrupting sleep in the fourth—or n-th—industrial revolution:

> One experiences pleasure at home when one consumes in the family circle (which has actually been transformed into a mere juxtaposition of individuals) the program that millions of other consumers are consuming at the same time, in an equally "private" way. (AM II, 82)

This is how the culture industry works. At the same time, intellectuals and scholars and savvy individuals overall tend to regard this working as by and large benign or else to suppose themselves somehow unaffected by the effects of this industry. It is the theme of Anders' as of Adorno's/Horkheimer's analysis of the culture industry that this effected a de-individualization matched from the start with the illusion of choice and above all, as Anders emphasizes this, the illusion of privacy. The point is easily extended to Facebook/Snapchat/Instagram/Twitter, and Anders reflects on its efficacy precisely as something we do to ourselves and as ourselves, on our own dime—Anders liked to emphasize that we pay for this—and, this is a psychoanalytic truism, with the result that what we subscribe/submit to, is completely volitional:

> For this conditioning is disguised as "fun"; the victim is not told that he is asked to sacrifice a thing; and since the procedure leaves him with the delusion of his privacy or at least of his private home, it remains perfectly discreet. (PM, 16)

Thus, Anders' "solo performers," referring to the consumer as he/she creates himself or herself as such, surfing the net, "designs" his own advertising experience. "Modern mass consumption is a sum of solo performances, each consumer an unpaid homeworker employed in the production of the mass man" PM, 14).

Thus, as Anders describes a mid-twentieth-century living room interior arranged around a television—note how constant that feature remains, even as it has migrated to our cellphones and tablets as well[45]—his description matches current mediatic self-inversion. The scene is familiar to the philosopher—it is Plato's cave:

> The seats in front of the screen are so arranged that the members of the family no longer face each other; they can see or look at each other only at the price of missing something on the screen; they converse (if they still can or want to talk with each other) only by accident. They are no longer together, they are merely placed one beside the other, as mere spectators. (PM, 17)

Speaking of radio, we noted earlier with reference to Kittler that gramophones and phonographs went together with the loudspeaker technology developed along with mass-programmed Baudrillardian speech-without-response, that is, radio broadcasting. In this way, Anders can observe, just to quote this once again:

> The stage-managing of masses in the Hitler style has become superfluous to transform a man into a nobody . . . it is no longer necessary to drown him in the mass or to enlist him as an actual member of a mass organization. Today's Hitler would need no access to radio and not even mass-party demonstrations. (PM, 16)

If, today, Donald Trump and every other influencer seem to have a Twitter account (now handily suppressed, to popular applause, in the case of Trump), it may also be argued that,[46] quite in the spirit of Gil Scott-Heron's 1971 *The Revolution Will Not Be Televised*,[47] any revolution to come will require attention to the limitations that follow from our confinement to our screens, assuming we can attend to such limitations, including the same Twitter that remains sufficient to effect control as well as the illusion of participation.

Anders' concern, as we have seen, is with the "unilaterality" of social media. This is curated self-creation, tweaking our own feeds, all the better for data miners as we today may well observe. Thus, Anders emphasizes: "Soft totalitarianism likes nothing better than allowing its victims to have the illusion of autonomy or even to engender this illusion" (AM II, 238). But what is expropriated from the average American is what had previously been most properly, most intimately, his own. To this same extent, what is taken is never property: relations of capital are not to be put in question (banks are always to be "bailed out"). Instead:

The "sole" thing that must be taken from him is his "particularity" [*Eigentümlichkeit*], his personality, his individuality, and his privacy: solely himself. In contrast with routine socialization, which involves what the person has, we are here concerned "only" with a socialization of that which the human being is. (AM II, 239)

In his own anticipation of Baudrillard's 1988 reflection on *America*, itself a twist on de Tocqueville, Anders reflects that "America uses psychoanalysis for the sake of the establishment of conformism" (AM II, 237). Today's neuropsychology follows along, as we, absorbed in our eternally new, eternally data-mined, "surveilled," and curated bubbles of attention, create and re-create ourselves in the current remix—again, n-th industrial revolution—in the age of digital reproduction.[48] For Anders, the fact that you appropriate something *not yours* as yours, does not make it yours. The point (and this is why one must take Anders' distancing of himself from Heidegger with more than just a grain of salt) is a matter of *Eigentlichkeit*.

9

Being-in-Music

Music Critique and Musical "Situation"

It is famously impossible to "please" critics. Beyond Georg Kreisler's satirically close-minded music critic in his song of the same name, *Der Musikkritiker*, the description seems to fit critical theorists, both Anders and the yet more obstreperous Adorno, including his criticisms of jazz (and swing). The "material" basis for Anders' thinking on music goes back to the 1920s and 1930s, part of which time Anders worked as an international music correspondent, writing reports quite on the contemporary musical scene from Paris.[1] The concern was not as such an incidental one. Thus, I have already emphasized that it matters that Anders sought to habilitate with a 1930–1 thesis: *Philosophical Investigations Concerning Musical Situations*, a project Anders continues to pursue throughout the following decade and the decade after that, all without success, including an application, failed, for a Guggenheim, letters to Arendt and so on.[2]

A full account of what Anders' called the "musical situation" and his later research on music sociology compels attention to Frankfurt school critical theory, traditional philosophical aesthetics, and the enduring challenge of writing between philosophy and music. All of this I can only outline in the current context.

Thus, in Adorno's 1944 lecture presented at Columbia University and addressed, *nota bene*, not to the philosophers but the sociologists, he underlines what he calls the "decultivation of the German middle classes demonstrated in the field of music but noticeable in every aspect of German life."[3] The disciplinary distinction between sociology and philosophy is essential to underline as Adorno recalls a class he himself had given more than a decade earlier,

> In the winter term of 1932-33, immediately before Hitler took over, I had to conduct at Frankfurt University a seminar on Hanslick's treatise *On the Musically Beautiful*—which is essentially a defence of musical formalism against the doctrine of Wagner and the programmatic school. Although the seminar was focussed on philosophical issues, the participants, about thirty, were mostly musicologists. In the first meeting I asked who was capable of writing the Siegfried motif, the most famous of all Wagnerian leitmotifs, on the blackboard. Nobody was.[4]

The vignette could not be more classic "Adorno," and it speaks to the very distinctions Anders makes as these retain a contemporary edge: concerning the "sociality" of the "complex" that is music. To repeat Anders' critical array of questions already cited at the outset: "Who musicizes, who may, who may not? Who composes?" Anders thus proceeds to ask: "Is there authorhood in a sharp sense?" (MS, 178), raising a question that in a very different sense is explored (in an analytic mode) in Lydia Goehr's *The Imaginary Museum of Musical Works*, and although Anders is hardly Goehr's reference,[5] the theme is a convergent one. Further, it bears on my own question of the musical cover and the range of issues involved whenever technology is involved with music along with what Anders emphasizes as performance practice as its own kind of musical situation *qua* performative or "execution situation" [*Mitvollzugssituation*].

Anders develops in his reflections on music sociology as well as his phenomenology of listening a more than Heideggerian catena of questioning. Thus, Anders maintains his earlier insight into the question just as Heidegger had drawn the question as such into focus in terms of the situation of both questioning and questionability. At stake for Anders, as he writes in his "Philosophical Investigations Concerning Musical Situations," is to follow the Heideggerian imperative to put the issue itself in question, not as a simple object of theoretical investigation but the predisposition always already entailed, *qua* pre-decision already made for this or that Heideggerian preunderstanding of what, in advance, counts as music (MS, 21), difficult for Anders, who sought to explore music as it is its own world and at the same time as it is in the world, differently accessed according to cultivation but not less class and privilege.

What is clear, perhaps is that be it a musical cultural history of "situation," or a music sociology, or indeed a phenomenology of hearing music, listening-to, a philosophical aesthetics of music, even enriched with the resources of hermeneutic phenomenology will require a path between music theory, music history, which for non-musicians, and even enthusiasts, can be closer to "music appreciation"—quite where Adorno dedicates two chapters to the last[6]—which may sometimes presuppose familiarity with the referent but often not even that. Thus, the demand, beyond the suggestion that one hum a few bars, set to a class of university-level musicologists that at least one of their number be able to write out the Siegfried motif can be overwhelming. Perhaps, so one is told, Adorno smiled when he related their lack of competence.[7]

This tension is compounded by academic philosophy's explicit *habitus* when it comes to aesthetics and its respective ingroups. Inasmuch as the philosophy of music reproduces *grosso modo* the prime division of professional or university philosophy— that is, the same discipline that did not make room for Anders—namely, into "analytic" (the great majority must be accounted as such, including Jerrold Levinson and the already-mentioned Goehr) and continental, the last being a very small set, inhabited, in addition to thinkers like Adorno and Anders, by Pierre Bourdieu as well as more esoteric others,[8] names philosophers of music either ignore or else, as in the case of the late Sir Roger Scruton, actively refuse—thus Scruton, with contempt, pronounced Adorno a "dead duck" at a 2012 London conference.[9] If there are exceptions, I have noted the musician-philosopher Max Paddison on Adorno,[10] it remains difficult in philosophy of music to invoke critical theorists like Adorno or like Anders quite because of the institutional dominion of analytic-style philosophy as this continues, formation and appointment, to characterize most university academics to date. To just this extent, critical theory can be subject to exclusion (just leaving out names can be enough for this), philosophical reservations of sundry kinds along with objecting to Adorno on jazz. There will be similar objections in the case of Anders, who remains even less discussed.

In the same way, Anders himself may fall between two stools. The late Ludger Lütkehaus, brilliant as ever—and he was exceptionally sharp—argues that Anders fails to overcome Adorno,[11] an observation attesting to the level of the bar to be met. Anders himself will note, as Lütkehaus underlines, that he does fail short in comparison with Adorno, but, and this is to the point of the current study, without yielding his own concerns in favour of Adorno's.

Situational Phenomenology: Underway to a Hermeneutics of Music

In the book that would have counted, a counterfactual book as it were, the kind of book that can haunt a scholar's lifetime had things been other than they were, as the young Günther Stern's *Habilitationschrift*, had he gone on to a position as professor of philosophy and/or cultural sociology of music as his thesis patently provides ample basis for the same, Stern/Anders notes the conundrum built into the word "*Musikphilosophie*." For Anders—and here again the influence of Heidegger's explicit phenomenology of questioning is evident—from the start, one must reflect on the subject matter of a given investigation, in this case into the musical situation, while also raising the question *why*. The "why-question" (MS, 15) is key; it is to be understood in connection typically with the *what* question and, for Anders especially, with the *that* question, crucial to any post-Rilkean aesthetic question, quite as Gadamer also for his own part will emphasize, and as is also differently underlined by Adorno. But, hermeneutico-phenomenologically, preliminary questions have to be posed from the start; this is an always-already presupposition of reflection necessary before one can raise a "why" question in the philosophy of music, to which, class-conscious as he is, from a sociological vantage point, Anders adds the *when*- and the *where*-questions as these bear both on society and history (MS, 51). Intriguingly, Anders begins with a reflection on Augustine's *Confessions*, referring thereby not only to the *locus classicus* of phenomenology (intentionality for Husserl) but also sensual desire—*voluptas*, a range of various and all-too-human *tentationes* and, in a lovely poise, an articulation of delectation: "that is to say, the current incarnation and the express physical-sensual realization of music, in contrast to the textual content intended in music-making, the '*res*', visible" (MS, 51), thence, to read this a little further as "becoming-song."

Anders' theme is "musical situation":

> this situation consists in the necessary *being-simultaneous* [*Zugleich-sein*] of being-in-the-world and of Being-in-music [*In-Musik-seins*] in *an* existence, as *an* existence; consists in that one lives in the world, on the one hand, in the medium of one's own historical life, that one understands world and life comparatively (or tends effectively thematically towards this understanding—i.e., philosophizing);

that one is not in the world, on the other hand, but rather "in music," whereby the word music does not indicate a piece of the world that can be found in the world, in short: that one lives in determinations that shatter the average ontological fundamental character of human existence, even overcome it, and in turn indicate your kind of existence. (MS, 16)

As Anders explicitly distinguishes the question, "the *musical situation*, the in each case *being in-music*, turns out from its inception to be a negative insular situation within the historical life of the human being" (MS, 23). The terminology recalls Anders' 1924 doctoral dissertation, *Die Rolle der Situationskategorie bei den logische Satzen*, written under Husserl and which Anders begins with a reference to Heidegger's conception of 'world,' on the very first page. Speaking thus generally of circumstance and historical standpoint, where Anders/Stern uses the term to refer to the concrete question-worthy situation [*die konkrete Fraglichkeits-Situation*],[12] not unrelated to Karl Jasper's language of limit-situation [*Grenzsituation*] and arguably also, especially given the complex guilt-shame for Anders, related to Carl Schmitt's terminology of "definition" as Schmitt expresses this in his own doctoral dissertation on guilt, and thence to, noting typical kinds and common instances, the very idea of "situation."[13] Lütkehaus, for his part, explains the same constellation with Adorno (and Eisler) circa 1931, but the language of "situation" in this era is patently overdetermined.[14] Thus, for Anders, the musical "situation" is, *qua* situation, question worthy as such but not less intrinsically an affair that is inherently exceptional, as he states specifically, not only as a typical discounting or criticism ("Sunday music") but of what it is to be in-music. The constellation includes a transformative, transfiguring effect which can therefore be counted as having a certain druglike efficacy (here Anders invokes Nietzsche) not only because Anders distinguishes between the musical performer/composer/expert listener and amateur or indeed extern, just as Maurice Halbwachs does,[15] distinguishing accidental from casual listening as determinative of the situation (and Anders, as noted at the start, is quite specific about the social circumstances inherent to music in every case but which he also connects, and this is the hermeneutic-phenomenological and not less existential element, with, for the most part, and to speak with Heidegger, the most proximate or intimate character of the human being who does not live *historically*, an oblique circumstance that makes the musical situation *and* its exceptionality in its capacity to engender such, even to the height of the numinous, thus for deity as Anders writes citing Augustine) and not less crucially: "Human being in the musical situation is neither life within the continuity of one's own life nor life in the world" (MS, 23). The "neither-nor" (MS, 20) brings Anders to a generally Kantian reflection on music aesthetics, not merely, as he says, the alternation subject-object, but action and work, and thus he has recourse to the "Aristotelian concept of ἐνέργεια," not on the basis of subjective attunement but the musical situation and thus as "including the ἔργον in itself" (MS, 20). Here, the focus on performance/execution has to do with *play*: engagement, otherwise directedness. Fundamentally, the "musical situation is a cognitive situation," which, expressed as a kind of "Schellingesque 'organon of art," does not differ from "love," Anders writes, noting the triumvirate: Augustine,

Pascal, Scheler (MS, 24, Cf. 33 ff.). From this juncture to an explicit *via negativa*, Anders clarifies music as coordinating the unhistoricality—note that one may also speak of this in temporal terms, being as it were outside of, transported out of the specific place and time—of human existence. This yields a reading of Dilthey's brief discussion of musical understanding and the "tonal world." For Dilthey, and this must be understood in terms not only of his analysis of experience and alongside his famous lived world, this is the world of music, a world apart, "with its endless possibilities of tonal beauties and their meanings."[16] Dilthey is emphatic on this apartness *qua* submersion: "there is no duality of lived experience and music, no double world, no carry-over from the one into the other."[17] Citing sections from Dilthey's next line: "Genius is simply living in this tonal sphere, as though this alone existed; all fate and suffering is forgotten in this tonal world, but nevertheless in such a manner that all this remains within it" (39),[18] Anders glosses what is key to the sublimating retention/overcoming of this same "foregone life," such that this is ultimately no "lost destiny [*das Verlorene Schicksal*]." The key for Anders is Dilthey's recognition that "the *object* [*Gegenstand*] of music" is "life itself" (39). For Dilthey, "every relation to a musical work is interpretive." The emphasis as important for Adorno as it is for Gadamer: "Its concern is with something objective."[19] For Dilthey, hermeneutically minded as he was, muses on nothing less Kantian (Copernican), than Kepler's method of inductive inference leading to the discovery of the "elliptical path of the planet Mars,"[20] and thereby on the question of determinate-indeterminate reciprocities in terms of sentence and syntax—and here there is a crucial resonance with the Hilbert school and thus with Husserl and with Heidegger that would not have been, could not have been, lost on Anders, who wrote a doctoral dissertation on sentences:

> The sequence of words is given. Each of these words is determinate-indeterminate [*bestimmt-unbestimmt*]. It contains in itself a variability of meaning. The means of syntactically relating each word to the others are also ambiguous within fixed limits; thus the sense arises, since the indeterminate is determined by the construction.[21]

Dilthey adds an Appendix on "Musical Understanding," articulated on the basis of the unifying Kantian reflection that "Understanding rests on the retention in memory of that which is immediately past, and its participation in the intuition of that which follows."[22]

Dilthey emphasizes the tonal world that is the world of music:

> absorbed by the musician from his childhood on and which is always there for him. Everything he meets is transformed into it and comes forward out of the depths of the soul in order to express what was there; and for the artist fate, suffering and blessedness are above all present in his melodies. Here, memory again asserts its role in producing meaning.[23]

In the previous chapter on radio echoes/ghosts, we saw the importance of memory and apperception. But here Anders, glossing Dilthey, writes:

Human being in the musical situation is neither life within the continuity of his life, nor life in the world. And so far as human existence not only occasionally lives in this insular situation but fundamentally, his existence as a whole is as unhistoric [*unhistorisch*] as the kinds of musical movements. (MS, 23)

In the case of musical situation, there is an essential reference to the social—thus, one sees not only the importance of history and hermeneutics but also what will become the project of a musical sociology as well as dynamics or *energeia* and along with this an ecstasy, almost religious in this text not only as Anders draws on Augustine but, at the same time, an anticipation of what Eugen Fink will characterize as an "oasis" of play,[24] more communally/sociologically expressed in Anders as "Enclave" (MS, 44f).

Given the reference to Dilthey and Heidegger, Anders' reflections on the "situation" of music can also be illuminated via Gadamer's history of effects. Indeed, in his reflections on *The Relevance of the Beautiful*, Gadamer presupposes the same anthropo-phenomenological context of the musical situation, in every sense of the word, literally programme music, in a very different sense from that in which Dilthey claims that "program music is the death of true instrumental music" and the challenge that remains, in some cases even to the current day, for the inclusion of contemporary works of the kind Anders and Adorno and Gadamer would have named the "new" music.[25]

For Anders, the "musical situation" emerges (and this too is related to Gadamer's Rilkean emphasis: "that such a thing" stood among human beings),[26] in that the "that sentence [*Dass-Sätze*]" or proposition, the same "that" that is at work in the music as *energeia*, expresses an "incompatibility, better: a non-unity between ordinary life 'in-the-world' and life in-music" (MS, 21). Later, Anders goes on to apostrophize Mahler's astonishment that in spite of the events of music, the ordinary banality of the everyday continues. For Anders,

> As long as the musical situation lasts, the human being is in spite of all immanent referentiality of beforehand and afterwards in the music him or herself always in the *now*; but at once not in the now as the present within historical time, but an expressly unhistorical now. To this it makes no difference whether the now already frequently and earlier was, the now of a music, if the now is an unicum and in the realm of life unsituatable occurrence. (MS, 38)

And, already in an emphasis on what Adorno will also observe with respect to the "being" of the work of art, the question of meaning is constitutively irrelevant. For Anders, this constelled being is what is effected in (being-in) music, the sheer "that" [*Dass*] of the original linguistic form and note here that this can relate to musical composition for the composer:

> that (in the music) one falls out of the «world» that one is still somewhere; that one is always torn out of the continuum of one's life; that in each case, one still remains in this hiatus quite in the medium of time; that one's own (personal-historical) life becomes imaginary in the face of the other reality; that there will only be a gap between the situations of being-in-music; that one no longer remains "one self";

that one is transformed; that one must return to oneself; that music says something in every note and nevertheless—in the sentential sense—articulates nothing; appears, seemingly to go somewhere, whilst concealing what is opened up, what opens up; that one understands something—and yet ἄνευ λόγου. (MS, 17–18)

Taken by Heidegger's reflection on the question as such, just as Gadamer would likewise be taken, Anders reflects, "A question is thus already the dissolution of wonder quite as the answer dissolves the question" (MS, 18). The observation clarifies a methodical petrification of the situation modelled in the case on the question situation which can settle into a "rigid and disposable *problem*" (MS, 18). To this extent, in addition to Adorno, Anders' "situational" reflections on the relevance of sociology to music highlights his philosophical conception of and attention to "Being-in-music," underscoring the formative dimensionality of Anders' relation to Heidegger, despite his lifelong reservations contra Heidegger. For Anders intended his own focus specifically to highlight the *active* musical listener in-music; this is not quite any listening and thus Heidegger's language of world and being-in can be useful, typically heard in general terms of being-in-the-world but which, for Heidegger, also conveyed a reference to disposition and affect but also a particular intentionality. The world is not the same for everyone; the lifeworld we all share is also differently given to us depending on our preconceptions, cultural and theoretical, the projects in which we participate and which are specific to us, and the focus we have on the world around us, dependent in turn on the aims or purposes we hope to attain. For Anders, what mattered derived from Husserl, and to that extent it was a matter of musical consciousness, not mere perception but resonance. Key for Anders is the notion of music *for* an attuned listener.

The theme, broadly conceived, has found expression in a number of studies of listening to music, such as F. Joseph Smith's *Listening to Music*, and in *The Hallelujah Effect* I try to connect this very practically to H. Stith Bennett's phenomenological sociology of specifically performative music.[27] What is typically missing is express reference to Anders, and it can only be hoped that a now-growing trend to attend to Anders will change this state of affairs.[28] For his part and very differently—note that Anders refers more conventionally to Schopenhauer, which is also an exceptional focus—Nietzsche raises the question of music in his own thinking, philologically minded as that was in his first book, *The Birth of Tragedy Out of the Spirit of Music*. Nietzsche begins by making an *active* distinction, setting the creator's or performer's aesthetic, that is: "the artist's aesthetic" in opposition to the "spectator's aesthetic." The gendered contrast Nietzsche drew from Hölderlin's poetic dynamic of the contest between the sexes, as Nietzsche described it as everlasting in the first line of *The Birth of Tragedy*, punctuated with the reconciliations that Hölderlin could transfigure into beauty. Nietzsche's own imagery was crucial given the focus on the audience in Nietzsche's era (not less in our own), which artist's *creative* aesthetic Nietzsche named a "masculine" as opposed to the latter audience-oriented, reactive or "feminine" aesthetic.[29]

The resonance can seem unmistakable. There we read the same dialectical tension: "*Die Situation des Zugleich*," the situation of the at-the-same-time, simultaneous, at once. Anders' specific concern is to reflect the circumstance of being not sheerly "in-world [*in Welt*], but 'in Music'" (MS, 16). Here, to be specific, and this is for Anders a reference to Busoni as much as to Schopenhauer, the "word music does not signify any worldly thing encounterable in the world" (MS, 16). In the context of literature, the same claim to voice, and what is thereby spoken in an immediate address, testifies to a direct encounter, spiritual and bodily, traced in Rilke's faintingly uncanny perishing "from his stronger existence" [*von seinem stärkeren Dasein*], explained by Anders/Arendt (if here we might have reason to take the emphasis to derive more from Anders than Anders himself will later concede):

> This "something" consists in an in-stance of hearing, being-in-hearing [*Inständigkeit des Hörens, Im-Hören-Sein*]. Today there has to be a condition and an occasion for being-in-hearing. In place of complete objectlessness, for which our heart is no longer adequate, the condition for being-in-hearing becomes the disappearance of the object, which we pursue with our ears.[30]

The phenomenological attunement of hearing is key; this is attending, hearkening where one is inclined, the vanishing object "pursued with our ears." The theological element subtends the entire point: hear! as the prayer commands. At the same time, Anders' explication of both the culture industry and the culinary includes an Adornian reference to breath (and suffocation). Anders writes:

> What the *background music* demands of us (ninety-nine percent of the music played on the radio and television is, or is in the process of becoming, *background music*, for *c'est la situation qui fait la musique*) is no longer that we should listen to it, but only that it should be there, because without it there would be an unbearable vacuum. (Ibid.)

We have noted that Anders' reflections must be read not only in an Augustinian mode but also in correspondence with Adorno, and Husserl as well as, despite or because of his criticism, Heidegger. At the same time, the most important influence for the question of music and sociology will be the concerns of the Frankfurt School, and if this is true of his reflections in the 1930s, this only continues throughout his critical work as the world itself comes to embody the worst elements anticipated by critical music sociology. Here, to cite this once again, Anders echoes Adorno's charged reflections in his *Minima Moralia*:

> The supplied commodity is, for the listener, "air"; and this in a double sense: 1) it is something he takes for granted; but 2) without it he cannot breathe. This kind of destruction, the self-liquidation of the object, is not a special feature limited to radio and television but is characteristic of all current production as such. (AM II, 254)

Anders thus analyses the way we find ourselves today, complete with our current commitments to distraction via multitasking and displacement.[31] If the later Anders seems to draw on Adorno, he offers a more Heideggerian perspective in a 1959 reflection on the "antiquatedness" of space and time, combined with a Frankfurt School–style critique of the culture industry as this industry commandeers the space of our lives and our minds. Once again, we recall Anders' reference to the 'schizotopic' project of listening to music on the 'parallel track' as it were. Thus, to cite again:

> we often look for something to do while listening to music to offset the unbearable character of the enjoyment which is good for nothing apart from its own existence. "I just can't enjoy Beethoven without doing my knitting" [*in English, in original*], is not the expression of an eccentric woman, but the epoch's confession of faith. (AM II, 346)

In addition to the increasingly consummate deployment of the human being on automatic as creator of himself/herself as "consumer," curating himself/herself for others, that is to say, for transmission or "broadcast" for similar consumption (this is the point of social media) and as himself/herself reproducer of himself/herself as a homeworker on himself/herself as consumer (in the scheme of industrial consolidation), there is the particular relation to where we find ourselves. This is our relation to the earth, including what Anders speaks of, here not unlike Hannah Arendt (and thereby echoing the parodic spirit not merely of nineteenth-century cosmology advancing to the claims of mid-twentieth-century cosmonauts, but dating back already to Lucian's second-century cosmology), as "the view from the moon," in particular including the potential exploitation or "development" of the same.[32]

Positive Attunement: Sociological Reflections on the Musical "Situation"

The colloquial focus of the musical situation has been with us for some time, ever since Plato, as what Anders highlights as the Greek conventionality that is relation of *ethos* and music, which we learn already by way of Plato's *Republic* and his discussion not only of a certain sense of decadence or decline but by way of a catalogue of modes and ethical-political sometimes assumed to be pathic effects. This Anders refers to in his own discussion of tonality (and rhythm and *melos*) which is indeed the means that for Plato literally and not metaphorically changes or transforms the human soul.[33]

This is esoteric, the so-called *acroamatic*, as this also recurs in Kayser's *Akroasis*. But quite by definition, the acroamatic is not typically understood. Exoterically typically, what is assumed is that some ancient and suitably cantankerous character denounces the music of the young, complaining that the good old days of yore are no more: this then can be assumed to be the "musical situation," as Adorno writes in 1932.[34] The same sort of languishing nostalgia can serve double duty for any other sort of golden age complaint. There are many problems with this (quite apart from the eso-exoteric distinction noted) at the start of Adorno's "On the Fetish

Character of Music."³⁵ Characterized as "musical sermonizing," Adorno observes, in a phenomenologically relevant note typically unnoted even by the most exigent Straussian, that it remains unclear "to this day why the philosopher ascribes these [weakening or 'soft' making] characteristics to the mixolodian, Lydian, hypolydian and Ionian modes."³⁶ Including most of Western music, as Adorno points out, both the "flute and the 'panharmonic' stringed instruments also fall under the ban."³⁷ For Adorno, there is patent irony in Plato, but there is also a serious and political edge, given the circumstances of the Athenian human condition in its socio-historical moment, far from an ideal utopia an urgent disciplinary effort: an "Attic purge in Spartan style."³⁸

It can go overlooked even in discussions of Adorno and music, especially with respect to claims concerning his assessments of popular movements in music and, to be sure, jazz, that Adorno himself was a composer, a student of Alban Berg, and, similarly key here, that that same, now fairly long in the tooth, 'new' music has been subject to a range of negative stricture on the side of popular reception, a tendency that has yet to abate. Here it is important to note that Julian Johnson, and here not unlike many commentators, both invokes and avoids this constellation, with the best of intentions, beginning no less by citing Schlegel's apt observation that "What is called philosophy of art usually lacks one of two things: the philosophy or the art."³⁹ Johnson, like many scholars already knows better, even as he cites Adorno's claim that a crisis in music analysis is a crisis in music composition, we already know better and have already written ourselves into the locus, contested as Anders will tell us, of those empowered to have views on music and those entitled to be heard as such.

Anders' attention to music, in terms of both listening to and performing music, as well as the circumstance in which music is made, composed and rehearsed and performed, was affected by Anders' phenomenological formation as well as his instrumental training and perhaps most intriguingly his attention to what was requisite for hearing music, listening as an attentive participant. To this extent, Anders refused any simple division between musician and audience.⁴⁰

Key to Anders' understanding here was also the embodied, situation, circumstantial, disposition, attunement of the audience member who could be merely present or who could, a frequent theme, *fail* to listen or be otherwise preoccupied. For Anders, the key, and here there are some elements in common with Leo Treitler, the only thing at issue would be the constitution, as such, of musical listening as such.

Here to speak of such a constitution in Anders' writing, it is necessary to refer to Heidegger's discussion of listening as this is always already in a world for us, and thus able to be heard for us: we do not hear, we cannot hearken to, we cannot listen to tones as tones, Heidegger maintains, as he thematizes this phenomenologically. For Anders, this hearing differs for the one for whom music is his or her life, the player/performer/conductor, the composer, the music enthusiast, and another listener.

Because we are primarily under the sway of analytic philosophy, that is, including the legacy of the Vienna Circle as of Carnap and to the extent that we are as Anders was fascinated by this legacy, including Heidegger's analysis of express articulation [*Aussage*] which he articulates in *Being and Time* in technical terms of indications and predications and—crucial for media philosophers, more incidental or casual, here we think not only of Anders but also of Kittler, who was (as he himself maintained) Heidegger's student as well, and I have sought to bring in Baudrillard—"communication," as such, "speaking forth," as well as affirmative or positive (the very meaning of the word "positive" in logical positivism refers to this affirmative character) *validity*. Heidegger here refers to Lotze, and the questions concerning logical validity exceed the present theme but would have been understood by Anders, such that it is at least essential to point to it. But most significantly for our attention might be to also take note of Heidegger's nigh-onto Nietzschean expression of these "logical" terms as so many "Word idols [*Wortgötzen*]" (SZ, 156) as he names them, a good way to characterize the terminology of logical validity as such.

The reference that now follows offers its own reprise of Heidegger's hammer example, and the context of hammering as a referential context of meaning and expression offers an explicit allusion to logical positivism as such along with a variety of ways of speaking of a physical object in its different referential insertions, as a table, as a door, as a bridge, and so on. The point (Heidegger is speaking of hammers, and their properties, and in this case heaviness), concerns categorial assumption but more significantly, as Heidegger underlines, the presumptions that go along with this. Thus, where one can refer to a range of language games, Heidegger reminds us of the world already at stake. In this case, "Unadverted [*Unbesehen*] is the 'meaning' of the sentence already assumed in advance as: the hammer thing has the property of heaviness [*das Hammerding hat die Eigenschaft der Schwere*]" (SZ, 157).

For Heidegger, this inadvertence has everything to do with the kind of *Umsicht* that is engagement with the hammer as such, in use.[41] Thus, Heidegger goes on to explain the comparative, also in Wittgenstein's examples, too heavy, or the other hammer.

Here with reference to Aristotle, to the *logos*, to judgement as such, indicated by "the phenomenon of the copula," turns out to have a determinative "interpretive function." All of this is background, and I am concerned here with Heidegger's reflections on "*Da-sein and Speech. Language.*" Here speaking and hearing and above all heeding (the German is *hörchen*) is crucial, and this is already to be seen in Anders and Arendt's articulation of Rilke's *Duino Elegies*.

Heidegger exemplifies the fundamental use of the negative (silence, but also as he emphasizes, misspeaking), negative listening, unattending, overhearing, mishearing all of this is for Heidegger, despite its negativity, revelatory and foundational: "Listening to . . . is Dasein's existential way of Being-open as Being-with for others" (SZ 163). In each case, Heidegger contends that we speak as we hear: already in a world with pregiven concerns, attunements, significations—because creatures of the word as we are, we are also born into, thrown into language as well.

Thus, as I referred earlier (because in Heidegger this has a negative expression) to the distinction Heidegger makes between his own analysis of sound and that of cognitive psychology as such, contrasting psychological research into perception studies into "the sensation of tones and the perception of sounds [*Empfinden von Tönen und das Vernehmen von Lauten*]" which is for Heidegger thus a *derivative* phenomenon,[42] quite as listening and hearing and hearkening presuppose language and communicative attention. This Heideggerian distinction would be key for Anders and not less for Arendt, as we read already in their focus on the hearing addressed in a religious context in their reading of Rilke. In addition to such "observance," there is also a Hölderlinian reference to "following" as such: "hearing constitutes the primary and authentic [*eigentliche*] way in which Dasein is open for/to its ownmost potentiality for being [*Seinkönnen*]" (SZ, 163).⁴³

Thus, we recall that Heidegger emphasizes:

> "Proximally" [*Zunächst*] we never, ever [*nie und nimmer*] hear noises or complexes of sounds, but rather the creaking wagon, the motorcycle. One hears the column on the march, the north wind, the woodpecker tapping, the crackling fire. (SZ, 163)⁴⁴

Indeed, and this Anders could not but have noted for his part, Heidegger emphasizes that a specific kind of practised, scientific, theoretical attention is required in order to make the scientific world of the cognitive psychologist possible in the first place: "It requires a very artificial and complicated frame of mind in order to 'hear' a 'pure sound'" (SZ, 164). By contrast, Heidegger tells us, we *hear*, we do not deduce or infer that we "hear," motorcycles and cars—and today we can certainly add the little notification tone or chirp that can tell us that we have—and now we can think again of Anders in his diner, as this information is also communicated to the someone, the anyone next to us, be it at the dinner table or as a stranger on the street—just received a text (or tweet or what have you), which "notification" we perceive exactly *directly*, as this capacity for immediate, instantaneous perception is part of the way we are in the world: "Dasein is as essentially understanding proximally alongside what is understood" (SZ, 164). As Heidegger explains, when we listen to a foreign language we do not listen to random "tone-data" but "*unintelligible* words." We take this for granted; this will be the basis for a good deal of French existentialism and not less *Daseinsanalyse*, when we seek out a companion for a specific conversation (as opposed to other possible interlocutors) but in each case, and the example of linguistic comprehension is crucial when it comes to music, such that this is key to what Anders seeks to theorize in terms of the listener who is in-music as opposed to others. "Only one," Heidegger writes, and Gadamer will repeat this as the cornerstone of his hermeneutics, "who *already* understands, can listen [*Nur wer schon versteht, kann zuhören*]" (SZ, 164, emphasis added). When it comes to music, hermeneutically, phenomenologically speaking, Adorno and Anders will agree with Heidegger, whatever other differences they may have.

Transformation and Transfiguration

Schopenhauer makes a range of claims for music, philosophically, metaphysically regarded that Anders finds both appealing and, on an object level, untenable, despite the advantage of distinguishing music from the other (representative) arts by describing it as a "direct objectivation of the world will." The hierarchical trouble enters in for Anders with Schopenhauer's own fourfold objectivization of the will, "Mineral, Plant, Animal, Human," which is, if only owing to a certain overzealous coordination with Schopenhauer's own schematism, subject to a certain "artificiality" (MS, 73). Bracketing the musical scale via the evolutionary hierarchy of being in this way, for musical reasons as Schopenhauer's schema, is inadequate to explain all musical kinds (as Schopenhauer contends that it does) retains a certain appeal. For Schopenhauer's cosmological notion of the world will as a constellation of the "musical situation of humanity" (MS, 73) together with his notion of "Being-in-music," taken in Schopenhauer's sense, yields a *participation* in nothing less than that same world will, including a being together with beings exceeding human provenance. The result is almost as cosmic a level for philosophy as Pythagoreanism, and this, for Anders, would be alluring were it not for its already-noted musical inadequacy. But what is above all, for Anders, problematic—Anders is no Pythagorean and also no Platonist—music as such is, quite like language, quite like the tool, quite like truth for Heidegger (and Nietzsche), an inherently, even essentially human affair. "Music is music of human beings. Without the modulation of this facticity, no music philosophy can function" (MS, 74). Here is the nucleus of Anders' Goethean reflection on enchantment, transformation, transfiguration:

> The fact that music, like every magic that makes the human inhuman, is made by humans, that the transformation that cuts to the human quick, also derives from him, is on systematic grounds forgotten and not counted into his explication of music [*Musikdeutung*]. (MS, 74)

For Anders, the transfigurative capacity of music testifies to an extraordinarily human capacity: "the possibility of enchantment, and of the situation in which he conjures himself into the music, lies within himself" (MS, 74–5). This coincidence entails that any effort to understand this musical transformative situation [*Verwandlungssituation*] remains an anthropology of music. A good deal of the attention to world music that goes along with, in many ways also being the heart of so-called postmodern musicology, attests to Anders' insight, offered to be sure, *avant la lettre*.

That Anders goes into a reflection on Schopenhauer's and musical tonalities, especially the bass, suggesting that "that also belongs, to the world subjected to the force of gravity," (MS, 74) thus loses what can seem its metaphysical risks, as a human capacity that is not only human in this same sense. It is here that Anders can bring in what he calls dimension, illuminated with reference to the music of Tristan but also, as one of his frequently invoked examples, the kind of *incarnate* song called for by Augustine. Here, the phenomenological force of Anders' musical anthropological vision entails an utter reciprocity, the situational transforms of mutuality whereby one is with respect to another, always also at the same time, an other posed with regard to yet another and so on.

> here he lives as son, there as citizen, thus these distinctions of "as" are however always yet modii of being-in-the-world [*In-der-Welt-seins*]. But the human being is not only in the world, he is for example as functioning body, for example as vegative eventuality itself *world*. (MS, 75)

One can read this via Aristotle, from whence the Linnean schema stems, or via Goethe or Rousseau or Rudolf Steiner but today's reflections on the second gut or the microbiome that makes a difference, physiologically, hormonally, neurologically, and so on, may make this point a bit clearer for us. Anders' point is precisely this as subliminal, as transpiring beneath the level of cognitive apperception, "apart from all human freedom and supervision, "*am* Menschen," in the human as it were, as hormones and nutrition and growth and disease and death are in the human. "Life lives in itself as vegetative passage" (MS, 75), not in one but alongside, passing beyond, out of oneself. This image at once beautiful and uncanny, takes Anders to a reflection on the ontologico-ontic multidimensionality of the human being, including one of his rare but quite on point references to Nietzsche with respect to Nietzsche's imperative call for "the necessity of a 'physiology of music'" (MS, 76).

Critical Sociology of Music

Anders' phenomenological sociology of music is specifically and at the same time a critical theoretic sociology of music owing to the nature of music as such, and not less owing to a certain correspondence, if one may so speak, and Anders does, of the relation between the one who calls [*Rufendem*] and the one who hears; this includes, as he emphasizes, a specific spatial dimensional extension [*Raumweit*] for the caller and a specific listening tendency for the one attuned, that is to say, inclined to hearken [*Hörgeneigtheit des Lauschenden*]. It is no accident that the example here is prayer, and this same spatial dimensionality is also in play in Nietzsche's discussion of rhythm and prayer, and so on and not less, and the coincidence with the early Nietzsche is also to be observed—although Anders himself connects Weber to be sure—"with a social or religious event" (MS, 181).

Anders proceeds in accord with attention to media and mediation to reflect on a parallel, as this cuts two ways, with film, never forgetting its particular genesis out of (and reciprocal constitution of) the culture industry as such, in this case out of the very political and market-driven film industry, which makes of the filmgoing public a filmgoing public: attuned to "the stars," reading magazines devoted to the stars or watching television programmes on the same theme and following the same figures on social media as if these were members of their immediate family as they are to be sure key to their affective lives, which fact takes nothing away from the decisive detail that everything is fiction.

If Anders attends to space as he does, the reach of the voice, the attentive response of the listener, the material basis for the loudspeaker, the radio receiver, all elements of Adorno's better-known radio theory,[45] it is significant that at the same time Anders is also concerned with time, understood musically and understood in terms of history as such as this fact is always part of music history and appreciation and a complex component to understanding what we mean by music. Just as Heidegger can reflect that a historical object, an artefact found in a museum, say, is as such, *qua present*, an object that has survived the passage of time and is thus at once an object of its time and an object of today. If Benjamin's aura makes all the difference here, it can obscure the historical point of coincidence that Anders draws out with respect to music—this is also why we can speak of efforts to recreate historical performances, or at least our enthusiasm for musical period instruments/practices: "*Music*," Anders emphasizes, "*is ever indeed [zwar stets] the music of its time, but not every time is the time of its music*" (MS, 184).There is the world of music, and this in itself, and thus the appeal of Schopenhauer, is also the sense of "another world," and the challenge for Anders to explain this as however much there is a social dimension of music as such, music "in its facticity" can never be explicated on the basis of the social character of humanity.

Being-In

In a summary assessment of the whole of Heidegger's project, Anders remarks that "Heidegger's intensity, called 'existence [*Dasein*],' lacks all Dionysiac connotations" (PC, 364). But what is Dasein (for Anders? for Heidegger) and what is the Dionysiac (for Nietzsche? for Anders, for Heidegger)?.[46]

In general, accessing Anders here is elusive, and part of the reason is not Heidegger but Husserl and, to a greater extent, the concern that animated overall Anders' reflections on music as this also animates Adorno for his own part with his own concern with music and space—and time.

For Anders, what matters is volume and voluminosities, density, porosity. Indeed, in his 1958 essay, "The Acoustic Stereoscope," Anders articulates a phenomenology of stereo that could be useful for understanding the (internet and sometimes cult) phenomenon of ASMR[47]:

> We distinguish the voluminosities of sound-complexes; their different "massivities"; the "density" or "porosity" of musical texture; we call one tone higher than the other; voices of an orchestral piece seem to be "in front" while others are supposed to operate as "background"; there is a "continuum" between tones, even different types of continua—the chromatic scale and the legato; there are "jumps' from one tone to the other which do not "touch" the in-between-tones, (f.i. the "sext") and so on.[48]

For Anders, there are only "spatial acoustic objects or 'events' which, in themselves, have space-properties or -structures."[49] Here, Anders seeks to distinguish a certain kind of being-in, and absorbedness in music, cosmically, almost as if he were anticipating a certain new-age movement, sometimes allied to the postmodern, a kind of *Hearts of Space* effect, already in effect, thus Anders invokes the most romantic musical "cliché's":

> E.T. Hoffmann's panegyric words about Beethoven, Heine's description of Berlioz, Wagner's self-interpretations—these all seem to agree on this point: the limitless space of music which they describe in pseudo-cosmological or -religious terms.[50]

The "limitless space" in question is meant to be absolute or all-encompassing: "Whoever listens to this kind of music, is not supposed to hear the musical pieces as being at this or that distinct point, but rather to feel 'surrounded' by them, to be 'in music,' to be drowned by them."[51]

In 1958, Anders wants to describe a particular stereophonic phenomenon, at that time innovative enough, as he demonstrates a performative way to create the same, and thus one may match this account, which can otherwise seem merely occasional, with his earlier and later experimental and phenomenological analysis of radio. The key is inbuilt into the nature of listening as opposed to the visual, phenomenologically speaking. As Anders distinguishes the point towards his conclusion here, he reflects:

> It is the fact that in the visible world we have the freedom of movement, while in the acoustic world we are unfree: we are always led by the strings of the musical object itself, for the object is a "process." Facing a painting, we have the freedom of looking first to the right, then to the left corner; faced with a musical composition, we are carried by the stream inherent in the music itself; we do not move ourselves within the musical "object," but we are being moved, led along by it.

Phenomenologically speaking, the acoustic stereoscope takes away no kind of aesthetic freedom from the listener, as Anders remarks, quite by contrast, with the "sensomotoric" deprivations inherent in using the child's toy example of the optical stereoscope (like a View-Master),[52] with reference to which same Anders begins his original essay, arguing that where the former is a toy and limited to amusement value, in the case of the acoustic stereoscope different access is offered to the music for the listening ear, and unlike the constraints on the eye imposed by the rightly named View-Master, in the latter instance, when it comes to listening, we are not "deprived of such a freedom by an acoustical gadget, since we lack this freedom anyhow."[53]

To this extent, Anders is invaluable, quite as Ellensohn emphasizes in his own study but also, and in this case as van Dijk had earlier suggested with respect to dynamic efficacy or working effects—this is what I call the "Hallelujah Effect"—of the culture industry in practice, both in the sense of critical theory and in its analysis of industrial and commercial or corporate culture, but also for what Anders called the musical situation, for the musicians, the composers, the listeners who are amateurs or who are themselves musicians.

There is much more to say, but the conversation remains to be had, and for that there must be interlocutors.

10

Transistor Radios and Media "Überveillance": From Anders' "Radio Leash" to Tracing

"Ground Control to Major Tom"

The monolithic power of radio and television media—war of the worlds style, *Voice of America* style—is seemingly long gone. But, were we aware of just how trackable and how tracked we are,[1] we would, perhaps, have cause to worry. We are as oblivious as we are owing to what Adorno called *standardized ubiquity*: thus we are constantly, always and already, connected via our phones and tablets, and today rather than listening in Anders' antiquated sense to the radio, automobile trips—when we can make them in an era where Lockdowns come and go—are monitored and supervised not via an Orwellian Big Brother but just the GPS lady (it is rarely a male voice). Once again, note that the driver as well as the passengers in the car can make every effort to ensure she speaks without interruption, even if/when she says the same thing over again: we don't want to miss a thing and so risk losing our virtual tether.

Not wanting to miss *anything* begins as Nietzsche already reflected in the newspaper age of the nineteenth century, read, as Nietzsche observes, "watch in hand." Thus, Adorno referring to the doubling phenomenon, Anders noted in his essay on radio via acoustic signals. Thus accommodated, Adorno writes in a chapter he entitles "Musical Situation," "No one will experience the phenomenon of the continuation of one piece from different houses as a sort of ghost like apparition any longer."[2] Now, it turns out that Anders' text is more influential than may be assumed if one only attends to Adorno's focus on Heideggerian *Dasein* rather than, as Anders writes, about broadcast musical phenomena. To be sure, part of the reason may be that Adorno adopts some of Anders' insights as his own. Thus, Adorno cites/criticizes Anders at length while emphasizing that "it would be superficial to dismiss [Anders'] assertions about the haunting character of radio."[3] Characteristic of Anders' philosophy of technology is that technology is more than a mere extension of our senses but of our desires. Importantly this conceals an appropriating pride and a "shame," as we have seen that Anders speaks of it. We know what the technology, the tool, the gadget can do for us, and to get it in our power and at our service, whatever the tech in question might be—information, "notifications," digital inscriptions, we pay, as Anders unlined this, for digital connectivity in our homes, and we dutifully charge our similarly self-financed cell phones, and so on. Thus, our psychological involvement with, and complicity with, our machines is part of the ongoing millenarianism of our techno-cargo cult era.

Political Philosophies of Technologies

i. Left and Right Philosophers of Technology and 9/11

With Adorno and Anders, but also Herbert Marcuse and Lewis Mumford, along with other less well-known names in the philosophy of technology, including technology and political theory in the case of the Canadian Gilbert Germain and the US political theorist Langdon Winner and John Street in the United Kingdom, and even van Dijk's undervalued study of Anders in his *Anthropology in the Age of Technology*, in addition to Baudrillard and Virilio, Ellul, and Illich, many theorists of technology could be located towards the so-called "left" rather in the sense in which Alasdair MacIntyre could call Tracy Strong a "left Nietzschean," if not always quite as "left," say, as Marxist theorists or, indeed, *critical* theorists of technology.[4] By contrast, those who write on Jünger and technology tend to be imagined on the right,[5] and it is significant that Heidegger would be what Karl Jaspers named a "limit" case between right and left.

Still for many scholars, the political turn in the philosophy of technology can appear to be ambiguous. The turn to the political can indicate what is sometimes described as public policy philosophy, one that would emphasize practical viability, making common cause with corporate interests and government interests, including the military.[6]

In the interim, a great deal has changed with respect to privacy as Anders is already concerned with this in 1958 in terms of wiretapping and the like, pointing out that surveillance devices as such "are totalitarian" (AM II, 216ff.). As Baudrillard analysed: one sees only what is afforded via the eponymous window of one's browser, and one should assume that one is thereby already tracked and otherwise monitored, with or without one's consent, as internet privacy laws turn out to be ways of assigning authority over one's activities to government supervision and not otherwise. The "see something, say something ethos" patent in the war on terror and expanded in the era of Coronavirus implies treating oneself and one's neighbour as a suspect, guilty until proven otherwise. Thus the idea that one has a virus one might transmit to others even if one is oneself not ill, means that one may treat the healthy as ill, and as a precautionary means, maintain social distancing, mandate vaccines and the wearing of masks, and lockdown or quarantine, etc.

The idea is to err on the safe side; false accusations and untoward medical side-effects (including death) or psychological damage may be written off, ignored, as part of the price for increased security. There are moral concerns, though these tend to be brushed aside. If McCarthyism was vulnerable to the dramatic challenge, "At long last, have you left no sense of decency?" it is not at all clear that the question could have any purchase in today's era of governmental overreach: lockdown, masks, and plans for compulsory vaccination and bodily tracking/tracers.[7] The threat of terrorism covered all bases, and the threat of a fatal virus seems still more efficacious. Nor is one meant, and this would have been the most challenging constraint for Anders, to ask provocative questions of the kind raised in the wake not of Hiroshima or Chernobyl but 9/11, questions concerning the technical details of vaporizing New York City skyscrapers.[8] Like the two towers themselves, the third building, the so-called "building 7," collapsed into itself, just as the twin towers did, in seconds: imploding into choking dust, the kind of dust that persisted for weeks and months afterwards, along with the fires burning, likewise for months, at the site of the former World Trade Center in lower Manhattan, an ongoing contamination of air quality inspiring the business community on Wall Street to commission independent science surveys of air safety as the government immediately declared the air perfectly safe. The experience of those who lived in the area contravened this on the empirical level, and people delighted to own downtown lofts gave up this advantage often citing, if not perhaps to their buyers, air quality as the reason why.

ii. Technology and War: On Maintaining the State of Exception

Rosa Brooks' popular account, written from a lawyer's (as opposed to a hacker's) perspective, *How Everything Became War and the Military Became Everything: Tales from the Pentagon*,[9] relates a very old story which somehow always needs retelling. Technology development and proliferation, beginning with the popular computer and thence to what becomes the internet, is related to military innovation from planes to bombs to drones. If the scheme for a global library goes back to a Belgian librarian, Paul Otlet,[10] himself developing or elaborating interior architectural schemes for the world on a Corbusier scale, namely as modelled, virtualized rotating manuscript wheels dating from the Italian renaissance—as these would ultimately play into what became the World Wide Web and the ideal of a "databank,"[11] including the so-called Memex, support for deployment on a national and international scale goes back to Vannevar Bush's 1945 "As We May Think."[12]

Media *archaeological* analysis is essential. Yet such an archaeology has to do more than today's conventionalities do, not only where Friedrich Kittler reminds us of the complexity of the development of portable typewriters and the gender transformation of clerks from male to female personnel in his *Gramophone, Film, Typewriter*,[13] but psychology in all its forms, as Kittler emphasizes citing Don Gordon's *Electronic Warfare: Element of Strategy and Multiplier of Combat Power*.[14] As Kittler writes, alluding to Clausewitz to do so:

> *War on the Mind* is the title of an account of the psychological strategies hatched by the Pentagon. It reports that the staffs planning the electronic war which merely continues the Battle of the Atlantic, have already compiled a list of the propitious and unpropitious days in other cultures. This enables the U.S. Airforce "to time [its] bombing campaigns to coincide with unpropitious days, thus 'confirming' the forecasts of local gods." As well, the voices of these gods have been recorded on tape to be broadcast from helicopters "to keep tribes in their villages."[15]

The point is more material than one may suppose from the language of psychological strategies. Hence, in *The Hallelujah Effect*, I cite Goodman's *Sonic Warfare*[16] just to the extent to which, like Kittler, Goodman underscores this long-standing habit of American military inventiveness. Mind control in this literal or material sense is more straightforward than one may think. It goes back to Edward Bernays' 1923 *Crystallizing Opinion*.[17] From the beginning (what Horkheimer and Adorno call the "culture industry"), the calculated military deployment of this crystallization technique was deployed by corporate culture.

Here the hermeneutico-historical suggestion is to consider not only the day before yesterday but the day before that. Thus, I recommend Paul Forman's discussion of the complex cultural context of quantum mechanics in a historical and necessarily hermeneutic (this would not be Forman's word choice but it is indispensable for a discussion for which perhaps only the late Patrick Aidan Heelan would have been an ideal interlocutor: "*Kausalität, Anschaulichkeit*, and *Individualität*, or How Cultural Values Prescribed the Character and the Lessons Ascribed to Quantum Mechanics"[18] but maybe more obviously "Behind Quantum Electronics: National Security as Basis for Physical Research in the United States, 1940–1960").[19]

iii. Hacking Hackers: Philosophy of Technology in the Wake of the Political Turn

In 1992, John Street published *Politics and Technology*, this last not a book that mentions either Anders or, indeed, Heidegger, but which does do what very few studies of political theory do, namely to raise the question of technology and to ask how "technical and political change might be linked."[20] How, today, may one link technological change as we live it, largely via social media as much as anything, and the political, given current events on a global level? If post-Brexit is one thing, post-Covid-19 seems overstated: are we there yet? And how, if so, might we be sure? Such a "pandemic" orchestrated at the level of the 2020/2021 pandemic seems inherently repeatable, like a seasonal virus or a cold, and not something one might get through or past. To this are to be added the questions of misogyny, of agism and disablism and overall intolerance, all in the midst of unrest around the long-standing themes associated with Black Lives Matter. How, especially given the current changes to police and municipal and civic tensions, is this related to the question concerning "security"? Does it matter that this is likely to be connected to a "medical" imperative insisting that everyone in the world be "vaxxed," new technology of the kind Virilio warned about, of the kind Anders warned about, indeed even Heidegger warned about: turning human beings into a kind of capital resource of themselves quite by way of the new vaccination technology, including new biochemical ventures. Here the question is not a matter of distrusting a certain technology but assuming that a health protocol that dates back centuries, namely vaccination, in this case of an utterly new variety, itself with only discrete efficacy, should be imposed on every man, woman, and child on the earth, except for those who have the privilege that affords them the option, be it hidden or overt, of opting out.

Thus, some argue, naively as it can seem, that what is at stake is a matter of getting funding and support for the philosophy of technology, whereby philosophers of technology might somehow get a voice in policy.[21] The field of the sociology of knowledge may (or may not) offer help with this question. This is related to larger problems in philosophy of science (but that would be another book), and at the same time the philosophy of technology exemplifies the complexities that haunt the political. It could be, as already suggested, all to be laid to Don Ihde's account—and perhaps there is more here than meets the eye, so to speak. As Patrick Heelan pointed out with some wry humour, Ihde seemed to be especially gifted in taking up thoughts from others, be it Husserl or Heidegger or Heelan, and express them accessibly and in the process, as part of the process, obscuring the original constellation.[22] Instantiating this point, one of the more creative ventures of Ihde's career was *Listening and Voice: Phenomenologies of Sound*,[23] and Ihde had not yet talked himself into what would be the key to his own success as a name figure in his philosophical niche, eked out of the closing of continental philosophy on behalf of mainstream analytic philosophy.[24]

It is relevant in a context of Gadamerian *Wirkungsgeschichte* that certain texts constitute a common reference point, and thus that it makes a difference that a translation of Anders' 1956 book on the antiquatedness/outdatedness of human beings would be "vetoed" as Don Ihde would confirm. Preventing an English version of Anders was unfortunate but it is the resultant gap, the lack of *Wirkungsgeschichte*, that is ultimately decisive and cannot be corrected. Silenced as book of its era, Anders' 1956 study of the "antiquatedness" of the human being was addressed to its times, even more radical than Marcuse's later *One-Dimensional Man*,[25] just as Anders' second volume published in 1980 addressed the 1960s and 1970s. Thus, the fact that Anders was not read in English (quite as opposed to the French and German traditions) is a historical detail that also renders Anders' voice more of a hermeneutic challenge than it should be. It is true, quite in accord with Ihde's assessment, that Anders is "negative."

But the negativity is no personal predilection of Anders, no more than it was in the case of Jacques Ellul, similarly dismissed by many of his contemporary readers in philosophy of technology as *too* negative.[26] In this case, the caution goes back to Greco-Syrian Antiquity. Thus, in the second, 1980 volume of his *The Antiquatedness of Humanity*, thematizing the "destruction of life in the age of the third industrial revolution," Anders reclaims Lucian's all-too-human insight concerning the temptations of techno-sorcery. Now Lucian, important for Nietzsche, patent perhaps in his epithet, as the celebrated "lover of lies,"[27] and no less for David Hume and Erasmus and Jonathan Swift, also served as the original source for Goethe's 1779, *Der Zauberlehrling*, "The Sorcerer's Apprentice."(Figure 9)[28]

Old as the cautionary tale may be, no amount of repetition of this same ancient point would seem to permit insight much less "political change," to quote Street's formula. And what Goethe's sorcerer's apprentice did with his water and mops, we are currently doing with geoengineering the sun, flooding the atmosphere with particulate matter of all kinds, acknowledged and unacknowledged for the sake of weather and climate control, along with depth charges in the ocean, killing species, especially cetacean life, at rates that have already led and will continue to lead to mass extinctions. The lesson

Figure 9 Ferdinand Barth (1842–92), *Der Zauberlehrling* (1797). *Goethe's Werke*, 1882, Ink drawing. Wikicommons. Public domain.

from Apuleius and Lucian to Goethe and Anders is that we human beings will not stop until, having set incorrigible disaster in motion, we cannot continue.

Note here what should be a massive brake on the historical and current technopolitical situation, namely that we human beings consistently fail to anticipate the negative. Thus Robert K. Merton sought to give a sociological reflection on the matter, and he does this subtly via an inversion (Jacques Lacan would tell us why this seems self-occluding) of Goethe's formula: "*Die Kraft, die stets das Güte will und stets das Böse schafft*" [That power that constantly wills the good and ever fashions evil].[29] In philosophy, Peter Singer can give utilitarian reasons why this is so, supposing that Lucian and Goethe and Anders are not your conceptual cup of tea, such that evil seems to be wreaked, even when the negatives of the practice are tragedy-of-the-commons-simplistically *obvious*. Oblivious to danger, we frack, we burn forests in Indonesia, Africa, and South America for the sake of agro-commerce, spray poison pesticides, devise 5G systems with measurably deadly effects at the cellular level, and run pipelines, with 100 percent failure guaranteed somewhere along the line, at repeated loci (we just *underreport the effects*, as Anders writes in *Gewalt, Ja oder Nein?*,[30] a point to which we return later in the current text), as if what none of us see, cannot hurt us; thus, we diffused rather than plugged the Deepwater Horizon gulf oil leak, now repeated July 2021 in a ocean fire geyser, counting on, as polluters always count on, the sheer size of the ocean and the general absence of witnesses at the bottom of the sea, to let the news of the disaster fade. By contrast, the advantage of programming constant media access shapes consumer belief. For this reason, beyond "postphenomenology," more dangerously "colonial" power threats are to be countenanced.[31]

Postphenomenology, Post Critique: From Surveillance to Überveillance

Already in 1958, surveillance was, for Anders, well advanced. In the title of the same chapter included in his 1980 collection, Anders, who was gifted at collimating his titles like so many aphorisms, ponders *The Antiquatedness of Privacy*. The decades that have since transpired seem not to have made a decisive difference, as Google and Amazon, Twitter and Facebook make more than patent: "Not only is it the case that 'the world is delivered to your home,' but also" (AM II, 210)—and we can think of the concerns raised and immediately forgotten concerning "smart" appliances but which of course dominate the current day along with the surveillance capacities built into smartphones, tablets, television monitors, quite in addition to Siri and Alexa, such that Anders' parallel today lacks any kind of shock value—"Your home is delivered over to the world" (AM II, 210). Anders refers to his earlier essay on the world, complete with his appropriation and extension of Heidegger's terminology of being-in-the-world, herewith simultaneously taking over Husserl's notion of world for his own purposes, writing of "phenomenon" and "matrix" but Anders also explicitly alludes to Arendt's totalitarianism. The already-widespread phenomenon of devices used for surveillance [*Abhör-Apparate*] are for Anders from inception, as noted earlier, explicitly "totalitarian" (AM II, 216). At issue is not only a matter of the determination and restriction of cultural and social possibilities but control. Thus, with respect to the vaunted neutrality of technology, "*inventions are never merely technical inventions*" (AM II, 216). That is to say: "every device is already its application" (AM II, 217).[32]

At issue is nothing less than world control, "planetary" in the language of the day, global standardization and manipulation. Anders touches issues that we have in a sense neutralized by speaking of issues of "*privacy.*" Anders thus emphasizes:

> "Integral presumption" and "integral shamelessness" is the corrective means requisite for the state aspiring to be totalitarian to realize its ideal of perfected integrity. (AM II, 220)

This is to be understood with reference to totalitarianism properly conceived, "Whenever totalitarianism is at stake," Anders writes, "the individual is the first 'occupied territory'" (AM II, 220). To this extent, as apt for the current moment as at the time of Anders' writing (just to consider as he surely considered the then-wars in Indochina), "Expansionism begins at home" (AM II, 220). But the means to this end is already at hand, consummately so, via surveillance technologies. There is, for Anders, no technology that one could possibly imagine that would produce both the brazen obtrusiveness of the totalitarian state and the shamelessness of the individual under surveillance—and here Anders goes beyond Foucault's reflections on the panopticon, with a clarity that only Virilio might have shared—than surveillance technology as such. As Anders observes, as if he had a direct connection with the current day, "Where surveillance devices are used as a matter of course, the main prerequisite for totalitarianism is created; and thereby totalitarianism itself" (AM II, 221).

As "example," Anders reports on the installation, for legal purposes, of such surveillance devices in the United States itself of no less than 1,000 cases in buildings limited to Los Angeles between 1957 and 1958 (AM II, 221), noting to be sure that such details are more public in the case of the United States "than in other countries." Adding as his second example, from 1952, taken from the notes of the Supreme Court Justice William O. Douglas that no less than "58,000" persons were then under surveillance (AM II, 221). At issue at the time were debates concerning tapping or bugging telephones, debates including, as Anders writes, issues of privacy and, as always, John Locke seems to be the prophet here, issues of personal ownership of personal articulation, in this case the words spoken on the telephone as property of the one who "owns" the telephone (AM II, 221). One may, if one cares to do so in the context of material or object or media archaeology, reflect on the tendency of telephone and cable and internet providers to "rent" the instruments for a monthly fee and a certain or patent legal advantage, but perhaps the issue we now call content may be more relevant. Facebook owns all one's posts and photos, for legal purposes, so to speak, likewise other social media platforms, and if we do not object, it is because we are, by now, well used to it.

Überveillance Überall

One of the points made in the 2013 edited collection, *Überveillance and the Social Implications of Microchip Implants* is that surveillance, in order to work as such, must be obvious to those surveilled.[33] This is the morphological excellence of Bentham's Panopticon. The effect is pure Foucault, discipline and punish: and we do submit—*no tickee, no laundry*—No Scan/No Vax/No Fly.

In an essay from 1966, *On the Antiquatedness of Evil*, Anders reflects that "technology obeys too well" (AM II, 396f). Thus, for the sake of air travel, a hectic affair as it is, unless we are privileged enough to walk through a mere metal detector, we expose ourselves to either backscatter technology airport scanning devices or millimetre-wave technology units.[34] To be sure, this pales in the face of both the reduction of personal air travel owing to government restrictions, PCR testing requirements, masking and vaccine mandates, and the current deployment of 5G/6G, etc.

What motivated Anders with respect to post-1945 radiation in Japan and the failures associated with nuclear reactors like Chernobyl is invisibility over time, the same reflections set into the centre of first book in his Beckett chapter, "Being Without Time." In addition to this, the chronic effects of radiation as indeed of vaccines, especially the current 'experimental' variety, as these are called, do not make an appearance (this is part of the point of Anders' reflection) until quite far down the road. Compounding matters, health effects vary from person to person, that is, one person's skin differs in absorption, and one's person's blood plasma likewise differs in penetration and response, meaning that for some people it is much worse.

The Skies Down to Earth: Being Without Time

Anders' pop concern with Honolulu crooners regaling one in the wilderness, the scoping involved in airport security not to mention GPS and Facebook and Twitter tracking etc., may be variously compared with Adorno's reflections as the "stars down to earth" (again to refer to his *Minima Moralia*) on technological determinism. At stake is a reflection on the consequences of having the things and capacities we have (and *know that we have*, which is one of the key reasons that despite its charms, the idea of a technological sabbath does not work as we also know exactly how long it will be until we can be online again), and note the claim, as Anders' point here may help to illuminate some aspects of Adorno's related arguments, again, for Anders: "the phonograph and radio have robbed us of live music performed in our homes" (PM, 18).

And, ignoring the factive accuracy of Anders' remark, as for the most part: we do not have "live music performed in our homes," it's easy to raise the idealizing objection: *Well, we could*. If Adorno can argue that one can scarcely grasp the point of talking about "ghosts" as Anders does, the reference is "outdated," "antiquated" in a conventional sense. But this is not quite correct, and in Anders' centre chapter, "Being Without Time" (AM I, 213f), he makes it clear using Beckett's "Waiting for Godot," tacking through Heidegger to do so, as he says, all animosity temporarily reserved, at issue there in a complicated inversion of negative theology where once "was merely the absence of attributes that was being used to define God, here," and there is an uncanny parallel with the dogma of asymptomatic transmission in the era of Coronavirus:

> *God's absence itself* is made into a proof of his being. That this is true of Rilke and Kafka is undeniable; likewise that Heidegger's dictum which he borrows from Hölderlin—"where danger is growing, rescue is growing, too"—belongs to the same type of "proof *ex absentia*."[35]

Anders' reflections on Beckett's Godot (on waiting) have acquired a greater plausibility in 2020 than they could ever have had in 1956. Extended under Lockdown, it has become plain that we understand all too well what Anders means by "Being without Time," as he explains that "time appears to be standing still and becomes in analogy to Hegel's 'bad infinity') a 'bad eternity.'"[36] The ghosting language is a matter of a pairing, an echoing, a double-effect, and if one might wish to argue that Kittler heard this doubling, Adorno, despite his musical and theatrical and literary sophistication in reading Anders' "radio ghosts," did not.

For Anders, we ourselves, we modern, antiquated, outmoded, "leftover" human beings represent, or mirror the pair on Beckett's stage. Thus, we are all of us not only 'sons of Eichmann,' but Estragon/Vladimir. I need to quote here at length, Anders reading Beckett in a chapter on privacy and security concerns, on the internet, on life lived on the terms of, including lockdown orders taken via, social media and so on:

> The pitiful struggle they are waging to keep up the semblance of action is probably so impressive only because it mirrors our own fate, that of modern mass man.

Since, through the mechanization of labour, the worker is deprived of the chance to recognize what he is doing, and of seeing the objectives of his work, his working too has become something like a sham activity. Real work and the most absurd pseudo work differ in no way, neither structurally nor psychologically. On the other hand, by this kind of work, man has become so thoroughly unbalanced that he now feels the urge to restore his equilibrium during his leisure time by engaging in substitute activities and hobbies, and by inventing pseudo-objectives with which he can identify himself and which he actually wishes to reach: thus *it is precisely during his leisure time and while playing that he seems to be doing real work*—for instance by resuming obsolete forms of production such as cultivating his balcony garden or do-it-yourself carpentering, etc.³⁷

This is Anders on Beckett and today this is life on social media, checking Facebook, Twitter, email, and doing this repeatedly, again and again, partly for its own sake, partly idly or to pass the time. Thus, Anders can seem to be anticipating today's social media and "being-on-line," when he writes that the mass human being [mass-man], has today "been deprived so completely of his initiative and of his ability to shape his leisure time himself that he now depends upon the ceaselessly running conveyor belt of radio and television to make time pass."³⁸

It is in this sense that we must understand the resonance of Anders' "ghosts" as he speaks of these in 1930 to discuss a certain phenomenology of musical experience, hearing a broadcast from discrete space to discrete space, recharged, by different radios playing in different loci (like a cover, *avant la lettre*, of an imaginary Nono performance), as the encounter with such spectres does not "haunt" the current world as much as it is the ongoing and presupposed backdrop of the supposedly "connected" lives we live today.

> The inverted meaning of the scene in which Estragon plays "shoe off, shoe on" reads: "Our playing of games is a shoe off, shoe on, too, a ghostly activity meant only to produce the false appearance of activity." And, in the last analysis: "Our real shoe on, shoe off—that is: our everyday existence—is nothing but a playing of games, downlike without real consequences, springing solely from the vain hope that it will make time pass." And: "We are their brothers—only that the two clowns *know* that they are playing, while we do not." Thus it is not they but we who are the actors in the farce. And this is the triumph of Beckett's inversion.³⁹

Anders' point in his reading of Beckett depends upon *standardization*, the same standardization Adorno will invoke, standardizing what we count as play and leisure time. His point then is that the bathos of Vladimir/Estragon, two "metaphysical clowns" as he characterizes them, metaphysically incapable as they are of distinguishing "between being and non-being,"⁴⁰ is quite that

> they do not possess yet, as we do, recognized and stereotyped forms of leisure pastimes, neither sport nor Mozart Sonatas, and are, therefore, forced to improvise and invent their games on the spot, to take activities from the vast store of everyday

actions and transform them into play in order to pass the time. In those situations in which we, the more fortunate ones, play football and, once we have finished, can start all over again, Estragon plays the *da capo* game "shoe off, shoe on"; and not in order to exhibit himself as a fool, but to exhibit *us* as fools.[41]

Time is at issue for both Anders and Adorno, both of whom reflect on the musical "situation," both of whom write about "listening to" music, both of whom write about radio, quite in the context of what Adorno names "Space Ubiquity."[42] For both Adorno and Anders, there is a difference between "live" music and radio music, and Adorno critically insists, with himself as clear referent—this is the same Adorno, let us not forget, who could ask his music sociology students on the first day of class to write the Siegfried motif on the blackboard—that

> A man with musically trained ears, who is walking along outside a restaurant and hearing music inside, will almost always be able to determine whether this music is really being played in the restaurant or if it is being transmitted by radio.[43]

At once Adorno observes, in no small part because just these elements are the concern of his radio theory, that "Of course, this partly depends on the specific modifications of sound which any music undergoes by radio."[44]

Taking over Anders' "ghostly" language, Adorno adds that "radio music always seems to be an *echo* of music coming from a distant place."[45] The language is repeated to be sure in Kittler's more uncanny discussions, which Kittler, rather in the spirit of Anders, also reads in connection with Goethe.[46] The contrast between the radio "voice" and indeed recorded sound as such, as Adorno continues to analyse in his own writing, and the performative possibility, the lived world possibility of lived live music, as it were, is, on Anders' phenomenological and sociological reflection, displaced from the home by radio and recorded forms: this today is YouTube, and so on.

Note that as already suggested earlier, the issue is not that "we" mourn the absence of "live music" in the home, as if most of us might simply organize a string quartet in our spare time, although we may fancy that we could, with only homes of high conspicuous consumption (or prestige-signalling) feature pianos much less cellos and so on (this is not the same as digital keyboards or guitars).[47] Thus, music history and music sociology detail how the place of the piano in the home is given over to the phonograph, and, eventually, this is a matter of living room staging, to the television set.[48]

Anders explains the interior architectural evolution of the living room as an evolution, sociologically speaking, which would appear to depend upon the elimination of the formerly "massive table in the center of the living room" (PM, 17). Anders' point echoes the earlier-noted negativity:

> the television set, a piece of furniture whose social symbolism and persuasive power can measure against those of the former table. This does not mean, however, that the television set has become the family center; on the contrary, what the set embodies is rather the decentralization of the family, its ex-centricity: it is, so to speak, the negative family table. (PM, 17)

Today, we can wonder what Anders is talking about—this is already Adorno's tactic as we have seen. These days, we arrange via a variety of streaming services to have music with us anywhere we like, screwed into our ears, synchronizing our thoughts on nearly any occasion, all on demand, on auto play.

Anders' point here is the point George Orwell articulates in his dystopian novel: "Intimate conversation is eliminated in advance" (PM, 17). Thus, earlier we noted that people driving *to destinations they know* can elect to listen to GPS guidance, even when someone else is in the car with them, quite as opposed to speaking with one another. Thus, we transform ourselves of ourselves into passive, rather than *active*, listeners.

Anders offers another description of the same phenomenon in connection with the über- or super-veillence of surveillance so very ubiquitous today, especially post-Covid, that most were unmoved by the report that one's "smart" televisions are smart by virtue of the capacity to listen in to conversation, the ubiquitous phenomenon of tapping or recording phone conversations,[49] now "unleashed" from or "untethered" to the phone, to use a term Anders was fond of employing. In the interim, Facebook and other apps can opt to hear one's conversations as they like—and Siri does like—and so we are prompted to purchase this or that, with an ad for whatever we might have been talking about.[50]

As audiences, "online" as we live our lives, we are caught in the ghostly phantom of what Anders called "sham reality."[51] Thus, we relate to the world as "spectacle" as Guy Debord contended in the 1960s, or via what Illich in the 1970s and 1980s named the "Age of the Show." And like Adorno, Anders emphasized the transformed experience and the space–time of the same for the lifeworld of the human being.

Hence, in a section entitled "*The return of the soloists*," reflecting in 1961 on the general "Obsolescence" or outdatedness or antique convention that is to speak of the "masses," Anders offers a phenomenological sociological analysis whereby what is abolished is the external world, namely the community: the public world. At the same time, as in the case of Nietzsche's twilight reflection *How the True World at Last Became a Fable*, what is lost is the private world.

In place of "presence," there is self-absorption and some part of the surround sound, and the headphone-driven music experience is part of this. Thus, these days we can be hard-pressed to imagine Anders' American drugstore diner jukebox experience. Indeed, the proximity of cultural programming, given iPhones and headphones or earphones along with the culture of surveillance capitalism, has already intensified, and the current and ongoing "pandemic" crisis is driven by nothing less. The issue concerning music in public spaces, as this is dramatized by the Orwellian feature of the loudspeaker, is that the agent ambulating in public can still retain a sense of freedom and choice, programming his or her own soundtrack. But this also means as Anders analysed this issue that the music one chooses to play publicly tends to be the music one plays privately: "The customers behave *in publico as they do at home*" (AM II, 84).

Today, social disattention to the sensibilities of those around one is common. But at the time, as Anders wrote this in the 1960s, the American habit of playing the music of one's choice at one's table at a drugstore fountain shocked Anders because by simply sitting at the next booth, he was a victim of his neighbour's taste. To this extent, the "privacy or isolation of the receiver disguises the mass character of the commodity and the mass character of its transmission" (AM II, 80).

We similarly tune others out when we attend to our cell phones in family gatherings, office meetings, and so on. This habitus made the recent lockdown less of a shock for many who simply continued this focus in private, without practising it in the open world, on public transportation, in public places, and so on. And in almost each case, individuals report a self-sufficiency they take, for some reason, to be unique to themselves. The games we play, the music we listen to, permit us to participate, on our own terms, in this 'mass character' as Anders uses the terms.

Frighteningly, given the omnipresence of Zoom instruction, the same constraints apply as the very possibility of online learning requires that one have access to zoom and other platforms, accessed at a distance:

each of us is supplied personally not only with our de-individualization and our form of mass existence, but also, at the same time, with the illusion of privacy (insofar as it is generally a question of a dual conditioning). (AM II, 81)

Thus, something so neutral, benign, as the experience of hearing the music one likes, quite when one likes, becomes, as both Anders and Adorno argued, a means of de-individualization. Given the ubiquity of surveillance, part of the illusion of privacy, Anders is uncompromising about the consequences for what is increasingly outdated or antiquated in our human condition. And who would want to be *out of place*, beyond the *timeliness* of the current mode of human being, that is, Heideggerian *Dasein* as Anders understood this more concretely as he claimed than Heidegger himself, that is, in-the-world? Perhaps it is Anders who can best help us negotiate modern social media and what it does to our social being in the world as one uniquely placed to think about technology in its most intimate expression and not less in terms of its capacity to change the world and both directly and indirectly, as Anders underscores as well, to destroy the world.

11

Pop Music (and Jazz), and Covers (and Copies)

Transitioning: From Anders' Radio "Leash" to Cohen's *Hallelujah*

Some accounts describe Anders as an essayist or "journalist," writing as he did on the Paris music scene as on Vietnam and Hiroshima as on American pop music in the 1950s and 1960s. The description is misleading, and Anders always assumed the authority of the professional academic, reflecting not less a performative perspective on musical sociology (this is the theme of what would have been his Habilitation thesis, becalmed as it was, among other things, by its timing, coincident with the Nazi regime).

Anders' terminology (as we have seen that this is drawn from Heidegger, and also appears in Adorno) of *Zuhören* corresponds in Anders' usage to a specific musical attunement, *listening to*. Anders is thus not describing careful attention: listening very hard or very carefully. Instead, at stake for Anders is a phenomenology and thus epistemic. And, as Gadamer observes, "To understand something, I must be able to identify it."[1] Adorno makes a similar claim, compounded with reference to music and to what is specified in context, and thereby to what may count as musical surprise and as variable:

> You have to know exactly *if* something sounds and only to a certain extent *how* it sounds. This leaves room for surprises, those that are desired as well as those that require correction: what made its precocious appearance as *imprévue* in Berlioz is a surprise not only for the listener but objectively as well, and yet the ear can anticipate it.[2]

For Anders, the informed musician hears or "is in-the-music" differently quite by comparison with someone lacking a musical formation. At issue is a matter of being-in.

To illustrate the phenomenology of such "listening-to," Anders uses a classical phenomenological tactic, that of free variation, listening, as he says, to "anything at all" from Debussy, "be it the few tacts of the 'Pelléas et Mélisande'-Overture, one of the 'Études' or some third thing" (MS 14). At issue is a separate question regarding musical experience in a social context. The point, at least on the surface, simply owing to its ubiquity, may be compared with Cohen's *Hallelujah*—which was at the end of Cohen's life so very *everywhere* that even Cohen jokingly called for a "moratorium." The ubiquity of pop music permits us to understand Anders' denunciation of a certain song maintaining, as Anders writes in 1958, that "love is only in Honolulu."[3] For Anders:

> The thing here is that this music was inescapable for hours. Why this agitated me, remained incomprehensible to my companions. For them, this circumstance was not only not unpleasant but explicitly delightful. (AM II, 241)

Anders, expert as he was on the phenomenology of "listening to," has no trouble understanding the reasons why, and he extends the point to the contemporary American scene, hiking Mount Washington in the company of a transistor radio:

> Patently, they enjoyed a certain feeling of security, as long as they could still hear the music transmission and find themselves "in" it: the feeling of still being "there" ["*da*"], down below. Like pilots, who like maintaining reliable contact with their base. They had not yet gone out of range. The acoustic leash, which connected them to the valley, had not yet been broken. (AM II, 241)

Like Adorno, Anders has little patience with the *words* of the song, which he, like Adorno, also cannot help hearing (like an Anglophone listening to the words *Je vais, je vais et je viens*, written in 1967 by Serge Gainsborough for Brigitte Bardot and illustrating the substance of the song, recorded in 1969 with Jane Birkin, *Je t'aime… moi non plus*, or Herbert Grönemeyer's 1984 pop song, *Männer*) the assertion "that love is only in Honolulu" (AM II, 241). The problem as suggested in the previous chapter was what Adorno calls *ubiquity*. Thus, for Anders, master of mobile phenomenological *variation* in space and over time,[4] the music in question "was inescapable for hours." Elsewhere, I offer a discussion of the experience of playing radio in the car and the perception, the hearing of proximity and range as we enter and leave the range of certain radio stations: to this extent, very phenomenologically speaking, as Anders describes the attention of acoustic intentionality, we *listen-to* not merely the what of what we are hearing on the radio but the thatness of radio contact, weak or strong. We don't need the car example; anyone who finds themselves in an area with weak mobile smartphone reception will know exactly what is meant as will someone seeking a WiFi/WLAN connection.

If Anders' Honolulu song has been (perhaps rightly) forgotten, Honolulu or Hawaii more generally, were popular references in the 1950s[5]—this was and this is no coincidence, and we will come back to this a bit later, in the setting and the subject of Elvis Presley's *Blue Hawaii*, and his cover of *Le Plaisir d'Amour* [*Fools Rush In*].

Anders' focus is on the loss of privacy (note again a reference to Arendt), and we today associate this with the omnipresence of social media, as Anders seems already to be pointing to this, as indeed and explicitly he was. Thus, and just as Anders begins his 1930–1 "Philosophical Investigations Concerning Musical Situations," with Saint Augustine's reflections on "music as *tentatio*" in his *Confessions*, (MS, 14), he reprises the seductive claim on the working power of technological reproducibility in the case of the work of art and social culture:

As everyone knows, Saint Augustine portrayed the *"memoria"* that reproduces the past, recollection, as the vehicle for the discovery of sins, of conscience and repentance. Today, this reproduction is unnecessary; in its place a device of reproduction has been installed that transforms the past into something reproducible and present at every moment, into something that properly does not belong to the past. *The* **peccavi** *of the conscience, has become the* **pecco** *of the magnetophone tape play back.* (AM II, 232m, boldface in the original)

We are familiar with this in its different adumbrations through Benjamin and of course Adorno.

In my own earlier hermeneutic and phenomenological reflections on Cohen's *Hallelujah*, the point, as I sought to underline this, was attuned to what Anders describes as "something reproducible and present at every moment," that is, what Adorno named *ubiquity* in the case of music, and which today may be arraigned under the rubric of what we call the "cover" and therewith, the culture of remix, replay. As Anders writes, "Human sociality is drilled into unilaterality" (AM II, 253).

Death and Taxes

For whatever complex actuarial or statistical or cosmological reasons, 2016 marked a death boom, especially of the famous, of actors, artists, musicians. The harvest of the dead included Alan Rickman[6] and Leonard Cohen[7] and the deaths also included David Bowie, Prince, and so on. In the entertainment industry, as the Frankfurt School has already observed, the *death* of the actor provides Hollywood with yet another opportunity to use star capital for the sake of the same industry. And we do the same when we tweet songs thereby including ourselves in the process, as Anders reflects, "as producers also always already the 'products of our own products'" (AM II, 249). At issue is dominion even in matters of taste. The focus of the culture industry, fitting for an industry, is the control of taste and thereby of culture itself. What is crucial is that by the same token we are "*robbed of experience and the capacity to take a position*" (AM, 251)[8] or to have an opinion of our own. Here Anders remains quite close to Heidegger's denunciation of the "they" that set up and determined what one believes, feels, or holds as true. Everyone is not only a consumer but a producer/advertiser/promoter of themselves, but, even more importantly, there is no place where this new world does not extend, where the individual can, as it were, opt out. One "has to be" a cheerful participant (AM II, 427).[9]

To confirm Adorno's culinary complaint, not dissimilar as we have seen to the eating/consuming language Anders employs, regarding the commodification of music, the "Hallelujah Effect" turns out not to be a result of singing *Hallelujah*, no matter how stunningly sung. The effect, phenomenologically speaking, is one that works retrospectively, this backwards working dimensionality is the imaginary causality (and for Nietzsche, as for David Hume, all perceived causality is imaginary) that Nietzsche in *Götzen-Dämmerung*, in two sections of *Die vier grossen Irrthümer* (looking at the subject's capacity to will his own thoughts, the very transcendental unity of apperception, but not less the same subject's apprehension of the causality of external events) attributed to a certain *Nachträglichkeit*, whereby and via such backwards working of our awareness of causality is revealed as a seemingly reverse temporalization.

What we do in dream time we do, for Nietzsche, in our waking hours. Modern marketing psychologists call it branding or priming, a phenomenon Freud's nephew, Edward Bernays, deployed to great effect, and it is still with us in commercials and not less in politics on the world stage, where, to be sure, it always was: Adorno named it "programming," which is what we still call it. Thus, the *Hallelujah Effect*, the "Honolulu effect," is a matter of what we today call triggers. These triggers are both acoustic and visual: one needs the music video, *and* one needs repetition. The contemporary example of ASMR recordings is a recent exemplification of this phenomenon. The effect is self-induced somnolence, self-manufactured, subscribers become 'patrons' of producers, directly and this is very different from YouTube influencers (though of course these can sometimes overlap or coincide).

Pop as Contemporary Music: Fortunes and Futures

Even the pop of yesteryear, even pop music before one's own time, including, for example, Anders' extended reference to his "fresh-air mountain advertisement for Honolulu" (AM II, 245), our parent's pop music, or, worse, our grandparent's pop music, the musicologist's version of Tin Pan Alley included, can be heard, and this is part of the point Anders seeks to make concerning listening-to, as the music of a then-contemporary, a then-when. If you are German, you can listen to the Comedian Harmonists for this; if you are anyone else, you can listen to Kurt Weill; and if you are American, perhaps, you can listen to Liza Minelli, via Bob Fosse's 1972 film, *Cabaret*.

Thus, one could speak of Elvis, the king "forever" as some still say, and this is true of Bob Dylan as well as of Frankie Laine,[10] of Lesley Gore, she of various teen *ressentiment*-angst pop songs, *It's My Party and I'll Cry if I Want To*, *She's a Fool*, *Judy's Turn to Cry* (all related themes, all from 1963), and *You Don't Own Me* (1964, covered by Bette Midler, Diane Keaton, Goldie Hawn in the 1996 film *First Wives Club*), and so on.

Note that the Elvis Presley reference dates from the mid-1950s, and Lesley Gore the early 1960s: Ought one not mention Beyoncé or Adele or, indeed, someone much newer? This is a limitation of any instantiation or example. But pop music, like bubble gum, illustrates "speculation," fancying and favouring, disinterested interest, with or without the promise of a return, another name for interest, across a range of context and contexts, experience and recognition, the uses and disadvantages, the very real *Nachtheile*, of history for life.

What is at issue for Anders, as he writes about radio, portable and otherwise, and about television, and juke boxes, but also canned music playing in the background in restaurants but also projected into the sidewalk to lure customers, is transmission. This includes streaming, engineering, digital mediatization, and this is the essence of modern technology that is, as Heidegger says, *nothing technological*.[11] Anders, as we have seen, has a conspicuously different take on this, and thus he foregrounds the nothing in Heidegger's formulaic word on technology. For Anders, and contra Heidegger's own claim for the distinction of his philosophy from life philosophies, be they Bergson's from whom Anders points out, Heidegger is at pains to distinguish himself or Jaspers or, especially, Sartre, Heidegger offers a life philosophy inimical to life. The definition is steeped in Nietzsche's own terminology, and Anders concludes by quoting Nietzsche directly indirectly, with just a closing riff on Ernst Lubitsch/Jack Benny:

> rather the result of the "Self's" will to power, of its omnivorous urge to appropriate everything. Nietzsche's words: "If there were a God, how could I bear not to be God?" seem to be transformed into "If there is History, how could I bear not to be History?" The desperate motto of all active desperados, "all or nothing," changed under the hands of the existential one into an "all and nothing," which makes it well understandable that the book that continues his work is not entitled *To be or not to be*, but *Etre et Néant*.[12]

Here it may also be worth raising the question of the non-aesthetics that Nietzsche claims as his own, beginning with his first book, first sentence no less, name that tune, where Nietzsche talks about *die aesthetische Wissenschaft*, "the science of aesthetics," and where he begins to analyse the art of the artist, a concern he will maintain for the rest of his life.

But by framing the question of the artist (as opposed to aesthetics, as opposed to the spectator), one stumbles into Nietzsche's gender issue: it is not for him a question. For Nietzsche distinguishes aesthetics as such as an art for spectators not only as an art before witnesses, calculatedly dedicated to performance, the art of the *Hallelujah Effect* where the subjectivity of the aesthetic subject, *qua* spectator, plays with and as the effect in effect: all taste, good or bad, because all judgement, *qua* judgement, this is another spin on *ressentiment*, is *feminine*, receptive, reactive. By contrast, for Nietzsche, "an art for artists only" is not merely monological, as he says, that is, unconcerned with *either* spectators *or* effects, but expressly, explicitly *masculine*.

In *The Hallelujah Effect*, I look at male and female desire, pointing out that the one remaining prohibition even after de Beauvoir and Irigaray, a permanently aesthetic *Bilderverbot* remains. This, and debate on race and gender, only underscores this is a ban on representations of the male as object (of desire) for a female subject (of desire). Thus, and this is why in Judy Collin's 2015 cover of Cohen's *Hallelujah*, featured on her album *Strangers Again*, Collins did not opt to sing the line, *Her beauty and the moonlight overthrew ya*, but gave the verse to the male singer, Bhi Bhiman; *she tied you to a kitchen chair, she broke your throne and she cut your hair, and from your lips she drew the Hallelujah*.

One may not, *qua* female subject, desire or objectify which is to be sure worse, just to the extent that judgement is involved, the male object. It is for this reason that Cohen can sing, not only because this is the song of a king, a priest, singing to his god, *It doesn't matter which you heard / The holy or the broken Hallelujah*. Every word is a word of prayer but, and this is not a little problematic, only for the male. The female, the woman, both Anders and Adorno will speak of the "girl"—using English—

so addressed and so objectified in this context is already and always excluded: *But you don't really care for music do ya?*[13] Anything that might begin to look like female desire *I've seen your flag on the marble arch*, is dismissed out of hand: *our love is not a victory march, It's a cold and it's a broken Hallelujah.*

Key to the *Hallelujah Effect* is that in his own first book on tragedy, Nietzsche in fact focuses less on Wagner than we are 'primed' to suppose (Wagner thought so, the original review by Ulrich von Wilamowitz-Möllendorff, self-published, said so, and in translation Walter Kaufmann repeated the claim to which ever since Anglophone scholars march in lockstep) but indeed the relative *absence* of Wagner. Thus although Nietzsche's first book is dedicated with a preface to Wagner, naming him an advance warrior (*Vorkämpfer*)—this is the kind of praise that is intended to dismiss a predecessor as a "pioneer"—Nietzsche's first section reads in, some detail, Beethoven's Schiller chorus, a reading subsequently further thematized in the book's succeeding sections, so much so that Nietzsche concludes with an elaboration of this spirit of reconciliation in Beethoven's ninth, invoking the transfigured time space of

> a region in whose joyous chords dissonance as well as the terrible image of the world fade away charmingly; both play with the sting of displeasure [*Stachel des Unlusts*]. (GT §25)

Nietzsche takes the reference further if this confounds most who rarely attend to this focus as he suggests that we then best understand tragedy, imagining "dissonance become human" (GT §25). To this same extent, Nietzsche's first book is not *about* Wagner, father-figure friend or not. It makes matters worse, but I am inclined to think that Nietzsche had the temerity to frame his first book to appeal to Wagner, who also had his own focus on Beethoven, to invite him to make common cause with Nietzsche for the sake of nothing less Wagnerian than Nietzsche's *own* vision of a future culture and a future musical work of art.

Indeed, and this should not surprise us, less than Wagner, and apart from his explicit references to Beethoven, *The Birth of Tragedy* was focused less on either musician than it was dedicated to an articulation of Nietzsche's own literally programmatic theme *Out of the Spirit of Music*, an exposition of Nietzsche's philological, quantitifying rhythmic discovery.[14]

Nietzsche, whose discovery "decodes," and we might say it is the basis for, our current pronunciation of ancient Greek, tells us that the tonal *ictus* of the ancient Greeks excluded our manner of emphasis or the stress ictus characteristic of Latin and all modern European languages. Instead, the Greek tone ictus (some think it more helpful to say pitch ictus, but tone is Nietzsche's term and hard to exclude inasmuch as Nietzsche specifically cites: *O Freunde, nicht dieser Töne*) was on Nietzsche's theoretical account of it to be understood in terms of quantity and time: that is, musically.[15]

In this sense, the culture of the cover has its philosophical expression, but such covers are self-covers, reissues, rearticulations, new envisionings; thus, we may think of Kant's *Prolegomenon* (to his first critique) Nietzsche's *Gay Science, Ecce Homo* (with respect to *The Birth of Tragedy*) as philosophical quasi-covers. Nietzsche wrote his own 'prolegomenon' to his first work, appending it as a new preface to a second edition, retitled as one might repackage an album, with a new disjunctive subtitle, *or Hellenism and Pessimism*, which new preface was entitled *Attempt at a Self- or Auto-Critique*, an attempt routinely misread by scholars as if Nietzsche there intended to withdraw or deny his first book (odd assumption given that he uses it to preface a republication), but where Nietzsche seems to castigate the book's voice for its errors, as if to match his junior colleague's critique of his first book, von Wilamowitz-Möllendorff's *Zukunfts-Philologie* (Future-Philology),[16] where this would be no kind of praise in the field of *Altphilologie*, arguably inspiring Nietzsche's reflections on then-contemporary styles of doing *Altphilologie* (his meditations on the then-popular attention to the historical Jesus)[17] and to history[18] in his *Unzeitgemässe Betrachtungen* (*Untimely Meditations*). Misreading Nietzsche's new preface as *retractio* to his newly reissued *The Birth of Tragedy or Hellenism and Pessimism* is easy to understand (misreadings tend to be more, not less, obvious), and scholars suppose Nietzsche to say that his text might have done better had he, and it is assumed to have been a metaphor, "sung and not spoken." Reading this way, we have forgotten Nietzsche's reflections on "Music and Word" to quote the lecture Carl Dahlhaus includes in his *Music and Romanticism* along with the challenging claims Nietzsche makes in his first Basel lecture reminding us, and this too is a word, contra Wagner, uttered before any falling out between them, that ancient Greek tragedy had exactly nothing in common with the way we moderns encounter tragedy, beginning with the way we actually encounter it: on stage, in a darkened theatre.

But repeating his texts did not do for Nietzsche what he had hoped (I don't think repetition ever does this for anyone but there is nothing for it) and reflecting in *Ecce Homo* on the futures (here we need the plural s) of his "Dionysian music," he refers to Wagner directly, pointing out that if anyone profited from his ventures, it would be not Nietzsche, but Wagner:

> In order that one may be fair to the *Birth of Tragedy* (1872) it is necessary to consign a few things to oblivion. It created a sensation, even fascination because of its defects— its application to Wagnerism as if the last were an index of a beginning. Solely on that account this treatise would be an event [*Ereignis*] in Wagner's life: from this point onwards great hopes would surround the name of Wagner. (EH, §GT 1)

Cover Culture

The "cover," the recognizable hit we know, exemplifies Ernest McClain's point concerning the mechanical calculations of *Summertime* and not less Adorno's culinary metaphor, made still worse with Anders' emphasis on the status of radio and television listeners in the United States as so many "noodled" geese, force-fed. In part, this productive (repetitive/receptive) status illuminates the irritation Anders expressed as noted earlier concerning the specific or uniquely amorous virtues of the song *Honolulu*, which is less important for its specific qualities than for the almost endless melody that is an earworm. To this extent, perhaps, Anna Kendrick's somewhat more recent *Cup Song* (which is not *her* song any more than it is the title of the song, but makes the point about covers, plastic cup pencil holder ad hoc convention of staged improvisation). Metonymy rules, displacing the title of the song, "originally" a Carter family song, *When I am Gone*, "originally" something else and to be sure because with this song one sees what a difference the age of what Walter Benjamin called technological (usually translated "mechanical") reproduction. Everything depends, as Adorno emphasized, in the reproducibility, the recorded factor of the music and not less its radio transmission—this is what Adorno means by the "radio face," a "physiological" reference which has less to do with the look of the radio, although as Anders says, the radio like everything does happen to have a face, than with the fact that sound from one surface, produced in one space, recorded as such whether in the radio studio or else in a very similar recording studio per se, might be reproduced in another dimensionality and space, localized in a speaker, however virtualized in two, as Anders analysed the audio stereoscope.

In the case of the Carter family hit, *When I am Gone*, the song *qua* song is an artifact of the recording process, possibility, and institution, very much in the sense of what Benjamin analysed in his reflections on the work of art and precisely technological reproducibility, a concern repeated with different degrees of esoteric accessibility in Adorno's *Current of Music*, and to be sure throughout Anders' own reflections on music performance, practice, and sociological situation is by virtue of having been recorded as such, an original as such.[19] Prior to that, who knows, and indeed the song in question was not invented or "written" by the Carters (which ever one likes) as it was a hobo song, as it was a gospel song. Like the *Hallelujah, I'm a bum, Hallelujah, bum again* with the same rhythm and repetition of the rails (we know this rhythm a bit better perhaps with Arlo Guthrie's cover of the late Steve Goodman's protest song: *City of New Orleans*).

The repetition of love and Honolulu goes together with the acoustic "leash" tethering Anders and his pop music aficionados hiking out of radio range. The element of distance and the function of filling time, thus the association of radio and car travel, is key. This has not changed in the modern era. Where Anders' companions had a battery-powered transistor radio, today's travellers listen to music not individually sourced to accompany their urban commute with such world-altering affects that a Coronavirus pandemic could (and did) piggyback on the same all-purpose media delivery system.[20]

It is the rail rhythm that is repeated, reprised, channelled by and with, we could say, the cups in the *Cup Song*. Perhaps some element of hobo affect is lost or sanitized—think of the Kendrick music video—but also lost is a sense of mortality, all Huck Finn mourning sensibility. The (originally interrogative) title of what we take to be the original (inasmuch as it was thus originally recorded and so set as the original via a recording session as a record and thus as something that could henceforth be played again and again and which on the basis of the same could be reprised or imitated or "covered") itself became, like the *Hallelujah* itself the title of a book on the Carter Family: *Will You Miss me When I'm Gone? The Carter Family & their Legacy in American Music*.[21]

To be sure, as songs respond to songs in the recording industry, think of Don Mclean's 1972, *American Pie*, there is a song about the specific challenges of recognizability. Rickie Martin's *Garden Party* (this is the Rickie of Ozzie and Harriet

Nelson American television fame), a rock song about covers and the need for even the originator of a particular song as covered by others to stick, as it were, to the programme he first set. So when Rickie, who had become a kind of prototypical teen idol at seventeen in 1957, dared at the age of thirty-one to veer from his signature "look" at a Madison Avenue Rock revival concert in 1971, by saying this I mean only that he looked rather like every other soft rock music star of the time, people were horrified enough to boo him off the stage. As music audiences will do. When there were music audiences.

In the same way, Cohen's *Hallelujah* sings the song of a song that has, so to speak, "been around" beginning with its biblical origins (the Hallel psalm) and a staple of hymns and spirituals, in masses and oratorios and continuing in popular music on every level. In my book, I offer a certain account of the historical horizon of influence of Cohen's *Hallelujah*, but I suspect a complete outline may well be impossible in all the details or registers of what Anders calls the musical "situation": Cohen's song took off only following the paradigmatic interpretation on John Cale's 1991 album, which he enthusiastically dedicated to Cohen, *I'm Your Fan*.

You Don't Really Care for Music, Do Ya?

Anders' objection to the exclusive association of love and Honolulu, rejecting his hiking companion's taste, attests to a culture clash. Speaking a second language, literally minded as we noted earlier—recall Nietzsche's reflection that at the end of the day, when it comes to Beethoven, the listener no longer hears Schiller's poem at all— Anders found the lyrics illogical (was it true that love was limited to Honolulu? Why nowhere else?), German expat clashing with LA music tastes.

Anders' objections exceed the content of the song he found unendurable—as music is Kant's example of sense imposition as it is difficult to shut or close one's ears to it. At the same time, linguistically hermeneutically speaking, it may be argued that one never listens to "the words" of a song more than when listening to music in a "second language." To this must be added, and this would apply to both Anders and his hiking companions, the tendency to insist on a constant programming (self- or other-imposed) of one's own life following the dissemination of radio ubiquity echoed or expanded by television ubiquity all the way to MTV and today to YouTube and TikTok too. Seeking such programming reinforcement, his companions opted to bring their transistor radio along with them; it was not issued to them as part of a permit for climbing that day.

Here the question is the question of musical taste and the musical subjects? Nietzsche's *The Birth of Tragedy* raises the question of the subject as subject by questioning what we think we know about the lyric. We all know that lyric poems tell what it is to say I, but Nietzsche raises the question of the subject, telling us in his first book that "The subjectivity of the lyricist is a deception." It gets complicated here, and it helps to go back to Cohen's *Hallelujah* because it is a singer's lyric song *about* a singer's lyric song. "I heard there was a secret chord."

The listener seeks the music, Anders suggest, but even more so, the secret of the secret, the best of the best. And bettering the best, to take an old song, as the Beetles wrote, a sad song and make it better, that would be what a cover is all about. Thus, and this is already eclipsed, think of Katy Perry's *Hey, Jude*, or Paul McCartney's performances of the same.

Here the point is that the putative "best" version cannot be heard. This is not because of some dispute regarding "taste," the same point Kant makes in his Critique to observe that where other nations speak of taste, the Germans speak of the "Aesthetic," and when it comes to covers, there is no limit. Thus, the ultimate version or, indeed, the question whether there will be any candidates for such, cannot be stipulated. Covers emerge and fade into oblivion, even where, think of the *Cup Song*—now already fading from pop consciousness—some pop group finds and resuscitates them.

What I call "cover culture" is what H. Stith Bennett, himself a phenomenological sociologist, calls "recording consciousness." It is this consciousness that Anders analyses, that Adorno analyses, in terms of the exemplification of the phenomenon of programming standards, self-creating oneself as a consumer, recognition being the most decisive factor, thus the need for repetition in determining just what we "like." Thus, Adorno will tell us, irking his reader in the process, aesthetic philosophers, and musicologists alike, well in advance of cognitive science, that our "liking" reduces to a matter of recognition: we like what we know; we like things we recognize. This seems to clash with the aforementioned reflection on covers (being all about a new variant or what have you) but meshes with the industry's efforts to get exposure for a performance.

For a phenomenological exemplification of this in practice, for performers, musico-sociologically speaking, I refer the reader to Bennett's recently reprinted 1980, *Becoming a Rock Musician.*[22]

In Bennett's study of the rock music phenomenon of what he describes as "getting" "the music," everything depends quite as Anders argues, although Bennett does not cite Anders, on the radio. Updated for our time, this is all about listening to a tape or a CD or YouTube video repeatedly, over and over again—as I recall David Darling complaining to me after a master class at the Juilliard School, hosted by the late Hugh Downes, that auditions were hopelessly tilted as students sought to replicate the CD. Today one might say the mp3. But what is absent is the musical culture that mattered for Adorno as it mattered for Anders and of course, as I argue elsewhere, as it mattered so much for Nietzsche, who also played piano and wrote music. The culture is the grammar of composition, the tradition that, quite technically as Adorno argues, makes "the new music" possible, the same tradition that induced Adorno to challenge the ubiquity standards of jazz and pop music as such, radio music, recorded music, and which also allowed him to manage to alienate readers high and low. It is certain that had Anders had other readers, as he did not have, in music sociology that he, like Adorno, might have done the same.

If both Adorno and Anders point to the sound *difference* this makes, the point at issue in Adorno's "hear stripe"[23] is that it is sociologically musicologically, phenomenologically, affectively oriented to and for the listener attuned to "the" music. Thus, it is as a phenomenological sociologist of music not unlike Anders in this convergent sense that Bennett can point out that the same medium, the radio, the loudspeaker, small and focused sound, enables a new phenomenon in the performative ambit of technologically mediated reproduction whereby the performer can learn by ear, even if he or she cannot read (or write) music. As Bennett points out without focusing on Anders: "the" music is not what it is as Anders describes such "being-in" music. Much rather what contemporary aspiring musicians seek to acquire as what Bennet calls "the music" is the sound as it is produced on the radio, via the CD, via the mp3, as we speak of it today, *qua reproduced*. The point here presupposes Benjamin on the work of musical art in the age of its technological reproducibility as much as we need Adorno, who wrote on the familiarity factor, the ubiquity key to the same.

Anders emphasizes the sociological elements key to any social access to music especially as performed. But in an age of technological reproducibility, what the aspiring performer, especially in the case of Bennett's rock musician, seeks to "learn" to "get," as Bennett says, the music in question is a specific recording. This can be very specific to the level of a given cut on a given album, but it is also a generic constant of what Bennett calls on the side of the listener, think here once again of Elvis Presley's *Only Fools Rush In*. Bennett's "recording consciousness"—it is this that David Darling was grousing about—which is the modulated equivalent for today's performers of Anders "being-in" music—entails that when one hears (or simply remembers) Elvis' first words, *Wise men say, only fools rush in*, the intonation, Elvis' musical voice is part of what one hears/remembers, which continues with Elvis: *Like a river flows surely to the sea, darling, so it goes, somethings are meant to be*.

Bennett's point seems patent: to "become" a rock musician is not the same as "becoming" a concert pianist. Given recordings, the distinction is less clear. Alfred Brendel is an artist precisely owing to our consumer's recording consciousness, and arguably, so too, Glenn Gould and so too, because esotericism does not get one out of this recording dependency, a sometimes-overlooked composer and performer, the Basel-born musicologist, composer and pianist Ernst Levy (1895–1981)—friend of the musicologists Siegmund Levarie and Ernest McClain. One can vary the argument for cellists (Casals, Ma) and quartets (Guarneri) and ensembles (Hilliard), as well as opera singers (Caruso and Callas) and so on.

Indeed, and this is to Anders' point concerning "being-in" music, Nietzsche, although a gifted improvisor on piano, was assessed by Hans von Bülow as a tyro composer. As von Bülow wrote to Nietzsche in reply to Nietzsche's request for an assessment of his work, on July 20, 1872:

> your *Manfred-Meditation* is the most extreme case of fantastic extravagance, the most unedifying and anti-musical instance of notes placed on music paper that I have come across in a long time. . . . Did you consciously flout all the rules of musical language, from the higher syntax to simple matters of correct notation? . . . .If you really have a passionate urge to express yourself in musical language, it is indispensable that you acquire the rudiments of this language[24]

For his part, von Bülow took barely a day to reply to Nietzsche's request mid-summer. For his reply, Nietzsche would take until mid-Autumn and would need to draft two versions of his letter to boot (circa October 29, 1872).

Today's readers read Nietzsche as prototypical genius, hardly in need of von Bülow's critique (commentators are unsparing: *how could* von Bülow, the discarded husband of Cosima Wagner, dare to be so unfeeling in response to Nietzsche's overtures in sending him his youthful scores, how cruel!). Indeed, as Giorgio Agamben has reminded us, philosophical and musical genius would seem to be "beyond" grammar, and Nietzsche himself urges us to overcome our fealty to grammar teachers (the first grammar books were written by the Alexandrians); our entire philological tradition is, Heidegger is right about this (if he is conspicuously wrong when he assimilates Nietzsche to the same), more Roman than Greek.

Nietzsche, who lived as much as possible, quite in Anders' sense "in-the-music," did not protest but replied, all good acolyte style, underlining that he was *not* unfamiliar with the rules of composition which he had studied since childhood, including a study of Albrechtsberger (and among the works Nietzsche would have known, as everyone did, including what he would have taken to be Beethoven's aka Albrechtsberger's theory of composition, Ignaz Ritter von Seyfried and Henry Hugh Pierson's *Ludwig van Beethoven's Studien im Generalbass und in der Compositionslehre*).[25]

If this exchange makes sense in its era and context, we can wonder if the same would hold for contemporary composers and contemporary music? To this same extent Anders' reflections on musical sociology, offering a phenomenological and sociological explanation, predate the more practical sociological reflections of Mike Roberts' *Tell Tchaikovsky the News*,[26] the last a reflection on the social lives of trade musicians, including economic details, the sheer numbers of those who work—who can work—and thus who live—and can live from music of all kinds, pop and rock and jazz but also new and modern and contemporary as all of this has been completely transformed with the invention of recording.

Not a matter of labour contracts (although Anders underlines the privilege that it is to be a performer and to be an educated listener), Anders was concerned with technological means for, that is, the digital mediation of, music in addition to the culinary factor, as Adorno would say, that is, the consumer's taste, at issue is also the artist's aesthetic disposition. One can compose with respect to—one can argue that one cannot but, inasmuch as the culture industry industriously continues apace, in obeisance to capital and such like (and this is no minor detail but perhaps more central than ever, thus I speak of the effect of the *Hallelujah Effect*). Alternately, one can compose with respect to the tradition and beyond, hence the pre-eminence of the name John Cage (1912–92) for Adorno and others, including Arthur Danto, although other composers also brought in silences and ambient and odd sounds and not less, or more promissory, than conceptual (and post conceptual) art.

Anders speaks of Romanticism and Impressionism. And although I have been talking about pop music, with the occasional reference to Nono, Anders discusses Busoni and Schoenberg. Here the focus remains a matter of the technological reproducibility, as Benjamin spoke of this, of the work of art in our techno-mechanical-digital era. Thus, for another example, Cage's "4'33'"" is not only performed, starting with its original performance art performance of a piece originally titled "Four Pieces," it became three movements, by David Tudor in 1952,[27] broadcast by the BBC in 2004—informatively, so relevant is the challenge of this performance to what radio engineers call dead air, in the case of the BBC radio transmission, including the requirement to switch "off their emergency back-up system—designed to cut in when there is an unexpected silence on air."[28] It has even, albeit in gallery spaces and museums, gone on tour, as it were, including a CD featuring 4'33".[29]

Darker Stars: Death and Silence

We are approaching the end of this study, and the aforementioned reflections have included death, with some severity, and, despite his brave words contra Heidegger on matters "concrete," Anders's own physical death would not be among the most graceful. As Illich has underlined, there is no factor of exception here, although and to be sure the wealthy pursue this with the support of endocrinology and surgery where this fails. Most death, as Illich underscores, is a "foul" affair, even if we have—this is what Illich names "expropriation"—sought in recent years to medicalize and thus and by various means as the current Coronavirus crisis only dramatizes, to conceal the fact, dressing numbers, spinning truth. And between hype and hope, some people have different access to different treatments. Illich, however, argued uncompromisingly that on a common level, for the common human being, medicalization has never saved the individual from the foulness that has historically always adhered to death but, and this has never been more true in history, not even when Illich first wrote this line, it "isolates" the one dying from view. Today this is not a metaphor, it is protocol, it is law, it is forbidden.[30] In Anders' case there would be considerable hardship, financial and otherwise, including physical pain and above all loneliness. There is a cost to being not merely misunderstood as Heidegger would complain that he was not understood but critically, crucially unreceived. All the things that Anders missed in his life were exacerbated, as things tend to be, at the end of his life.

Here I refer to one of Leonard Cohen's last recordings, given the earlier reference to Nietzsche's book on the death of tragedy, and silence, with Cage's 4'33". Thus, to change the tone, to talk about the music, I will do what I have been doing all along, emphasizing the words.

Cohen's *Hallelujah* is a lyric, an "I" poem, like *Bird on a Wire*, pluralizable, replaceable, interchangeable, almost as Anders presciently argued in is "The Antiquatedness of Privacy." As punctuated with the "who" of *Who by Fire*, Cohen's pronoun changes to what will always have to matter, as Ivan Illich writes: a matter of grace, *Umsonstigkeit*, that is what happens *for*, *with*, *to* the individual the I, when it is mediated by or via a distinctive other, a you.[31] Thus, even in the divine *You Want It Darker*, the poet speaks in the first-person plural, you and me, all of us: *we kill the flame*. Then we hear not Greek and not Saint Augustine's Latin, *ad sum*, but Hebrew—*hineni, hineni* [here I am, here I stand] *I'm ready, my Lord*. Cohen is a master of the pop song, he has the music industry in his pocket, and he repeats and plays an infinite melody, this time with the echoes of a cantor: *If thine is the glory, mine must be the same: You want it darker.*[32]

If, as Anders says, citing Max Weber, what is most important is always to be found in the notes, between the lines, "*Das Wichtigste steht natürlich in den Anmerkungen*" (AM II, 14), the most important things are the questions. These Anders invokes both to situate his own situational reflections on at once [*Zugleich*] the phenomenology of music with reference to Heidegger and to differentiate his own perspective from the same with reference to a very posed and dynamic *dass*, which, as Anders writes, "stands between situation and question" (MS, 16).

Here we can also recall Nietzsche's wonderful reflection on the "moment," the *Augenblick*, poised in and against history, useful for the same of understanding Anders "dass" reflection as this is positioned between the more Heideggerian and phenomenological poles of *So-Sein* and *Das-Sein*. For Nietzsche, who emphasizes "the value of feeling unhistorically through and through"—and precisely for the sake of life, which is also for Nietzsche always also to say for the sake of art and thence to science:

> Whoever cannot set down upon the threshold of moment, all pasts forgotten, whoever is incapable of standing on a point like a goddess of victory without dizziness and fear, such a one can never know what happiness is and still worse: he will never do anything that makes others happy.[33]

This Nietzsche writes on history and life in his *Untimely Meditations*. The task for the scholar, in his case: for the classical historian, is always "untimely" [*unzeitgemäß*] and thereby his title, to work "against the times and thereby on the times and hopefully for the sake of a coming era."[34] It is forgetting timeliness, one's timebound condition, that "makes happiness happiness"; in other words, as Nietzsche goes on to explain, the ideal is utterly unhistorical. As a result, all our joy is oblivion: the forgetting of the memory of past feeling gives us delight, one more time.

Intriguingly, just this untimely insight corresponds to the delight of recognition that is for Adorno, as we may recall in his own work on effective "work" of the culture industry, such recognition is all we need to like anything, especially in broadcast music, traditional or new, as in film, photographs, any work of art.

We may thus track Nietzsche's insight behind Anders' mordant words contra the miserably perfect *ascesis* of authentic Heideggerian being-in-the-world:

> When "Dasein" sleeps, it wakes itself up, if it wants to read the paper it tears this "tool of mediocrity and average-life" from its own hands. It excludes itself from leisure, friendship, friendliness, in short, from culture. Its *exercitia* fill the twenty-four hours of the day, its drudgery to march toward death lasts the whole life. (PC 362)

Part Three

Schizotopic Thought: Planetarism and Apocalypse Blindness

12

Political Media Theory, Hiroshima, and Nuclear Power Plants

From the Holocaust to Hiroshima: "Chernobyl is Everywhere"

I have argued that to Anders' reflections on genocide and Auschwitz must be added his pacificism along with the question of one's ownership of action, of "having been," of "having been done," regarding the atom bomb as this was twice deployed in Japan in August 1945. To this, nearly impossible to coordinate on the same level of analysis, must be set Anders' efforts at cultivating a future for kindness or humanism as we may read in his correspondence with the Hiroshima airman, Claude Eatherly, but also very literally his letter to Eichmann's son. For us at this juncture, given Anders' title: *Hiroshima ist überall*: "Hiroshima is everywhere,"[1] there would seem to be a continuity with "Chernobyl is everywhere."[2] Indeed, as Anders goes on to say, counting out Ten Theses to underscore his point as is his wont, "the real danger consists in the invisibility of the danger."[3]

The "everywhere," an obligatory "*überall*," is ubiquitous but precisely as such "invisible." Better said, opaque: precisely where radioactive contamination of one kind or another remains, however "real" it happens to be, however much it affects us as we are exposed at almost every turn to radiation, microwave, cell phone, broadcast radiation of all kinds, and not merely, as noted in the course of the last chapter, every time we go through federal buildings or airline security, an "out of sight, out of mind" affair.[4]

Other Than, One More Time

Earlier I called attention to the conventionality of Anders' name change, casually taking the nomination "other than," "something else," as we hear the apocryphal story, at the behest of an editor who complained that there were too many "Sterns" on the newspaper masthead. Thus, it was also noted that Anders, who apparently took this designation to heart, turns out to be less well known than he should be,[5] despite belonging to a group of famous names, by marriage, blood, and other affinities.

In *Gewalt, Ja oder Nein?*, Anders argued that the ideology of non-violent resistance meshed rather too conveniently with the *Nicht-können* that happens also to be the rule of law: a bourgeois, all-too-bourgeois, excuse for inaction out of respect for law. This, our cowardice, is exacerbated as it is an automatic consequence of our dependence on technology. Earlier, we also had cause to note that Anders' critique of technology is not to be separated from his phenomenological analysis of music broadcast- and reception studies, which themselves have a technical psychological component, as do radio and television programming.[6] The internet can seem to break that parallel until we consider bubbles and the particular constraints of our experience of social media, as noted earlier with reference in particular to Baudrillard's early analyses of this phenomenon of what he called "speech without response." Thus, our interactions are framed or limited in advance: we are fed certain posts, in response to which we can register a like (thumbs up on Facebook) or a love (a little red heart on Twitter). We are primed to respond to posts with such likes; it's a seemingly easy enough thing to do, and we seek such responses, reading these numbers, as much as we might read comments on our posts. Some users turn off comments all together, thus limiting themselves to the numeric satisfaction of likes. As noted at the start of this study, Seymour's recent book *The Twittering Machine*, focusing on social media and the very possibility of political and of active citizen engagement, explores some of these limitations with an array of dark consequences.

Anders, who used the language of the "homeworker" (the mass human as the self-made)—with some aid or inspiration from the argument Heidegger also makes with respect to the radio and in the ambit of Benjamin and Adorno as well—argued in a fashion more sustained than any of these, that the consumer himself or herself literally fabricates himself or herself in the image of mass media into a participant in mass-media by means of consumption. Thus, in his 1980 *Antiquatedness of Humanity*, where Anders meditates on the antiquatedness of this and of that, including the very notion of the mass human being, Anders writes on our new need, here again to cite this epigraph that still requires comprehensive attention despite the ubiquity of consumer attention, content for "consumption": "Give us this day our daily hunger." (AC II, 15) Only Illich and Baudrillard and more recently Sloterdijk come close to repeating the singularizing, isolating, point that Anders identifies.

For Anders, our preoccupation with (and by) technology is a preternaturally religious one, filling the void or wake of what Hegel in his *Phenomenology of Spirit* had already announced, well before Nietzsche, as the "Death of God." In this way, different from Nietzsche's mad man who breaks into churches to play his *Aeternam Deo*, calling the churches so many sepulchres for the deity, Anders' homeworker is a trans-modern "hermit," working tirelessly—like a monk in his cell. Anders repeatedly emphasizes the quality of being alone, even in company with others, conspicuously so—correspondent to the vocation of creating oneself, all as unpaid home labour, in as many hours of the conscious day that one has to oneself in order to produce the mass human being, capable of appreciating on command, on demand, mass media and to pay for the privilege, as Anders adds in the kind of stylistic dissonance and insight that was his watchword. Spoon-fed by the media, watching TV as one does all alone (even in the presence of others), TV is a quintessentially autistic medium. So too of course the internet (and Facetime and Zoom, etc., are not likely to change that).

So too, similarly, YouTube, on computer or on our cell phones and tablets, as I write in *The Halleluiah Effect*.[7] There, I already invoked Anders along with Adorno's more well-known involvement in the Princeton Radio Project,[8] to make the case that we might well understand Facebook, and other forms of surfing today's internet, Twitter and Instagram among other newer apps, as exemplifying this quasi-autistic effect just where this same isolation and self-reference may turn out to be its most subversive quality, and correspondingly, its lasting appeal.[9]

Already in the mid-1950s, Anders argued that there could be no possibility of democracy in a world with television (which would make the case that much worse, so we can now extrapolate, with the internet). Thus, as we saw earlier, contrary to popular belief, it turns out not to be the case that connectivity and cell phones enable revolutions, if it does turn out to be true that the internet and aforementioned cell phones make it possible to coordinate meetings with family and friends and not less encounters, erotic and otherwise, with friends but also with strangers.[10] Today's computer and cell phone connectivity (and internet bubbles and "filters" and algorithms) makes that same democracy less possible than ever.

Here, given that Anders wrote of Chernobyl—and since then, with Fukushima, we have now had a number of such disasters, disarmed of specifically "nuclear" associations as these are more generically designated as environmental catastrophes[11]—and that he also wrote of television and radio, we may ask what radio, even "spooky" radio, could possibly have to do with violence? Apart from the bomb, it seems that Chernobyl, Three Mile Island, Fukushima are incidents, accidents, things that happened to transpire with effects that are happening quite without the responsibility and certainly apart from the deliberate intention of anyone.

Even citing, as I above cited Steve Goodman's *Sonic Warfare*, to explain some of the points Anders makes with respect to music on the radio, or music as a television broadcast, like music on YouTube or Spotify, and so on, music as such seems, *qua* entertainment with the on-demand appellation that covers over the necessity of exposing oneself in the undertaking not merely to sound but all kinds of other ambient radiation, the antithesis of violence. Goodman himself, echoing Kittler, prefaces his book with a dissonant quote from Francis Ford Coppola's *Apocalypse Now*, referring to Wagner's *Ride of the Valkyries*, which Goodman, a Scots philosopher and performance artist (producer of bass weighted electronic dance music under the name "kode9"), attributes to one "General Kilgore," actually: *Lieutenant Colonel* Bill Kilgore: "We'll come in low out of the rising sun and about a mile out, we'll put on the music."[12]

That was Vietnam, Hollywood style. The reality today, as Goodman, himself a dubstep, electronic music sonic artist, demonstrates, is more pernicious: ongoing, ubiquitous (Adorno's favourite word when it came to the culture industry has never been more apt). For Anders, who writes on radio in 1930, the "spooky," uncanny, for our brains, as we are compelled to process this, would be the point that derives from his experience of his father's tone generator, hearing the same music from every window, continuing seamlessly as he remembered walking down an apartment building corridor.[13] For us today, there is no surprise, accommodated as we are to streaming, earbuds in our ears, there is no there there. We hear and overhear without remark and it can be hard to grasp what Anders was on about.

Adorno arguably borrows the image, if without attribution, in his *Current of Music*.[14] One year after Anders' "Spuk und Radio," the role of loudspeakers in Huxley's 1931 *Brave New World* recurs, playing, as noted earlier, the same part in Orwell's 1949 novel, *1984*.[15] Once again, loudspeakers were involved in—and were specifically developed for—party rallies.[16] This entrains and captivates us in our commercial jingles, priming or triggering our thinking (that is our brains at work again) in response to our attention to radio programmes, television programmes, all the way to YouTube playlists and streaming media, all programmed: on demand.

The social manipulation of Facebook in the recent US presidential election (and of course this "social engineering," as it is called, has been deployed for some time and must be, though it rarely is, considered along with direct and overt hacking, not necessarily via supposed Chinese or supposed Russian interference but routine party politics, as this is something that has been going on for some time now)[17] is the latest manifestation of what Anders called "force-feeding." You do not have a choice. As consumers, homeworking on ourselves as such, we do this ourselves to ourselves when we watch television or, as today there is no difference, when "go online" or use our cell phones, for texting, or via GPS when we navigate the city: Google earth is, of course, anything but neutral.[18] For Anders, if "democracy supposedly consists in that one disposes over the right to express an opinion of one's own, then democracy is through mass media rendered impossible."[19]

Anders is not here making a case about what one takes to be one's opinion one way or the other. Anders' point concerns whether *one is free*, to begin with, *to form such an opinion in the first place*. On Anders' account of it, judgements framed in accord with mass media are not because they cannot be your own, that is, no matter whether you were for Donald Trump, or Hilary Clinton or Joe Biden, as it transpires, because, so the claim went, *anything* would be better than Trump, and the DNC itself had already manipulated or hacked Bernie Sanders out of the running so that the only candidate to set against Trump would be Clinton (who duly lost) or, more recently, Biden (who duly won). Like the unconscious, publicity works in one direction as public opinion research since the early 1920s has underscored. Thus, repetition, so Adorno argued on behalf of the Office of Wartime Information in the Second World War, beginning with his work on the Princeton Radio Project in 1937, entails that the radio does not play works that happen to be popular (this would be a democratic hit parade) but makes them popular.[20] Thus, being played on the radio, as any professional musician will tell you, makes a hit a hit. The title of Adorno's "current" of music reflects this same point: radio makes popular hits popular simply by broadcasting them.

Here to be opposed to something is the same (in end effect) as to be *for* it. In either case, one is preoccupied by it. If the ancient Stoic thinker Epictetus took care to remind us that quite as one would be horrified to be *physically* handed over to the power of another, one ought to recoil from allowing one's *mind* to be affected by another's insulting word (or seduction or flattery) because just at that moment one has surrendered one's mind to that other.[21] Nietzsche reflected in *On the Genealogy of Morals* on the power of words when he explained that the slavely moraline ascendance of reactive thinking effects a revaluation of values simply by changing what things are called.

At issue here is not a matter of slavely moraline character or predisposition, good or evil. Much rather, it is the architecture of modern 'social' media that, as Baudrillard argues, "always prevents response."[22] These media forms, television, Facebook, Twitter, and so on constitute, for Baudrillard, an address we give ourselves over to without the possibility of actual or real response and so we post on our Facebook walls and tweet into the void, while being exposed to the assaults of the same, augmented by television, cable and network, radio, and internet feeds, as so much received "Speech Without Response."[23] And as Anders writes in the 'Antiquatedness of the Individual,' concerning the illusion of exchange in a section entitled "The Collective Monologue":

> [w]ith respect to most of our conversations, namely 'small talk' [Engl.] it is patent that the words and expressions that we exchange with our partner resemble tennis balls flying here and there between tennis players; i.e., that the 'balls' we 'serve up' by speaking are identical to those we have received by listening and that those we receive are identical to those we have been given. In brief, receiving and giving have become interchangeable. (AM II, 153)

Programmed to 'receive' speech to which one cannot respond, beginning with radio and television sets in the living room, but now on every media platform, 24/7, it is plausible to suggest that we are "primed" to look for this at every waking moment, as internet psychologists have studied and today, perhaps, perfected the phenomenon. For Anders, who already anticipated this, we are at work at every moment in the creation of ourselves in the image of mass media, now named social media.

If we surely suppose ourselves free to respond to someone on Facebook (or on Twitter), all we really "do," all we can do, is hit the like button, letting the sender (and Facebook) know that we have seen and received the message.[24] Better yet we can augment the same event of "speech without response" by reposting or retweeting, and if one writes a "reply," perhaps alongside "replies" already written, one generates, creating any range of possibilities for miscommunication, only what Baudrillard presciently named a 'simulacrum' of a conversation. The result never yields speech plus response, not for theorists of semiotics like Umberto Eco as Baudrillard cites him. For, speech *with* response always presupposes communicative exchange (not action as Habermas one-sidedly insists), including everything that constitutes what Gadamer calls conversation. Conversation, as Gadamer emphasized, and which in a surprising coincidence Illich also argued, is an emergent property of the spirit resistant to formalization and every kind of institutionalization, as Illich plainly emphasized this resistance in a passage that cannot but confound even his admirers.[25]

But how compare Anders and Illich, except to note that both are marginal names? After all, what commonality can there be between Anders, a Jew, with a classically German formation, and Illich, a mixed blood, Jewish-Catholic/German-Slav? The commonality would have been their dedication to humanity and to the sound of the sign, the musical note for Anders, for Illich, the "vineyard" of the text.

"Seit ein Gespräch wir sind"[26]

Hölderlin, in his poem *Friedensfeier*, names beings such as ourselves, Anders' antiquated human beings, a "conversation." By this the poet obviously is not talking about the capacity to "like" or to "reply" on Facebook or retweet a tweet or even the ordinary kinds of exchange. Hölderlin emphasizes *"hören von einander,"* key to the phrase that concludes the verse: *"bald sind wir aber Gesang."*[27] And here, and this point too connects with Heidegger, Anders foregrounds the echoing capacity on our part to be attuned to one another, to argue that what is at stake is precisely what Kant called *Mündigkeit*, that is, the ability to speak in our own mouths, with our own voices, for our own part. The media today entail however, as Anders has it, that *"Der Mensch ist kein 'mündiges' Wesen mehr."* Thus, to quote Anders, here writing in English:

> The human being is no longer an entity capable of speech, such that he might express his own opinion with his own mouth.[28]

Anders is concerned precisely with, and this is a technological dictate, the "unilaterality" of broadcast and social media. Driven by media, collimating our senses, we are "eye people," we are "ear people," and Nietzsche's Zarathustra (*On Redemption*) already predicts a human being who had become only a giant ear. Anders uses the (rather appalling) phrase *"Genudelte Gänse,"* as I referred to this earlier, *force-fed geese*, emphasizing that in all the years he lived in the United States he himself had not been free, just as he never met *anyone* who was free for and on their own part to express his or her own opinion. Anders insisted on this, and elements of this insight dominate both volumes of *The Antiquatedness of the Human Being*.

We are thereby constrained by the social media we use (notice that saying social media makes the claim easier than saying radio or television) to create ourselves in the image and likeness of the media. But there is no extern's tyranny in this constraint. This is curated self-creation of the self in the exact image that social media would have us make of ourselves, which, as Anders emphasizes, we do willingly. "Soft totalitarianism likes nothing better than permitting its victims the illusion of autonomy or even to engender this illusion in them" (AM II, 238). As Anders explains, what is expropriated from the average American as he writes is what had previously been, in the sense in which Rousseau speaks of this and, more recently, Derrida, and of course Heidegger, most *properly* his own: him or herself. Social media as we speak of it is in this sense, for Anders, quite specific. Nothing of what we ontically 'own' is taken from us, no part of the relations of capital are put in question (the banks must always be "bailed out"). Much rather:

> The "only" thing to be sacrificed is his *"own specificity"* [*Eigentümlichkeit*], his *personality*, his individuality, and his privacy: *solely himself*. By contrast with the ordinary kind of socialization, involving what the person has, what is at stake here is "only" a socialization of what the person is. (AM II, 239)

What is characteristic of the modern age is that it is of our own free will that we appropriate the task of creating ourselves as "the masses," as Anders writes, and in and as our online personas. And these are various, depending on the social media we access but big data (and Google) tend to cross such fine distinctions. What is crucial for Anders' analysis bears repeating as it has become even more accurate given the role of social psychology and programming in the current Coronavirus crisis, "America uses psychoanalysis *for the purpose* of establishing conformism" (AM II, 237). But this does not mean that we create ourselves as everyone, we think of ourselves as unique, as individuals, and certainly not as workers or as the proletariat. Thus, Anders' point is more Benjaminian, that is, we create ourselves as human beings technologically, as we do in the current age of technical reproduction, thus consuming mass products *en masse*: the same products immoderately, without measure.[29] In age of digital media, the mass products we consume are digitally conveyed, and thus we are attached to our phones and we do not tend to think about the effects of omnipresent WiFi/WLAN or indeed cell phone radiation, on the eve as we are of the acceleration of the same to 5G/6G.[30]

Baudrillard makes the point that "power belongs to the one who can give and cannot be repaid."[31] Thus, Baudrillard argues, "the revolution everywhere: the revolution *tout court*—lies in restoring this possibility of response."[32] To do this however would presuppose a complete transformation of the architecture of the media as such.

For Baudrillard,

All vague impulses to democratize content, subvert it, restore the "transparency of the code," control the information process, contrive a reversibility of circuits, or take power over media are hopeless—unless the monopoly of speech is broken; and one cannot break the monopoly of speech if one's goal is simply to distribute it equally to everyone.[33]

We need what Gadamer names conversation—this is what Habermas never managed to understand, because for this we need phenomenology and hermeneutics, we need the face to face (unmasked), that is to be able to speak with and not less to be able to hear from one another, and to do that, as Baudrillard emphasizes, so-called social media excludes us, as what is needed for conversation are all the elements that escape transmission even as there are new emergent communicative possibilities that grow out of our new modes of expression of storing, "liking"—such a minimal affect to minimal effect—sharing, retweeting. All of this is part of communication and media, but for conversation with one another we need the ironies and the hints that escape mediatized exchange.

Baudrillard thus highlights what may be named Hölderlinian, Gadamerian conversation when he writes in his "Requiem for Media" that "Speech must be able to exchange, give, and repay itself as is occasionally the case with looks and smiles." We need more than the possibility of registering a like or a love. We need the body, we need the look, we need the smile, faint or feigned, absent, or irrepressible. In other words, for Baudrillard, response, to be response cannot be "data," mined or otherwise; it "cannot simply be interrupted, congealed, stockpiled, and redistributed in some corner of the social process."[34]

Anders is even less optimistic as we may be sure, and less compromising, naming us "cosmic parvenus, usurpers of the apocalypse," contending, this is the dark side of his language in his open letter to Klaus Eichmann, *We Sons of Eichmann*, that we have our excuse at the ready, in our non-action by the impermissibility, the illegality of any counteraction. It's against the law. Thus, Occupy Wall Street was put down in New York City, way back in 2012, in good fascist fashion by "locking up" (it will do to remember the language Trump used to trump Clinton) as "terrorists," as the Occupy Wall Street protesters were designated.

To this day, violence *can* only be deployed by those in power. That is the violence of violence. Once set in motion, once deployed, violence constrains all acts that follow. Someone rebukes you, again to recall Epictetus on responding to insults, only let it trouble you and you place yourself in the thrall of your abuser. At issue is not an imperative to turn the other cheek, this is not Epictetus' claim but instead to underline that if your mind or unconscious mind is troubled, it is conquered forever. At best, as Nietzsche says, one can look away, but how close one's ears?

Like the gas chambers in Auschwitz, like the bomb deployed in Hiroshima, Anders argued that nuclear power plants present an ongoing and invisible circumstance of constant danger to the earth and everything on it. Thus, Anders spoke of "globicide [*Globozid*]" (AM II, 410). And in the face of such an extreme, global danger, he argued especially towards the end of his life that the stakes constituted an emergency, calling for defensive action. *Notstand und Notwehr*.

Yet he was already aware of this trajectory in his first writings on the bomb, which already forms what is in effect the second half of his first book, including the section: "*Über die Bombe und die Wurzeln unserer Apokalypse Blindheit*" (AM I, 233f.). In an excerpt published at the same time, Anders began by adverting to our displacement of ourselves into the position—this was an argument that begins in Nietzsche and can be tracked with some dexterity in Heidegger but very plainly in Sartre—of the divine. The argument Anders makes is the Promethean argument, and the reference is Aeschylean; we can create nothingness, absolute destruction, as he writes in *The Antiquatedness of Evil*, in a section entitled the "Theology of the Atomic Situation":

> Indeed "like unto God" ["*Gottgleich*"] only in the negative sense, as *creatio ex nihilo*, cannot come into question, instead we are now capable of a total *reductio ad nihil* as the mark of omnipotence as destroyers. As omnipotence we may really characterize this, that we (or more correctly, our broomsticks, instruments we have summoned) are capable of obliterating the entirety of humanity and the human world. (AM II, 404)

The 'broomstick' is the Goethean reference, today we have vaccines, PCR tests, titanic enough for Anders' Promethean vision, as we may again recall Ferdinand Barth's abjectly monumental and picture book nineteenth-century illustration of an adolescent sorcerer eking a spell from a book, broomstick propped against it with conjured spirits already eagerly swirling around (see Figure 12.9, Ch. Ten above). Our problem with such precociously radical prospects, with thinking exceptionality, is precisely what Anders, who speaks of the "emancipation of objects" (AM II, 406), calls our apocalypse blindness. In the midst of disaster, especially grievous disaster, we do not see ourselves, neither the where nor the how of what Anders calls the "situation." Lacan, in another context, speaks of this as the Real.

In his own day, Anders argued that nuclear preparedness requires a constant state of violence. For, even if and even as we manage not to see it, "the devil has moved into a new apartment [*Der Teufel hat einer neue Wohnung bezogen*]" (AM II, 410).

For Anders, even as we do not like to be lumped together, the point of invoking this "new apartment" together with our complacency with nuclear preparedness as with nuclear energy as with drone attacks at a distance—and Sloterdijk discusses the continuity of drone attacks and the history of bomb warfare from one world war to the nuclear climax of the other—cannot but entail that all of us must be counted as so many new "sons of Eichmann," implicated by nothing other than our inaction, even if we name that non-action by the fetish name nonviolence, peaceful resistance. Thus, we are complicit in a planned crime against humanity and not less against all other living beings, all plants and animals, and even the earth itself, that same Gaia having already been blown to bits on a regular basis by atomic testing in deserts, Pacific islands, and oceans, earth and sea and air contaminated already for millennia to come. In the face of the violence of violence, our complicity not only gives consent but perpetrates violence. There is no non-violent action.

Violence Contra Violence

Chernobyl is everywhere, more than ubiquitous, more than ongoing.[35] And, how, in the wake of Fukushima, are we not able to see this today? To be sure, part of the problem is the invisibility of radiation, the slowness of radiation. Its protractedness is the deep time that belongs to nuclear contamination.[36] If Erwin Chargaff, a trained chemist the author went out of her way to meet when she first arrived in New York three decades ago, could already criticize genetic engineering as "practicing biology without a license"—and it is to be noted that in Heidegger's *Gelassenheit* lecture it is chemistry and biology, the science of tweaking life that is for Heidegger the greatest danger of the day, even more than nuclear power or nuclear bombs.[37] To that end to be sure, we need to read Virilio and Illich in addition to Anders.

Reflecting on the near-and-present danger of globicide by specifically reflecting on "incidents" at nuclear power plants, just as the media suppresses such incidents by failing to mention them altogether or underreporting them,[38] Anders writes:

> Although I am very often regarded as a pacifist, I have in the interim come to the conclusion that nothing more can be attained by nonviolence.[39]

As we have seen: "Renunciation of action does not suffice as action."[40] Thus, Anders refers to the proportionate challenges of emergency, *Notstand*, and the need in such exceptional circumstances to defend oneself [*Notwehr*] by any means possible.[41]

The underreporting of the media is more than a problem of so-called and recently thematized "fake news." For the problem of the media is that it has been an organ of public opinion from time immemorial but specifically calculatedly so since the end of the First World War, when Bernays published his *Crystalizing Public Opinion* and that project drove the coordination of industry, media, government, and especially military interests. Thus, as David Bertolloti has argued in his chapter on the bomb in his *Culture and Technology*, US media censorship went hand in glove with national interest during the war, with nary a sense of this censorship, and it is noteworthy that discussion of this censorship, quite official, quite successful, remains unquestioned to this day.

> There were also times before the attack when the secret of the atomic bomb was in jeopardy, and in danger of being uncovered. In one instance, a reporter from the *El Paso (Texas) Herald* investigated reports of a huge explosion on July 16, 1945—the date of the test bombing. The official report stated that an underground ammunition dump had accidentally exploded. Such a mundane story was too bland for this enterprising reporter who went on to embellish his account with exaggerations of "the greatest fireworks show" which "illuminated whole mountain chains." (Apparently, Japanese spies did not read the *El Paso Herald*—circulation 27,046). Another near disclosure occurred when reporters asked what was being built in Oak Ridge, Tennessee in October of 1944. The response was that "35,000 workers are making Roosevelt campaign buttons." Even comic strips were censored to ensure that the slightest hint of disclosure would not occur. In an April 14, 1945 Superman comic strip, Superman was to be bombarded by a 3,000,000-volt charge generated by a cyclotron: "Even Superman can't take it." Superman had been warned.[42]

As Bertolotti adds, the comic book was told to cease and desist, and cease and desist it did. As Bertolotti cites *The Newsweek* article, "The Superman Way":

> Superman could take it and did. What he couldn't take was the office of censorship which asked McClure Newspaper Syndicate to discontinue references to atomic energy. A new series of strips, then in production, was cancelled, and Superman went into a sequence in which he played a baseball game single-handed.[43]

Significantly, and this bears on the question of the story, the official narrative, and engineering and scientific assessments, echoes of 9/11; thus, Bertolotti recounts the systematic efforts to resist or tweak eyewitness perceptions of this, spinning the sunset itself. Thus, and this is significant in the plain-talking Midwest:

> Chicago readers had to be assured that an exceptionally dramatic sunset in the west had no relation to the atomic bombing. Said a meteorological forecaster: "Any time you get smoke and moisture in the western sky you have a red sunset."[44]

Of course, the same tactics continue to this day when it comes to explaining the significance of what is otherwise visible in the sky as signs of weather manipulation and geoengineering, about which more in the final chapter.

What about Violence: Yes? or No?

Baudrillard's "Speech without Response" turns out to be a particularly effective way, along the lines of *The Man Who Shot Liberty Valance*, to print the "legend" rather than the fact.[45] Nevertheless, and this is the point that is also exemplified in John Ford's film, the problem remains the violence of violence. For violence does not leave one with the choice of non-violence in response. The Stoic counsel of *ataraxia* reminds one that one cannot respond *at all*, no #metoo, which means that one is bodily bound, even as one keeps one's spirit, which one does by acknowledging what is and what is not up to us. The engine of technology has its way with whatever is set in motion. For this reason, the Ancient Greek tragedian, Aeschylus reminds us: *by the sword you did your work, and by the sword you die* (*Agamemnon*, line 1558). And we read too as we may prefer, the Gospel, to cite the King James version: "Then said Jesus unto him: Put up again thy sword into his place: for all they that take the sword shall perish with the sword" (Matthew 26:52).

Jesus, who said, we are told, "I am the life," encourages non-violence—for the sake of life.

It is Anders who reminds us that the path of non-violence is not a means of resistance. For this reason, at the moment of violence incited by the High Priest, the bodily seizure of Jesus by the guards who would bring him to Pontius Pilate, before torturing him to death—the same torture common among Romans at the time, highly effective for "encouraging the *others*," as it was then supposed and as it was still in the last century, wreaked upon the British to compel them to leave Palestine (the so-called "Seargent's affair")[46]—Jesus responds to one of his followers who sprung to his defence to sheath his sword instead and *allow* the violence about to transpire to transpire. This practical syllogism entails the crucifixion.

Let me be clear, Anders, a perfectly good if atheist-minded Jew, is not himself talking about Jesus Christ, another perfectly good Jew, to vary Jacob Taubes' encomium of a "nice guy."[47] Anders much rather underlines that non-violence offers no defence against violence. In the parallel I draw here, the non-resistance on the part of Jesus's disciples in the Garden of Gethsemane, so little resists that it is central to the myth of redemption that is sacrifice, yielding Jesus bodily (again, we recall the stakes of the aphorism cited earlier from Epictetus' *Enchiridion*), to be carried off to a vapid conversation with the man in power (about truth), thence to a desultory sentence to death, and then, in short order, to whipping, piercing with thorns, heavy labour, including three falls, and, finally, as the antecedents are crucial to the quick fatality of the last, crucifixion.

Anders' reservations concern the impotence of the impotent and thus the culpability of recommending an ineffective form of resistance. Does one think, he asks Manfred Bissinger,[48] that anyone with power cares in the slightest about the sandwich [*Schinkenbrot*] one does or does not eat? The only thing done by one's hunger strike is suggest to you, the self-starving one, that you are doing something, which of course you are doing *to yourself*. Masks, as we know, especially as self-imposed, quarantine, as self-imposed, work the same way. This is the great achievement of the ascetic, as

Nietzsche points out. One takes what one *does not do*, not eating in the case of a hunger strike, turning the other cheek in the case of non-violence, keeping away from social contact, refraining from free breathing, *as if* it were an action.

If Anders was critical of the "happenings" of the 1960s, it was because unlike those who were young in the 1960s, Anders, who was already old at the time, saw non-violent actions, handing over bouquets of forget-me-nots to soldiers, as bits of "theatre," to which, like fascism, he said they were related. Beyond anything so phenomenologically ready to hand as the quotidian illustration that captured Heidegger's readers in 1927, of a broken tool, Anders recalled the violent, near invisibility of what had struck him some ten years earlier, at the age of fifteen, on his way home after the First World War, spent as a much too-young soldier/volunteer in France. This trauma we have already cited as it would stay with him:

> On my way back, at a train station, maybe it was in Liege, I saw a line of men, who strangely seemed as if they began at the hip. These were soldiers who had been set on the platform on their stumps, leaning them against the wall. Thus they waited for the train that would take them home[49]

Today's medical interventions entail that more shattered soldiers survive than ever before. Anders reminds us, this is the point of his "Promethean shame," that the violence continues.

For Anders: "Today, the real danger consists in the invisibility of the danger."[50]

13

"The Devil's New Apartment"

Apocalyptic Thinking

This is the last chapter but cannot be a conclusion. At stake is the current "situation," that is what is jargonistically—in the sense in which Adorno speaks of "jargon"—called "climate change," and at stake is the question of a range of personal freedoms such as the same freedoms Adorno himself took for granted, already, presciently feeling guilt for the givenness that once was the freedom to breathe when others had been gassed in camps and firestorms, the freedom to move in the world, freedoms which were never to be sure without complicity; we human beings deal death with every step, every breath we take. These are choices, not in the way popular culture invokes such issues, as if what is at stake turns between climate change "deniers" and virus deniers—the language is intended to punish those who do not think in accord with the mainstream, AIDs deniers and suchlike, as opposed to those who acknowledge (and whose virtue is rewarded for acknowledging) "climate change," specifically global warming, which warming is then urged may be combatted with geoengineering, already underway for decades, but due radical expansion.

Given today's health crisis of militarized governmental restrictions, lockdown and quarantine, mandatory masks and vaccines, ubiquitous surveillance via contact tracking/tracing, it is necessary to add a reflection on the positioning of attention and focus. For to date, the activist or crusader or protester might have assumed, as Anders did assume, that others need only be persuaded to see the dangers posed by human incursions on the world and that those others might, collectively, be moved to stop or to litigate or legislate against such incursions: less drilling, fracking, mining, fishing, logging, and so on.

Adorno reminded us in his broadsided attack against jargon as such (and not only Heideggerian authenticity) that the rhetoric of wellbeing now well-ensconced as mantra of the new capitalist world order is inherently empty. For Adorno, as co-author of the *Dialectic of Enlightenment*, points to the suspicion that we should not lose sight of,

> that, after all, the overpowering conditions of society really were made by men and can be undone by them. ... It was not Man who created the institutions but particular men in a particular constellation with nature and themselves. This

constellation forced the institutions on them in the same way that men erected those institutions, without consciousness.[1]

But that is only one side. From another perspective, things take on a frightening guise.

Nightmare

Sometime late in 2019, a terrible dream left me with the uncanny and unpleasant insight that, from a certain point of view, the great thundering concern with Thunberg, the rightful and important Dakota pipeline protests, agitation and despair in the face of the burning of the Amazon along with the murder of its peoples, concerns with Monsanto's destruction of the agricultural ecosphere on so many levels, protests against waterway poisoning consequent to feedlots, protests against the fracking industry that destroys aquifers and consumes pure water to suffuse it with chemical contaminants, because that is how fracking works, protests against the deployment of 5G, along with the 9/11 truth movement initiative on the part of engineers and architects (contra official, government narratives), grievous worries about industrial farming practices and the depletion of the soil, beach modification and the destruction of coastline, the bees, dolphins, trees, koalas and kangaroos, and now so many vanished beings after the fires in Australia, geo-"accelerated" as they were, a loss compounded these days by increased and heedless logging, all that and more in a single sentence that could be infinitely expanded, all of it could be regarded very differently indeed.

One can see all this as horrific and as compelling a programme of change or "reset": "we" must stop, "we" must reverse course, and so on.

But from another perspective, a perspective easy to consider as it is the perspective of government and corporate power, the problem is the protesters who gather to block the streets, the workers who strike, unwelcome emigrants from other nations.

This is the uncanny insight into "what is," to use Heidegger's language. This is the insight into the world of power and privilege, the view from the perspective of those Rilke, in the last century, could, once again, name without flinching:

> The kings of the world are old and feeble.
> They bring forth no heirs.
>
> Their sons are dying before they are men,
> and their pale daughters
> abandon themselves to the brokers of violence.

To be sure, as Anders would remind us, there are numerous ways to read this. Today, Heidegger tells us, these are the wealthy, again, as Rilke's verse continues:

> Their crowns are exchanged for money
> and melted down into machines,
> and there is no health in it.[2]

Today, so go the arguments since the zero population growth movements of the 1960s,³ arguments advanced by the wealthy, by the United Nations, the World Health Organization, the Bill and Melinda Gates Foundation, the World Economic Forum in Davos, what is needed, what would solve everything, is fewer people. This is the new blue *and* red pill question for a new *Matrix* reboot. But to be sure, and despite the dark value of the Netflix series, *Black Mirror* (2016–19), we find it hard to grasp the notion, relegate such considerations to conspiracy theory and react, overreact to alternative accounts such as this perspective shift.

The zero population growth movement has gotten more attention today given the renewed focus on Bill Gates, who advocates sterilization built into vaccinations distributed to the poor of not only the various countries of Africa but also India, Southern Asia, and so on. A lot fewer people, millions fewer, billions fewer. Thus, the roster listed earlier, the UN, WHO, the World Economic Forum in Davos, represents a range of clubs for the wealthy, thus the metonymic justification for citing Rilke's "kings of the world," a text Heidegger cites in *Wozu Dichter?* [*What are Poets For?*]. Heidegger reminds us that this poem is drawn from Rilke's beautifully moving *Stundenbuch*, *The Book of Hours*. These are from *The Book of Pilgrimage*, written just after the end of Rilke's love affair with Lou Salomé.⁴ Heidegger's text dates from 1946, and I have suggested that it would have mattered for Anders just owing to this engagement, as Heidegger quotes Rilke, were that not enough, as writing/citing: "*Gesang ist Dasein.*"⁵

I have twice cited verses Heidegger quotes—and this overlaps, despite manifest animosity, with Adorno's reflections on "humanity" already cited earlier—dedicated to what Rilke in a letter calls the "vibrations of money."⁶ Today we know Heidegger's references to the "open" in part owing to readings transmitted via Jacques Derrida and Giorgio Agamben, but fewer thinkers have drawn attention to Rilke's "kings" and the larger portion of extant scholarship focuses on the references to the child, valuable and important as these are in a world captivated by the nostalgia for the child, just to recall Simone de Beauvoir's chapter on "Childhood" in *The Second Sex*,⁷ or Gary Shapiro's reflections on Nietzsche's world-child, among others reading Rilke, perhaps to peak in the opening poem-film sequence echoing the filmic montages of a more elegiac era, Wim Wenders' *Der Himmel über Berlin*, Peter Handke's *Lied vom Kindsein*, "*Als das Kind Kind war* [When the child was a child]," in *Wings of Desire* (1987).⁸

Thinking of "what becomes questionable along with the thingness of things," at issue is "human material" for Heidegger who thereby observes that the "menace" of technology may not be ascribed to growing "Americanism" inasmuch as it had already ensconced its danger from some time ago, cites at length Rilke's letter of March 1, 1912, from the height of European culture, sugar Gothic and all, writing from Duino itself:

> The world shrinks into itself; for things, for their part, likewise do the same, shifting their existence [*Existenz*] ever more into the vibrations of money [*die Vibration des Geldes*], developing there for themselves a sort of ghostliness [*eine Art Geistigkeit*], which even now already surpasses their graspable reality [*greifbare Realität*].⁹

At issue is a focus on the thing *qua* thing, for Heidegger, materiality *per se*, including the human as material, and thus gold as such, when, as Rilke writes, "money was still gold, still metal, a beautiful thing [*eine schöne Sache*], the handiest [*die handlichste*], most understandable thing of all."[10]

Faced with scarcity, for those who understand economics, one has two choices. One can either modify one's course, embrace Schumacher style "small is beautiful" economics, adopt local resources for local needs, and so on, or, and this is the going alternative, one can eliminate other interests in the older and time-honoured tradition of aggression, intervention, direct action. If we want what others have, or if we want others to refrain from impeding our access to what we want, as Plato points out at the start of his *Republic*, we shall find that we are at war. But war can be conducted in many ways; thus, the aforementioned litany of transgressions against the natural world, including the 'harvesting' of human material, is a war.

War can also be waged by weather manipulation as such "weather wars" have been practised for some time. Peter Sloterdijk reminds us that this begins with the Battle of Ypres, Sloterdijk gives us the date, and in Dresden, and of course in Korea and Vietnam. Such weather wars via atmosphere, and Sloterdijk has texts on that as well, are best fought in the background, invisibly, via "acts of god," and the kind of thing insurance declines to cover, ongoing to the present day including chemtrail atmospheric contaminants, aluminium and other metals, theragrippers and nanomaterials and microplastics, bioweapons of various kinds, quantum dots and other tracer IDs—this is in part what may be found in such chemtrail contaminants—all with a long track record of scientific publications, government documents, press releases coupled with official denials, and general incredulity even among academics who suppose that "they" would never do such a thing, quite in the face of government publications attesting to the same plan to spray the world with poison just to block the sun, of course necessary, think global warming, in order to modify the weather, geoengineer the earth.

Despite this, one imagines that "terraforming," to the extent that we possess such techniques, would, despite the name, be deployed on other planets, supposing "we" ever get there rather than here on good old Terra.

And then there are bioweapons.

The United States has, this is a matter of public record, been working on bioweapons since the Second World War quite along with other combatant nations. It has used these as well, to certain effect, also a matter of record, in its wars since the 1950s and so on. Indeed, there are scholars who argue that two centuries ago, smallpox was deployed as a vector of deliberate depopulation in the United States. This is disputed and reargued in academic flurries that keep busy across the disciplines, following mainstream schools that flourish—"normal science" emphasis on the normativizing force of the same—from time to time, before fading. And a similar case may be made for introducing alcohol, devastating, poisonous in direct effect, to peoples who previously had no exposure to grain alcohol. Or refined (white) sugar, a variant/version of the same. But these are subtle arguments, and we are addicted to both so it is hard if not impossible to see these things as "poisons." Surely not. And if so, ubiquitous, and slow in any case, at least to us, acclimated as we are to both alcohol and white sugar. To cite Wilhelm Busch, the German physician's *Die fromme Helene*:

Es ist ein Brauch von alters her, Wer Sorgen hat, hat auch Likör!
[It is a custom from days of yore: Those with troubles, also have liquor (i.e., and thus the joke: "the cure")]. (Ch. 16)

The beauty of disease as an agent in war is that the agent, this is beautiful for criminal logic, is already at hand as scapegoat and it works invisibly. It works via the immunity of some, which would be the European settlers for whom smallpox was not a problem and the lack of immunity on the part of native peoples who die so catastrophically, so dramatically, that lands are cleared for expansion, in effect, *from sea to shining sea*. Afterwards, the entire academic community will occupy itself with denial, refusing the suggestion that the contamination was deliberate, because there are, after all, other hypotheses. Act of god, divine right, white man's burden, all that.

Genetic modification is part of this (GMOs are made using viruses), and what better vector than the common cold, that is, "Coronavirus," if one means to develop a viral pathogen to be used on an enemy population? Alternately, relatedly, vaccination is part of this because new, viral or mRNA nanotech is part of the novel vaccine. What better way to introduce that material into a population than by injection? In the olden days, one gave blankets away in cold winters. Today, we have vaccines. Thus, very directly, one can deploy such a means of population control on the population as a whole for the sake of reducing population. And primed to work in tandem with the virus not in vitro/in vivo (the distinction dissolves in this case), the virus in the wild, the fatal effects take months or years. And such outcomes, of whatever various kind, varying from person to person, underlying condition to underlying condition, is a 'side effect.'

The best way to kill, Plato tells us this, will be via those whose mission is otherwise dedicated to saving lives. For the sake of health, for your own good.

Thus, my realization from the end of last year, a lifetime ago: we need, I maintain, to change what we are doing to the world, the rapacious way we live, fishing, hunting, mining, logging. This kind of change is unlikely as all the powers that be are aligned against this.

What is however likely, nightmare or not, as there is massive support for this, popular as socially engineered, psychologically induced, and at the highest governmental and NGO levels, all "we" need to do is to reduce the number of people living on the earth who are doing these things to the world.

Do that and everything changes. The tragedy of the commons becomes a paradise if only you reduce the numbers of those with access to that same commons. Then, the ones still standing after the reset—those will be the wealthy, the true "kings of the world"—may then do whatever they like for as long as they like.

Counting Industrial Revolutions

I have called attention to Anders' habit of listing industrial, technological revolutions by number. The "technological" in question Anders borrows from Walter Benjamin but not less from the Heidegger, who antedates Benjamin on the question of the work of art. Heidegger looks to origins, Benjamin and Anders look to the transforms, the more Husserlian phenomenological variations on the work of art as it changes, both the work and, as Anders emphasizes, the "consumer" of the cultural work of art, the manufactured, delivered, imposed work of art in the age of the mechanization, industrialization; today we do not wish to simply say "technologization" but also "digitalization," "mediatization" in our own accelerationist intensification of Benjamin's original title quite as Anders himself accelerates the point and then revisits the same acceleration. Nor does it stop.

Anders writes in advance of the war on terror, of climate and weather control, and of the current pandemic. One might think that Anders would predate all conspiracy theories (as these are of recent vintage and it hardly helps to call Malthus a conspiracy theorist) but in fact the notion was deployed contra Anders as thinker of Prometheanism, so Christopher Müller argues, a subtle point repeated by Müller together with Christian Dries in their introduction to their jointly edited 2019 issue of *Thesis Eleven* dedicated to Anders and entitled *Inverted Utopias*.[11] And inasmuch as Anders is a thinker of broadly apocalyptic thought, the "inversion" that he attributes to Beckett in his central chapter, "Being Without Time," seems consummate.[12] The trouble is that it is difficult to parse such concepts: How can we think the unthinkable?

Prometheus, an Aeschylean figure, articulates an uncanny titanism, as Aeschylus tells us that mortal beings owe the sum of their technical arts to Prometheus only to add, as counterpoint it resounds in Seneca, that craft fails in the face of necessity: ανάγκη [*Prometheus Bound*, 511]. The contrast as it is echoed in Hölderlin's reflections on nature and art (the Greek is τέχνη) corresponds to a Nietzschean constellation of shame. Thus, and quite with respect to the notion of sin, good and evil, Nietzsche commissions a woodcut of a liberated Prometheus as frontispiece for his first book *The Birth of Tragedy*, in which Nietzsche contrasts Semitic and Aryan forms of sin or transgression—and shame, the same shame Nietzsche foregrounds in several books of *The Gay Science*. Technology, in its Aeschylean expression, corresponds to the titan's gift to us, as creatures of lightning and blood and titanic ash, in language taken from the alternate title of Mary Shelley's 1818 novel, *A Modern Prometheus*.[13] Thus, to this day technology is the engine of our ambition and fraught signifier of the future. Technology promises limitless possibility, even to the extent, as Anders wrote in a parallel with Adorno's reflections on breath in *Minima Moralia*, of "shaming us" by comparative contrast, leaving us to dream of a posthuman condition beyond the human just to the extent that we feel inadequate by comparison with the orderly *Ge-Stell* of the tool, any tool, the inveigled array that is part and parcel of *Zeug*, as Heidegger writes in *Being and Time*.

Anders' "Prometheus effect"[14] has long since been transmogrified into transhumanism and the cargo-cult aspirations of the same.[15] Like Adorno, who raised the question of our complicity in genocide,[16] Anders raises the question of our complicity in the ongoing violence of nuclear power plants as these are, quite as the political theorist Langdon Winner argues, and as we may rephrase with reference to Clausewitz, the continuation of bombs by other means.[17]

We are sure that Heidegger, when asked about technology, erred in his claim that only a god can save us. Today, we suppose ourselves in need of no god but and only the right tech, the right entrepreneur, cue Elon Musk or whoever the next guy might be. If

Anders, via Goethe, had already highlighted the problem of geoengineering with his discussion of the sorcerer's apprentice and his broomsticks, Sloterdijk clarifies the problem as human waste, spewn forth, into the wind, the ocean, everywhere:

> Nowadays what human beings meet in the weather are their own expectorations— become atmospherically objective—of their own industrial-chemotechnical, militaristic, locomotive, and tourist activities.[18]

Weather Talk: How to Do Things with Clouds

Perhaps better than most philosophers, even better arguably than Baudrillard, Sloterdijk knows how the names we give to things plays itself out in the media. And Nietzsche had already told us that everything depends on what things are called. In this way, the back story to all "fake news" concerns how what is "fit to print" gets into print and how what is silenced is silenced. Think of Harvey Weinstein over the years but think too of all the Harveys there have been in the entertainment industry, in academia, anywhere there is power, countless, unmentioned scandals.

If Latour has for some time been telling us that we have never been modern, his recent reflections concern the weather, *Facing Gaia*.[19] By contrast, Sloterdijk gives us the birthday of our real, all-too-real modernity, adumbrated via atmospheric expectorations, with *The Battle of Ypres*, including the why and the how, gas warfare in the First World War, born on:

> April 22, 1915, when a specially formed German "gas regiment" launched the first, large-scale operation against French-Canadian troops in the northern Ypres Salient using chlorine gas as their means of combat. (TA, 10)

This is not the place for it, but one might with profit return to Sloterdijk's rather detailed discussions of the colours—yellow— of destroyed lungs, the foaming of breath, and the ongoing respiratory damages effected from the poisoned air as it affected the lungs as this was also attributed to the media descriptions of Coronavirus.

For his part, Sloterdijk carries his question through two world wars, including the firebombing of Dresden but also deployment of (and the denial of) weather control in Vietnam as Anders likewise details the same weather control *and* the nuclear attacks on Hiroshima and Nagasaki. This kind of warfare has yet to cease, if the current Covid-19 crisis has worked to make it plain that weaponization may be the least of it. Indeed, as Steve Fuller argues, what is essential is "controlling"; this is the language of the social sciences to be sure, the "narrative."[20] Thus, trumping many political philosophy media analyses, Fuller assesses the importance of what things are called, to use the terms already introduced with reference to Anders, Heidegger, Rilke, and Nietzsche, in their more conventional branding practice. Fuller is interested in assessing the long game as it were, and in ensuring a certain militaristic process for research funding and although he plays both rhetorical sides in the process, he is not on the side of Anders or Agamben much less Adorno.

On the terms of Sloterdijk's own "spherical" analysis, "terror from the air" is the escalation of modern warfare as wars of action-at-a-distance, "the *de facto* norm for 'air battles,'" as "one-sided, irreciprocable air strikes" (TA, 51). Today's ongoing wars, whether declared and not, are "ex-plicated" in this way at a distance. Social distancing consequent upon the political measures undertaken in the wake of Covid-19 is to this extent only the most recent articulation. Sloterdijk is one step beyond the rhetorical question concerning wars that do or do not "take place," as Jean Baudrillard put it[21]: past, present, and future. In this way, Sloterdijk frames his discussion of the "militarization of weather," variously, in the third of his three volume *Spheres*: *Schäume* and *Luftbeben*, "Airquake," and *Terror from the Air*.[22] To this must be added the militarization of the air we breathe in proximity to those around us, reinhaling, to vary Sloterdijk's point for today, our "own exhalate," whilst creating the other as a mutually and reciprocally deadly danger to collective health. The benefits for totalitarian regimes remain unlimited.

Sloterdijk's invocation of Taubes and not less Gnosticism, especially with reference to Heidegger and Adorno reminds us of Marinetti's celebration of what the Italian Futurist calls the "beauty" of gasmasks, made still more clearly with Sloterdijk's discussion of the aesthetics of yellow foam characteristic of fatal lung damage.

These are/should be alarming topics—quite as awful as Anders' references to those US soldiers who returned with gold teeth extracted from the jawbones of Japanese soldiers—and Sloterdijk takes his points a little further than we are accustomed to seeing in professors of philosophy who are usually fast students of convention. To tell the story of war in the age of its technological reproduction, its escalation, as a "force multiplier" (to quote the Pentagon as Sloterdijk does),[23] Sloterdijk explains the technique involved at Ypres at some visceral length—thus, again, the colour of the foam that so uncannily maps on to the damage done to lungs by today's Covid-19, but, more technically, he goes on to describe the firebombing of Dresden as a "blast furnace effect":

> the attackers aimed to generate a fiery central vacuum by dropping a high concentration of incendiary bombs, to produce a hurricane-like suction effect—a so-called firestorm. (TA, 54)

The result of these "surgical" bombing effects was the production of

> a special atmosphere capable of burning, carbonizing, desiccating, and asphyxiating at least 35,000 people in the space of one night [which] constituted a radical innovation in the domain of rapid mass killings. (TA, 66)

Hiroshima and Nagasaki constitute force "multipliers" of the Dresden tactics deployed by Winston Churchill and Bomber Harris.

Here there is not just (to use a gaming metaphor) a "levelling up" (mere escalation) but *ex-plication*. Articulating *Ge-Stell*, Sloterdijk's explication corresponds to "the scandal of Being taken to its dark limits" (TA, 64). Here it is what we do not say that is key as all of this takes place against a backdrop of censorship. Sloterdijk's making "radioactivity explicit" contrasts with the expressly inexplicit: occupation censorship, as he explained, entailed that the very mention of even the deployment of the bombs would be denied in Japan until 1952. And if one can deny an atom bomb, as one did do, for some seven years after a fact trumpeted in lock step on the front page of every newspaper in the United States, to deny chemtrails overhead is a piece of proverbial cake. The same holds for the listing of numbers, as Fuller points out along with Agamben another analysts speaking of the death rates and the assembly of such rates as reported for Covid-19.

Such silencing accompanies explication as before; thus, this is a standard protocol—nor to be sure, do we, the consumers, worry about microwaves or about cell phone radiation and even 5G continues to concern few (despite the abundance of scientific and medical concerns) as consumers happily munch away at genetically crisped apples, and salmon, and apart from social media indignation are quite content to take their energy needs from pipelines and fracking or windfarms and solar panels as these fit into the same electrical grid provided by the same energy utilities.[24]

In consequence, we have a "radically new level of latency" (TA, 64). As Sloterdijk explains this latency, he means the term with uncomfortable precision because he is talking a metaphor for contamination, radiation, inflammation, sickness, and poisoning:

> The long concealed, the unknown, the unconscious, the never-known, the never-noticed and imperceptible, were forthwith forced to the level of the manifest becoming indirectly noticeable in the form of peeling skin and ulcers, as if they were the result of an invisible fire. (TA, 64)

Sloterdijk's "atmospheric explication," includes current weather manipulation (and it is routine for academics, especially academics, to deny as "conspiracies" aka "fake news" anything but the official story on anything from JFK to 9/11, or indeed the very idea of weather control, including HAARP).

Yet my speaking of chemtrails, my relatively frequent mention of 5G (mainstream academics do not mention this at all, even when writing about such thinkers as Anders), not to mention weather control, violates all standards of academic complicity. Only Illich, a priest and provocateur of another generation, himself born in the city in which Anders would die, could remind his listeners, very subtly of the original meaning of conspiracy: *Das Geschenk der Conspiratio*, the gift of conspiracy, a lecture he gave on being awarded the Bremen peace prize in 1998.[25] Illich begins his lecture by remembering other commemorative events at the same locus (in Bremen), disparate interests, research projects, friends from across the world, not all of whom would get along with one another after his death, such that even this tenuous collectivity would vanish, this is the way of the same spirit with which Illich kept his faith.

Very differently than Sloterdijk, speaking of air, of gas and vapours, Illich confesses that he speaks of "atmosphere," lacking a better option, "*faute de mieux*" and this 'spirit' is part of his notion of gratuitous or unearned grace or *Umsonstigkeit*:

> In Greek, the word is used for the emanation of a star, or for the constellation that governs a place; alchemists adopted it to speak of the layers around our planet.

Maurice Blondel reflects its much later French usage for *bouquet des esprits*, the scent those present contribute to a meeting. I use the word for something frail and often discounted, the air that weaves and wafts and evokes memories, like those attached to the Burgundy long after the bottle has been emptied.[26]

Illich's theme, as elaborated from *conspiratio*, is "the commingling of breaths,"[27] and in an era that has seen the closing of churches and the distancing of community members, including close family, Illich's emphasis on the complex history of the Christian greeting, the kiss of peace that is part of Catholic ritual[28]—it is this point that our current horror of death, care of and for the self, would seem to alienate. Contra the contract as such, as it were, Illich argues:

> Community in our European tradition is not the outcome of an act of authoritative foundation, nor a gift from nature or its gods, nor the result of management, planning, and design, but the consequence of a conspiracy, a deliberate, mutual, somatic, and gratuitous gift to one another. The prototype of that conspiracy lies in the celebration of the early Christian liturgy in which, no matter their origin, men and women, Greeks and Jews, slaves and citizens, engender a physical reality that transcends them. The shared breath, the *con-spiratio*, is peace, understood as the community that arises from it.[29]

If Illich is correct, the wearing of masks cannot but have political consequences, as the shortest path to ending community.

Just as much as today's academic denounces those who question the received view on Coronavirus, a faith which turns a complex shifting of official doctrine into the social media equivalent of Keystone Cops worthy turns and reverses, it is also typical to mock the notion of weather manipulation, chemtrails, 5G, what have you, as if there were or had to be assumed as unquestionable article of faith ("Faith"?—the very term should get some attention in this context) that the government could not, would not be involved in any such thing. Claims of government incompetence that vitiate intentionality loom large in such accounts.

Like HIV, Coronavirus has a standard story that must be observed. And gas warfare is something that Nazis do, thus deploying such weapons of mass destruction is for the likes of villains such as Saddam Hussein, and if Iraq, in truth, never had such weapons, this is the point of Fuller's rhetoric of "post-truth," as cited earlier, we are not deterred in assuming it must be true of Syria or some other foe *du jour*.

The language of concession compounds any issue of discussion. If scholars rarely invoke weather *weaponization*, it is also uncommon to speak of 5G, although some medical doctors have written about this, or else of "inventing" HIV (although virologists and molecular biologists like Luc Montagnier and Peter Duesberg do so, with career-killing consequences, consider the different cases of Franz Moewus, who crossed or otherwise somehow annoyed his Cold Spring co-researchers or Bruno Latour when he disturbed a certain status quo in writing about the Salk laboratory).[30]

If the weather comes up at all, it is named geoengineering as if we were in the middle of a sci-fi story and could "terraform" the world overnight in the hoary fashion of a *Star Trek* film, rather than doing the geoengineering we have always been doing (ordinary anthropocene slash and burn that is gardening, that is palm oil agriculture) and certainly as opposed to the explicit military application of such interventions. Hence, in the case of 9/11, the process theologian, who better to speak truth to power, David Ray Griffin did write about this, a range of books, some thirteen of these apparently. Here I cite just one: *The New Pearl Harbor Revisited: 9/11, the Cover-Up, and the Exposé*.[31]

If, in philosophy of science, one simply fails to cite outlier views, the practice works across the board, and the first chapter of this book began by detailing its effects in scholarship as a whole. Thus, at the beginning, I noted Don Ihde's pioneering tactic of silencing, a tactic by no means limited to Anders, used to great advantage in philosophy of technology and philosophy of science and overall. Thus, one may write essay after essay about Anders and related themes, one may raise questions about otherwise "damned" topics like the work of Ludwik Fleck and others in philosophy of science, and with respect to the sociology of models, social science, what have you. But this can be so much talking to the wind as one's colleagues do not read what one writes, and once they realize that one has written such things, they quickly cease to engage.

The tactic has a name in German, academically, and it is what happened to Nietzsche following his first book on tragedy, a silencing that is still in effect to this day in philosophy more generally, in Nietzsche studies specifically, and above all in Classics: *Todtschweigerei*. And all of this is done with a good conscience: no one thinks that their practices are unfair or indeed that they are silencing anyone, even as they reject applicants for academic positions and grants, refuse to publish their work, omit to mention them in their own research, and above all fail to include them in official conferences and speaker's programmes. The violent resurgence of mobbing and claims of pseudoscience and suppressed research in the current and ongoing Coronavirus pandemic is only the latest instantiation.

I have repeatedly noted that it is only Sloterdijk who cares to mention weather manipulation. And patently pointing to such a thing is problematic, given that, as Sloterdijk writes,

> Built-in to the premises of weather weapons research is a stable moral asymmetry between US acts of warfare and every potential act of warfare: under no other circumstances could there be any way to justify investing public funds in the construction of a technologically asymmetrical weapon of an evidently terrorist nature. Democratically legitimizing atmoterrorism in its advanced form requires a concept of the enemy that gives the use of means for the enemy's special ionospheric treatment an air of plausibility. (TA, 51)

Sloterdijk's point concerns HAARP, citing, likewise as already noted, the US Department of Defense 1996 publication, "Weather as a Force Multiplier: Owning the Weather in 2025" (TA, 64), naming the 1990s a decade of military escalation not only previously unthinkable but, "largely unbeknownst to the public, in the possibilities of atmoterrorist intervention" (TA, 64), including the logical implications of the use of drone warfare under Obama (and normalized as and *qua* propaganda, as already noted by way of a Hollywood movie, *Eye in the Sky* [2015]), which are quite "far from providing the antidote for terrorist practices, the stratification of weaponry works toward their systematization" (TA, 53). Indeed, as Agamben and Adorno remind us when it comes to the culture industry, montage is everything. Thus, in the aforementioned film, neither Helen Mirren nor Alan Rickman met the other during the filming. They seem to be in the same production, but they communicate virtually, just as the film itself was made, at a distance. Today's Zoom instruction replicates the illusion of communication.

It is in this "framed" sense that Baudrillard writes of the Gulf War. We see what we are shown, and since Hitler the movie industry has played several roles in various war efforts, all of them on the government side, one or the other, remember Steve Fuller's instructions on writing about pandemics for nationalist fun and glory. As Garrett Stewart writes as apothegmatic banality in his "Preface: Returns of Theory" to his *Closed Circuits: Screening Narrative Surveillance*, "All montage is espionage."[32]

Remaining on theme, and note that the Covid-19 virus still has us concerned with air and not less as Agamben has observed, with the conditions (if not the declared fact) of war, Sloterdijk observes:

> The fact that the dominant weapons systems since World War II, and particularly in post-1945 US war interventions, are those of the air force, merely betokens the state-terrorist habitus and the ecologization of warfare. (TA, 53)

Explaining that

> Air-design is the technological response to the phenomenological insight that human being-in-the-world is always and without exception present as a modification of "being-in-the-air." (TA, 93)

Sloterdijk highlights the difference between phenomenologists who "explicate human dwelling in its global atmospheric conditions" and Irigaray's material insight "that Heidegger's concept of *Lichtung* be bracketed and replaced by a meditation on air" (TA, 93). For her part, focused on breath, Irigaray reminds us that:

> It is not light that creates the clearing but light comes about only in virtue of the transparent levity of air. Light presupposes air.[33]

Confined to our homes for months, confined to our masks, rebreathing our own exhalations, forbidden fresh air, we could do well to read and reread Irigaray. And perhaps recall the ancient breath techniques, as she did, of yoga.

At stake here is the state of what Heidegger called "the question,"[34] as questioning is transformed *as a possibility* in the wake of technology. If we, as I argue, as Adorno and Anders articulated the basis for critical theory to recall this possibility, we are still trying to catch up to the intersection in thinking between Heidegger and Adorno, and even Heidegger and Anders, as well as Adorno and Anders, just to begin to be able to explicate Sloterdijk's "highly explicit procedures." Thinking Being, we can forget the "stars down to earth" such that, for Sloterdijk, "any thinking that stays phenomenological for too long turns into an internal water colour which in the best of cases fades in to non-technical contemplation" (TA, 934).

The "Modern Prometheus"

The allure of the titan's gift to us, creatures of lightning and blood and the titan's own vaporized ash, is the alternate title of Mary Shelley's *Frankenstein*.[35] The title, the reference, is part of Langdon Winner's concerns in his book, twice written on the same theme, first with a clear reference to Shelley's creature, *Autonomous Technology: Technics-out-of-Control as a Theme in Political Thought*, published in 1978 to virtually no response among political theorists and an if-possible-yet-more-becalmed reception among philosophers of technology. The California-born Winner updated the argument and refined the focus, to little resonance in a second book published with a more popular title ten years later, *The Whale and the Reactor*, now in its second edition.

Shelley's Dr. Frankenstein, her modern Prometheus fashioned a creature of cobbled together body parts, medical detritus, a creature, as a result, of "proud" flesh, insulted, inflamed: in stasis between necrotized tissue and still viable, still functioning organs. This condition of necrotization and inflammation is the condition of any transplant and the drugs one takes to prevent rejection of the organ are as much to prevent the body's reaction to decay in today's medical innovations, transplanted partly decayed hearts and kidneys, lungs and livers, from cadavers, human and not (xenotransplantation),[36] but above all skin, even faces,[37] and limbs. It is significant that, not unlike Shelley's sombre nineteenth-century vision, Ridley Scott's 1982 *Bladerunner* shows us a dark world of barely integrated cyborgs, filthy urban landscapes, complete with environmental catastrophe. Nanotech is invisible by contrast but the effects are just as real.

Even if we have not read Adorno, we *live* the culture industry: the consummate *Ge-Stell* of digital media including the all-encompassing Imaginary that is the screen. But even scholars focused on technology and sociology of knowledge, conversant with digital media, seem unaware of the rather more prosaic bubble in which we live—and on the terms of which we publish. It is not possible to buy anything one might desire in the supermarket market: rather, it is only possible to buy what is available there. Increasingly, after months of lockdown, consumers have adjusted their demands to suit what is available, when it is available. And count themselves lucky.

Geoforming, Masks, and Us

Adorno argued a painfully key bourgeois point in an aphorism entitled *Antithesis*, set into the contradiction that it is "to stand aloof" as he begins by writing for the very same reasons that Anders points to his own reflections on Kafka and in general. For Adorno, in *Minima Moralia*, the danger is that of the elitist in an impossible circumstance as one thereby "runs the risk of believing himself better than others and misusing his critique of society as an ideology for his private interest." The trouble is the dialectical fiasco that it is to write an autocritical sociology or phenomenological anthropology of one's own culture from within that culture: "Against such awareness, however, pulls the momentum of the bourgeois within him. The detached observer is as much entangled as the active participant."[38] Thus:

> The notion that every single person considers themselves better in their particular interest than all others, is as long-standing a piece of bourgeois ideology as the overestimation of others as higher than oneself, just because they are the community of all customers. Since the old bourgeois class has abdicated, both lead their afterlife in the Spirit [Geist] of intellectuals, who are at the same time the last enemies of the bourgeois, and the last bourgeois. By allowing themselves to still think at all vià-vis the naked reproduction of existence, they behave as the privileged; by leaving things in thought, they declare the nullity of their privilege.[39]

Uncannily, as we began this closing chapter by noting, it is Adorno's famous line on shame, the shame that connects his thinking with Anders, that is the shame of "still having air to breathe."[40] For those who still breathe freely, this holds more than ever.

We dedicate our minds, ignoring the possibilities that those minds can be subject to strictures of "control" beyond the "culture industry" of which both Heidegger and Adorno speak or else reviewing the use of music as a different kind of military "air-conditioning," in Kittler's words. Sloterdijk takes the point to reflect that

> infrasonic waves affect not only inorganic material but also living organisms—in particular the human brain, which operates in these low frequency zones—HAARP includes the prospect of developing a quasi-neurotelepathic weapon capable of destabilizing the human population with long-distance attacks on their cerebral functions. (TA, 68)

It is time to bring Heidegger and Anders/Adorno together, highlighting their common focus on phenomenology towards a critique of technological reason. Talking weather, daring to question events like "bomb cyclones," or everyday things like chemtrails and so on runs the risk of speaking of truth to power but also, and this is worse for academics, invites mockery, as what Sloterdijk calls "a form of incitement to blasphemy."

As our insurance policies spell it out for us, rather in the way that todays' emergency vaccine manufacturers are indemnified against any untoward side effects of their new vaccines: things like weather are not covered; these are "acts of God".[41] As Sloterdijk writes, "the principle of the weather is like that of birth and death: it comes from God and from Him alone" (TA, 88). Thus, we prefer to talk about climate change rather than weather manipulation.

And there is such a thing as climate change but, like Pogo looking for the enemy, it is us. Better said, we are the immediate cause and the intended target of the efforts underway with weather manipulation to contain and correct it. And if these efforts succeed, humans will die along with those beings humans already kill. If Anders, via Goethe, had already highlighted the problem of geoengineering with his discussion of the sorcerer's apprentice and his broomsticks, today's effective equivalent is nanotechnology. Sloterdijk clarifies that and buried in the listing of our "industrial-chemotechnical, militaristic, locomotive, and tourist activities," it is important to highlight the "militaristic." Thus toward the end of *Terror from the Air*, Sloterdijk explains that:

> In the age of atmospheric toxins, strategies, and hidden agendas all such quasi-religious consenting to place one's trust in one's primary surroundings—be it nature, the cosmos, creation, homeland, situation, etc.—takes on the guise of an invitation to self-harm. Advancing explication not only forces a semantic change in the meaning of naivety, it means that it becomes increasingly in-your-face, and even objectionable; the naïve, nowadays, is that which encourages sleepwalking in the midst of present danger. (TA, 108–109)

Beyond Heidegger, and Sloterdijk, we are still in the wake of modern technology and all its force multiplying effects, and we still need to ask after questioning.

But even there, we have a problem. We are currently living under lockdown, currently enduring mask mandates, enforced or constrained vaccination schemes, as if the science on vaccination had been settled as it has not and as only fiat could make it so. The silence of intellectuals on these matters is painful and telling, and for this reason alone, had we no other reason, and we have many, Anders is worth reading.

"Cosmic Parvenus"

Designating the citizens of his time—and I do not think he would exclude the citizens of 2020/2021—as "cosmic parvenus, usurpers of the apocalypse," Anders argues that quite in the face of the fact that we ourselves have caused this (think climate change, think susceptibility to social media manipulation), "sleepwalking," we exempt ourselves from fault in advance, excusing our inaction.

To this we should add what Virilio described as our "silence," which in the current crisis is endemic. There is complicity on the part of the average academic regarding social distancing and the oppression that is the wearing of masks, the blocking of breath, especially for children.[42]

Thus the new "Fat Man" (to take over the military codename for the bomb dropped on Nagasaki on August 9, 1945) is a metonymic signifier for an ultimate "trolley problem," as analytic philosophers fond of calculating ethical consequences name such things, considering the death of one of unlovely girth, deliberately thrown from a bridge, supposed thereby, according to projected calculations, so to offset the imminent or threatened death of a certain number of others (five or so, more or less) on a given track. Anders contends that even apart from such stochastic calculations, we imagine a constant state of vigilance, the go-to description for nuclear preparedness as entailing a constant state of violence. The parallel with the current Coronavirus crisis is obvious.

Thus, on the level of global endangerment, it is argued that climate change demands active or overt geoengineering, rather than the covert kind already in play, and thus insists that the only way to fight the poison we manufacture, what Sloterdijk rightly called our "expectorations" mentioned earlier, has to be not with *ceasing* our drilling, our fracking, the burning of forests, the mining of rare earths, the raising and killing of animals on an industrial scale, the trawling and seizing, repeatedly and massively, of all manner of ocean creatures on factory processing ships, and thereby the devastation of the world, land and air and sea, but and only to add yet more poison, that is the idea of geoengineering via atmospheric dispersants, coupled with radio frequencies, scalar transmitters, and so on.

To quote this again, Anders' old-fashioned and sardonic observation: "the devil has moved into a new apartment" (AM II, 410) seems bound to remain with us even as we dismiss it—we do not "believe" in "such things."[43] The dark star that followed Anders seems to be reflected in this notion, more than banal, of a new locus of evil, corresponding to our complacency with vaccine mandates (note how much more efficient these are than mere chemtrail dispersants could ever he), along with passports to ensure compliance, our support of nuclear technology (bombs and energy), our tolerance of not only weather manipulations, human-animal "chimbrids" (all as noted earlier), but also drone and other attacks; all this would seem to ensure that we must be counted, all of us, as so many new "sons of Eichmann," implicated by our inaction, even if we name that non-action by the fetish name: nonviolence, peaceful resistance, "sleepwalking."

The term "sleepwalking" is Sloterdijk's as cited above. Anders himself quotes Max Scheler who, unlike most modern thinkers (including Heidegger who explicitly denies a connection between technology and deviltry), emphasized that he maintained his 'belief' in the devil. Speaking of the devil's "new apartment," Anders' point is that the apartment is the one next door, which means that even as we seek to "smoke him out," as Anders writes, the devil is easier to miss than ever before.

In his Kafka book, and this is unsettling, Anders reminds us that it is imperative to recognize that it is not only war that can continue under another name. Thus, he writes, quite with Nietzsche in mind, that

> it is the particular stage of disbelief reached at any given time that is itself interpreted in a religious light. Nature, though neither created by God nor inhabited by any specifically divine presence, still awakens diffusely religious sentiments. ... or else the "twilight of the gods" has apocalyptic and therefore religious significance; or man "heroically confronting the ultimate nothingness" becomes the object of a mystique—the belief in a superman; or finally, unable to find God, the poet sings the mysterious lament of Rilke's poems in which the singing itself becomes an act of praise for the divine absence.[44]

This, as Anders concluded in the second volume of human antiquatedness, outdatedness, "is our doom" (AM II, 471). For Anders, who also had recourse to the figure of Benjamin's angel of history in his notion of "*historians turned toward the future*," everything would turn on whether we might be capable of a newly phenomenological hermeneutics, able to descry "in the machines of today the humanity coined by" the same, not to be part of this new future or to hack it, but for the sake of "interruption."

The problem is not simply climate change, but geoengineering, cloud seeding and 5G and soon, 6G, and so on, 'multiplied' to use the military expression of choice. The problem is the wearing of masks, the complicity with lockdown regimes and self-imposed quarantines, the susceptibility to social media manipulation. The problem is the shunning of the old and the ill by agreeing to prohibitions against hospital visits and care home visits, the problem is agreeing to zoom one's lectures, agreeing to a simulacrum of life, to virtually, as it were, *teach, love, theorize, socialize* online. But that is nowhere real or true, and this is not abrogated by the fact that it is all we (take ourselves to) have. This is the *softest* war against the all, that is against, in Adorno's words, they, the people, that has ever been undertaken, and it is soft because it can be: there is no resistance.

Under his dark star, Anders himself could not have been particularly sanguine about the success of any undertaking that might seek to challenge the consequence of the new social and world order under which we live and have lived thanks to modern media technology from radio to the internet. Nevertheless, Anders argued, and we should note his clarity and philosophical rigor, given that "impossibility has not been proven, it is morally impossible to renounce the attempt."

Notes

Introduction

1 *The New Shorter Oxford English Dictionary, Volume 2, N-Z.* ed. Lesley Brown (Oxford: Clarendon Press, 1993 [1973]), 2713.
2 Ernst Schraube attests to this breadth in his essay, "'Torturing Things Until They Confess,'" *Science as Culture*, 14 (2005): 77–85, underlining the relevance (and the challenge) of "a huge oeuvre that constitutes an extensive critique of technology: one meter of writings, more than 20 books and 100 articles, and 70% of his work yet to be published" (78). Yet the reasons are not simply subjective irrepressibility but what I am here noting as a negative constellation, that is a kind of black star, and consequent readerly "scotosis," just to use a technical, scholastic term. As Schraube goes on to observe, "Few other intellectuals share Günther Anders' way of combining theory and practice" (Ibid.). Just this combination and style requires attention to phenomenology, Husserl style, classically regarded, and Heidegger style, hermeneutically conceived.
3 Anders, "Sein ohne Zeit," AM I, 222.
4 The term "anthropological phenomenology" can refer to anthropology or ethnography per se and is still relevant in connection with Anders in contemporary research, cf. Ehgartner et al., "On the Obsolescence of Human Beings in Sustainable Development." *Global Discourse. An Interdisciplinary Journal of Current Affairs and Applied Contemporary Thought*, 7, no. 1: *After Sustainability—What?* (2017): 66–83. See here, among others, van Dijk, *Anthropology in the Age of Technology* (Amsterdam: Rodopi, 2000 [1998]). More broadly and with specific reference to Adorno, see Dennis Johannssen's "Toward a Negative Anthropology." *Anthropology and Materialism*, 1 (2013): 1–14.
5 This term also appears as the title of a collection edited by Helmut Reinalter and Andreas Oberprantacher, *Außenseiter der Philosophie* (Würzburg: Königishausen und Neumann, 2012). Predictably, the collection includes some authors who are, factically speaking and by contrast with Anders, quite mainstream, such as Montaigne, Pascal, William James, Franz Rosenzweig, and Hannah Arendt along with Simone Weil and Günther Anders. Christian Dries counts Hans Jonas among the excluded in his *Die Welt als Vernichtungslager. Eine kritische Theorie der Moderne im Anschluss an Günther Anders, Hannah Arendt und Hans Jonas* (Bielefeld: Transcript, 2012). See further in a related lineage, Édouard Jolly, *Étranger au monde: Essai sur la première philosophie de Günther Anders* (Paris: Classiques Garnier, 2019) as well as Adi Armon, "The Parochialism of Intellectual History: The Case of Günther Anders." *Leo Baeck Institute Yearbook*, 62 (2017): 225–41.
6 See for a discussion my several essays on Nietzsche, perhaps (most accessibly): Babich, "The Genealogy of Morals and Right Reading: On the Nietzschean Aphorism and the Art of the Polemic." *Nietzsche's On the Genealogy of Morals*, ed. Christa Davis Acampora (Lanham: Rowman & Littlefield, 2006), 171–90.
7 See, especially for its relevance for phenomenology, Theodor Adorno, *Negative Dialectics*, trans. E. B. Ashton (London: Routledge, 1973 [1966]).

8 Karl Marx and Friedrich Engels, "Manifesto of the Communist Party." *Collected Works, Volume 6 Marx and Engels, 1845–1848* (New York: International Publishers, 1976), 477–96.
9 Important sections of this text can be found together with valuable accompanying reflections in Müller's *Prometheanism: Technology, Digital Culture and Human Obsolescence* (Lanham: Rowman & Littlefield, 2016) in addition to chapters that were originally published contemporaneously in translation.
10 Marx and Engels, "Manifesto of the Communist Party," 487. Thus, it can be useful to cite the German original: "Die fortwährende Umwälzung der Produktion, die ununterbrochene Erschütterung aller gesellschaftlichen Zustände, die ewige Unsicherheit und Bewegung zeichnet die Bourgeoisepoche vor allen anderen aus. Alle festen eingerosteten Verhältnisse mit ihrem Gefolge von altehrwürdigen Vorstellungen und Anschauungen werden aufgelöst, alle neugebildeten veralten, ehe sie verknöchern können. Alles Ständische und Stehende verdampft, alles Heilige wird entweiht, und die Menschen sind endlich gezwungen, ihre Lebensstellung, ihre gegenseitigen Beziehungen mit nüchternen Augenanzusehen."
11 This is featured in Johann Wolfgang von Goethe's famous *Ausgabe letzter Hand. Band 1–40 in 20 Bänden* (Stuttgart and Tübingen: Cotta: 1830 [1827]), Vol. 1, 217–20. See overall Daniel Ogden, "The Apprentice's Sorcerer: Pancrates and his Powers in Context (Lucian, 'Philopseudes'), 33–6." *Acta Classica*, 47 (2004): 101–26. In the related framework of the need to control science/technology's potential overgrowth, see Anthony R. Michaelis' editorial of the same name as Ogden's essay, on the 1940's film *Fantasia*, "The Apprentice's Sorcerer," *Interdisciplinary Science Reviews*, Vol. 9, no. 1 (1984): 1–3.
12 The Manchester-based philosopher and gaming designer, Chris Bateman, uses this term in a dialogue with the current author, Babette Babich, and Christopher Bateman, "Touching Robots." Online https://onlyagame.typepad.com/only_a_game/2017/02/babich-and-bateman-touching-robots. Accessed September 1, 2019. on his blog, "The Last of the Continental Philosophers." *Only a Game*. November 29, 2016. Online. https://onlyagame.typepad.com/only_a_game/2016/11/babich-and-bateman-1.html. See further, for a fuller elaboration of Bateman's reflections here, and note that the publishers own name reflects this portable—and invisible—rubric, *The Virtuous Cyborg* (London: Eyewear Publishing, 2018).
13 Richard Seymour, *The Twittering Machine* (London: Indigo, 2019).
14 Ibid.
15 Anders, "The World as Phantom and Matrix." Norbert Gutterman, trans. *Dissent*, 3, no. 1 (1956): 14–24.
16 Ibid., 19, translation slightly altered.
17 See Tom Stoelker's January 2020 video interview with "Babette Babich on Love, Social Media, and Megxit." Video Interview. January 16, 2020. Online. https://www.youtube.com/watch?v=9Emj6JAEVKs.
18 See for references and discussion, Babich, "The Question of the Contemporary in Agamben, Nancy, Danto: Between Nietzsche's Artist and Nietzsche's Spectator." *Futures of the Contemporary*, ed. Paulo de Assis and Michael Schwab (Ithaca: Cornell University Press, 2019), 49–82.
19 The author of *Down Girl: The Logic of Misogyny* (Oxford: Oxford University Press, 2017), Kate Manne, uses the term, which she tells us she does not coin, to characterize the circumstance wherein a male, even a criminal, can elicit more sympathy than his (female) victim. In philosophy, we need only think of leading scholars accused of harassment whose careers continue and, in the plural and very recently as well, in the political sphere, of sitting Supreme Court judges.

20 See most recently, my contribution, "Good for Nothing: On Philosophy and Its Discontents." *Why Philosophy?*, eds. Diego Bubbio and Jeff Malpas (Berlin: de Gruyter, 2019), 123–50. It is to the point of the entitlement that drives "himpathy" that even mentioning the analytic continental divide can irritate colleagues, just talking about it seems sufficiently incendiary. See my "Philosophy or Love, Actually" a lecture for the faculty at Fordham University, February 1, 2019. The video version is available online: https://youtu.be/zCef5sEje2w. I included reference to the aftermath of such #metoo style aggression on 30 March 2019 in a lecture presented to the North Texas Philosophical Society, Dallas, Texas, March 30, 2019 and, more broadly on "James & Husserl & Analytic Philosophy: On Love, Actually." *Boston College Graduate Colloquium: Remembering Richard Cobb-Stevens*, September 20, 2019.

21 The quote by Jean-Pierre Dupuy is part of the publisher's features for Bischof, et al., (eds.), *The Life and Work of Günther Anders: Émigré, Iconoclast, Philosopher, Man of Letters* (London: Taylor & Francis, 2014). See too, Christian Fuchs, "Günther Anders' Undiscovered Critical Theory of Technology in the Age of Big Data Capitalism." *tripleC*, 15, no. 2 (2017): 582–611. Online: https://www.triple-c.at/index.php/tripleC/article/view/898.

22 Thus, Müller and Mellor write in their editorial to the 2019 issue of *Thesis Eleven*, dedicated to Anders, reflects that in the course of "the last 25 years, this has given rise to a now vast body of scholarly work in German, French, and Italian . . . and this special journal issue marks the growing interest in Anders's work in Anglophone research." Christopher John Müller and David Mellor, "Utopia Inverted: Günther Anders, Technology and the Social." *Thesis Eleven*, 153, no. 1 (2019): 3–8.

23 Fuchs, "Günther Anders' Undiscovered Critical Theory of Technology."

24 I discuss some of this in my "Are They Good? Are They Bad?"

25 Hannah Arendt, "Introduction." In Walter Benjamin (ed.), *Illuminations: Essay and Reflections*, 1–55 (New York: Schocken, 1969), 1.

26 Thus, although this continues to be repeated in commentary on Anders, it is important to underline that Anders was not Benjamin's "second cousin" or "distant cousin." As his nephew, David Michaelis replied to my query concerning their relation: "Yes, they were cousins, not distant. My grandfather and his father were brothers." Private communication with the author: December 12, 2019.

27 See James Dawsey, "The Life of a Rescuer: Eva Michaelis-Stern in Dark Times," June 25, 2019. Online. https://www.google.com/url?sa=t&rct=j&q=&esrc=s&source=web&cd=&cad=rja&uact=8&ved=2ahUKEwiOsZbCutHpAhWzqHEKHeBPC5cQFjAAegQIAhAB&url=https%3A%2F%2Fwww.nationalww2museum.org%2Fwar%2Fartic les%2Flife-rescuer-eva-michaelis-stern-dark-times&usg=AOvVaw3AodLtsezuxLID0 q3ojPZ0. David Michaelis wrote to say that Anders was given to tell his sister, that is, Michaelis' mother, "we wrote while you did."

28 See William Stern and Clara Stern, *Die Kindersprache. Eine psychologische und sprachtheoretische Untersuchung* (Leipzig: Barth, 1907) as well as their *Erinnerung, Aussage und Lüge in der ersten Kindheit*. For discussion, see Werner Deutsch and Christliebe El Mogharbel, "Clara and William Stern's Conception of a Developmental Science." *European Journal of Developmental Psychology*, 8, no. 2 (2011): 135–56. For Anders, called "Heinz" in his mother's recollections, Clara Stern. *Aus einer Kinderstube: Tagebuchblätter einer Mutter* (Leipzig: B.G. Teubner, 1914), and already represented in the studies co-authored with Anders' father, the lability of names may have been quasi-second nature.

29 Stern, *Die Rolle der Situationskategorie bei den "logischen Sätzen." Erster Teil einer Untersuchung über die Rolle der Situationskategorie* (Diss. Freiburg/Brsg. 1924).
30 See for a discussion of Anders and philosophical anthropology, beginning with a reflection on the difficulty of categorizing Anders on the terms of the same academy that excluded him, Marcel Müller, *Von der "Weltfremdheit" zur "Antiquiertheit." Philosophische Antrhopologie bei Günther Anders* (Marburg: Tectum, 2012). See too Christian Filk. "Der Mensch ist größer und kleiner als er selbst" Günther Anders' "Negative Anthropologie im Zeitalter der '(Medien-)Technokratie.'" *Medienpulse*, 50, no. 2 (2012): 1–19. as well as Filk, "'Frei sind die Dinge: Unfrei ist der Mensch'—Zur Entwicklung von Günther Anders Negativer Anthropologie im technisch-medialen Zeitalter." *MEDIENwissenschaft: Rezensionen/Reviews*, 23, no. 3, (2006): 277–91." Both of Filk's essays constitute contributions to media-anthropology, a field that has had an efflorescence that has also split into several different directions in many cases steered by the dominance of names other than Anders. Valuable here is the attention to the complexities of Anders' person in addition to the difficult—and still very immediate presence—of the theme of philosophical anthropology brought into connection with media, or what Anders understood under the notion of having and which we might better understand via the concept of the consumer as Adorno also understood this by way of the culinary. Filk also notes Nietzsche in conceiving the distinction between the human and the animal for Scheler/Plessner and Anders (this last a bit tendentious as we will see in the following text).
31 Cf. Anders, *Musikphilosophische Schriften. Texte und Dokumente*, ed. Reinhard Ellensohn, (Munich: Beck, 2017), 15–140.
32 By contrast, Adorno, for all his revolutionary sense and sensibility, was able to secure his own academic credentials, securing a *Habilitation* under Tillich by submitting, quite cannily and contrary to Adorno's own research interests, not a thesis on the same sociology of music Adorno also wrote on as he like Anders had a keen interest in the topic and would go on to write extensively on it, but and just at hand, matched to Tillich's interests, such is academic politics, on Kierkegaard. See: Adorno, *Kierkegaard: Construction of the Aesthetic* (Albany: State University of New York Press, 1962). Boer argues, quite conventionally, that the book is not sufficiently adverted to as the point of departure for his essay "A Totality of Ruins: Adorno on Kierkegaard." *Cultural Critique*, 83 (Winter 2013): 1–30. This is inaccurate as there are a number of books that directly engage this connection, especially via aesthetics, as may be seen in Hullot-Kentor's reception of Adorno and most conspicuously (in the literal sense of the conspicuous, *qua* Adorno Prize winner), in spite of the author's otherwise limited familiarity with Adorno's work (not that this is all that uncommon), Peter Gordon who wrote a monograph perpetuating the identification of Kierkegaard with existentialism as a philosophical rubric. On the question of academic facility, beyond questions of acknowledgement and privilege, Heidegger insists on naming Adorno's original teacher—Hans Cornelius, whom Adorno followed sufficiently dutifully that the former would refuse to direct the text submitted as a Habilitation (because lacking distinction from his own work). Cf., *inter alia*, Babette Babich, "Between Heidegger and Adorno: Airplanes, Radios, and Sloterdijk's Atmoterrorism." *Kronos Philosophical Journal*, VI (2017 [2018]): 133–58.
33 Stern/Anders, *Über das Haben*.
34 See for a discussion from an importantly Francophone perspective—including reference to Gabriel Marcel—Daglind Sonolet, *Günther Anders: phénoménologue de la technique* (Bourdeaux: Presses Universitaires de Bordeaux, 2006), here esp. p. 41ff.

35 Stern, *Über das Haben. Sieben Kapitel zur Ontologie der Erkenntnis* (Bonn: Cohen, 1928). See for a discussion of Anders in the phenomenological tradition, Laurent Perreau, *Günther Anders à l'école de la phénoménologie* (Paris: Kimé, 2007), especially foregrounding Husserl in addition to: Édouard Jolly, *Günther Anders. Une Politique de la Technique* (Paris: Decitre, 2017). Contemporary approaches foreground the ongoing turn to religious approaches, including Jason W. Alvis, "Transcendence of the Negative: Günther Anders' Apocalyptic Phenomenology." *Religions*, 8, no. 59 (2017): 1–16. https://www.mdpi.com/2077-1444/8/4/59, and see too Eckhard Wittulski, "Der tanzende Phänomenologe." *Günther Anders kontrovers*, ed., Konrad Paul Liessmann (Munich: Beck, 1992), 17–33. But see also the always-excellent Lütkehaus, most recently his *Philosophieren nach Hiroshima. Über Günther Anders* (Berlin: Fischer, 2018 [1992]), in addition to his *Schwarze Ontologie. Über Günther Anders* (Lüneberg: Zu Klampen, 2002). Anders himself also published in French in the mid-1930s, which explains some of the attention of French scholarship, including his acknowledged influence on Jean-Paul Sartre, "Une interprétation de l'*a posteriori*." *Recherches philosophiques*, IV (1934–1935): 65–80 et "Pathologie de la liberté. Essai sur la non-identification." *Recherches Philosophiques*, VI (1936–1937): 22–54.

36 F. Joseph Smith, the musicologist and philosopher, argues that Husserl himself was influenced by Wilhelm Stern's tone variator for his own analysis of tone: "In discussing tonal apprehension [*Auffassung*], Husserl follows W. Stern, emphasizing the fact that apprehension is not instantaneous, a thing of the moment but that it builds up gradually. . . . A series of musical tones builds a successive unity which is apprehended as such." See Smith, *The Experiencing of Musical Sound: Prelude to a Phenomenology of Music* (New York/Montreux: Gordon and Breach, 1979); here Smith goes on to characterize the perception of a symphonic form in the case of Sibelius, from "motives enunciated by the cello in the opening bars." Synthesizing the opposition, Heidegger poses contra Husserl, indirectly, saying that we never hear tones, Smith writes: "Tones are in themselves, as it were, dead entities, regarded acoustically, easily identifiable as the same continuous tones; but in phenomenal flux, as they appear in immanent time, tones are alive and fluid, and so are the forms they generate."

37 A related distinction informed the heard/lived-experience of tones in Anders' various writings on philosophic music aesthetics and history and musical sociology.

38 In addition to several sections that appeared in English simultaneously, the first part of this *Die Weltfremdheit des Menschen. Schriften zur philosophischen Anthropologie* (Munich: Beck, 2018), "Über prometheischer Sham," is included along with valuable commentary in Müller's *Prometheanism: Technology, Digital Culture and Human Obsolescence* (Lanham: Rowman & Littlefield, 2016), 23–95.

39 This first section is available in English translation as "On Promethean Shame" as cited above.

40 An outstanding translation by Martin Esslin has been available since 1965 in collaboration with Anders himself. It is intriguing to note how excellent this reading is, especially as Anders points to an observation that makes his own what can otherwise seem to be a clear recognition of a kind of Kojèvian shift of thinking, noted not by name but only from the period in which Anders found himself, "Since the early thirties when Hegel's dialectic and Marx's theory of the class struggle began to interest the younger generation in France, the famous image of the pair 'master and servant' from Hegel's *Phaenomenologie des Geistes* so deeply engraved itself into the

consciousness of those intellectuals born around 1900 that it occupies today the place which the image of Prometheus held in the nineteenth century." Anders, "On Beckett's Play, Waiting for Godot," trans. Martin Esslin and revised by Anders, *Samuel Beckett: A Collection of Critical Essays*, ed. Martin Esslin (Englewood Cliffs: Prentice Hall, 1965) , 140–51, here 149.

41 Anders, "Reflections On The H Bomb," *Dissent*, 3, no. 2 (Spring 1956): 146–55. See too, with Claude Eatherly, *Burning Conscience: The case of the Hiroshima Pilot, Claude Eatherly, Told in His Letters to Günther Anders*. Preface by Bertrand Russell; Foreword by Robert Jungk (New York: Paragon, 1961). Later concerns include Anders, "Ten Theses on Chernobyl." Sixth World Congress of the International Physicians for the Prevention of Nuclear War in 1986. Online. https://libcom.org/library/ten-theses-chernobyl-%E2%80%93-g%C3%BCnther-anders. Accessed March 8, 2018," as well as, as Harold Marcuse notes, a translation from a translation, of "We, Sons of Eichmann: An Open Letter to Klaus Eichmann." Jordan Levinson, trans. 2015. Online. http://anticoncept.phpnet.us/eichmann.htm.

42 See Jean Baudrillard, *The System of Objects*, trans. James Benedict (London: Verso, 1996).

43 Thus, Marco Marian connects Anders with the Situationists, Marian, "Günther Anders and the Modification of Reality." *Journal of Historical Archaeology & Anthropological Sciences*, 3, no. 6 (2018): 789–92.

44 In English since 2017, as Anders, "The Obsolescence of Privacy." Christopher John Müller, trans. *CounterText*, 3, no. 1 (2017): 20–46.

45 At Stony Brook in the 1970s, Ihde sidestepped any mention of Anders in his courses on the philosophy of technology (an opposition he would later, and repeatedly, brag about). The refusal was deliberate and, in a recent autobiographical reflection, still seeking to highlight what others lack, Ihde associates his refusal of Anders' contributions (along with Hannah Arendt) in: Don Ihde, *Husserl's Missing Technologies* (Oxford: Oxford University Press, 2016), xiv.

46 van Dijk, *Anthropology in the Age of Technology*. See for a recent, Ihde-influenced reading, conceived as counterpoint, Dominic Smith, *Exceptional Technologies: A Continental Philosophy of Technology* (London: Bloomsbury, 2018).

47 See overall but especially the first chapters of Neil Postman, *Technopoly: The Surrender of Culture to Technology* (New York: Vintage Books, 1993).

48 See the contributions to Yunus Tuncel (ed.), *Nietzsche and Transhumanism: Precursor or Enemy?* (Cambridge: Cambridge Scholars, 2017). See too Steve Fuller's many recent books on the theme of what he calls the proactionary stance.

49 See on this Tim Adams' interview (with) "Sherry Turkle: 'I Am Not Anti-technology, I Am Pro-Conversation.'" *The Guardian*. October 18, 2015.

50 Sherry Turkle, *Alone Together: Why We Expect More from Technology and Less from Each Other* (New York: Basic Books, 2011) and see too Nicholas Carr, *The Shallows: What the Internet Is Doing to Our Brains* (New York: W. W. Norton & Co., 2011) restating/updating Postman's *Amusing Ourselves to Death in the Age of Show Business* (New York: Penguin, 1985). Dallas Smythe makes this point in "Communications: Blindspot of Western Marxism." *Canadian Journal of Political and Social Theory/ Revue canadienne de theorie politique et sociale*, 1, no. 3 (Fall/Automne 1977): 1–27. Drawing on the history (and politics of technology) and writing with reference to Anders' notion of obsolescence, see Slade's *Big Disconnect: The Story of Technology and Loneliness* (Amherst, NY: Prometheus, 2012) and *Made to Break: Technology and Obsolescence in America* (Cambridge: Harvard University Press, 2006). Cf. too

Langdon Winner. *The Whale and the Reactor. Search for Limits in an Age of High Technology* (Chicago: University of Chicago Press, 1988).
51 Anders, "The World as Phantom and Matrix," see, using an analysis of the evolution of interior design, 17.
52 Ibid., 15. A helpful discussion, for those interested in the economics of political economy, is offered in the chapter, "The Audience Commodity and its Work." In Smythe's *Dependency Road: Communications, Capitalism, Consciousness, and Canada* (Norwood, NJ: Ablex, 1931), 22–51.
53 It is useful in this connection to read the provocative online essay by Franz Schandl, "Work Will Not Set You Free: Notes on Günther Anders."
54 A recent and critical mainstream discussion appears in Seymour, *The Twittering Machine* noted above but see too, more conventionally, Tim Wu, *The Attention Merchants* (London: Penguin, 2016), amid a number of other discussions including, Shoshana Zuboff's *The Age of Surveillance Capitalism: The Fight for a Human Future at the New Frontier of Power* (London: Profile, 2019), and her "Big Other: Surveillance Capitalism and the Prospects of an Information Civilization." *Journal of Information Technology*, 30, no. 1 (2015): 75–89 as well as the contributions to the security-minded collection edited by Elmer et al., *Compromised Data: From Social Media to Big Data* (London: Bloomsbury, 2015), but which must be matched to the mediatic phenomenology of simulation in Jean Baudrillard, *For a Critique of the Political Economy of the Sign* (St. Louis: Telos Press, 1981), in addition to his *The Consumer Society: Myths and Structures* (Paris: Gallimard, 1970).
55 I refer to the title of Morozov's *To Save Everything Click Here: Technology, Solutionism, and the Urge to Fix Problems that Don't Exist* (London: Penguin, 2014). In addition to Baudrillard, Friedrich Kittler also sought to make related point. See also David Berry, *Critical Theory and the Digital* (London: Bloomsbury, 2014). See the chapter on "Television and the Art of Symbolic Exchange." *Baudrillard and the Media: A Critical Introduction*, ed. William Merrin (London: Polity, 2005), 10–27, for a useful discussion in Merrin, *Baudrillard, and the Media* and for an early overview of Baudrillard, see Mark Nunes, "Jean Baudrillard in Cyberspace: Internet, Virtuality, and Postmodernity." *Style*, 29, no. 2, *From Possible Worlds to Virtual Realities: Approaches to Postmodernism* (Summer 1995): 314–27.
56 See, for example, my "Musical 'Covers' and the Culture Industry: From Antiquity to the Age of Digital Reproducibility." *Research in Phenomenology*, 48, no. 3 (2018): 385–407.
57 See, for example, on the socio-theoretical but no less literally material notion of 'entanglement,' Babette Babich, "Heidegger on *Verfallenheit*." *Foundations of Science*, 22, no. 2 (2017): 261–4.
58 Anders, "Einleitung. Die drei industriellen Revolutionen (1979)." In *Die Antiquiertheit des Menschen 2*, 16. See for a discussion of "unsalararied" work: Müller, "Die Unangestellten. Ein Blick in die Zukunft der Arbeit."
59 Arendt, "Introduction," 5.
60 Ibid., 6.
61 (Concerning honour without fame/concerning greatness without glory/concerning dignity without compensation). Benjamin published *Deutsche Menschen. Von Ehre ohne Ruhm. Von Grösse ohne Glanz. Von Würde ohne Sold. Eine Folge von Briefen* (Lucerne: Vita Nova, 1936) using the name Detlef Holz.
62 Cf. Arendt, "Introduction," 16.
63 Ibid., 17.

64 Cited in Ibid.
65 Walter Benjamin, "Franz Kafka." In Hannah Arendt (ed.), *Illuminations: Essay and Reflections* (New York: Schocken, 1969), 117.
66 Babich, "Musical 'Covers' and the Culture Industry."
67 As I show in *The Hallelujah Effect: Music, Performance Practice and Technology* (London: Routledge, 2016 [2013]), Jeff Buckley is more commonly associated with covering Leonard Cohen's song, *Hallelujah*, even more than Cohen himself, certainly more than k. d. lang and so on.
68 Cf. here H. Stith Bennett differently minded but still phenomenologically articulated sociology of music, *Becoming a Rock Musician* (New York: Columbia University Press, 2018 [1980]). The new edition features a preface by Howard Becker.
69 See the first chapter of Babich, *Nietzsches Antike* (Berlin: Academia, 2020), 17–48.
70 Michael Theunissen, *The Other*.
71 See Ivan Illich, *Deschooling Society* (London: Marion Boyars Publishers Ltd, 2000 [1971]). See for a discussion of Illich, including reflections on the crisis of the university today, Babette Babich, "Getting to Hogwarts: Michael Oakeshott, Ivan Illich, and J.K. Rowling on 'School.'" *Education and Conversation: Exploring Oakeshott's Legacy*, ed. David Bakhurst and Paul Fairfield (London: Bloomsbury, 2016), 199–218, as well as my "Tools for Subversion: Illich and Žižek on Changing the World." *Making Communism Hermeneutical: Reading Vattimo and Zabala*, ed. Sylvie Mazzinie and Owen Glyn-Williams (Frankfurt am Main: Metzler, 2017). 95–111.
72 See for a discussion of Illich with specific reference to medical technology, Babich, "Ivan Illich's *Medical Nemesis* and the 'Age of the Show': On the Expropriation of Death," *Nursing Philosophy*, 19, no. 1 (2018): 1–13.
73 Ivan Illich, "An Address to 'Master Jacques.'" *The Ellul Forum*, 13 (July 1994):16–17.
74 Heidegger, *What Is Philosophy*, trans. William Kluback and Jean T. Wilde (New Haven: College and University Press, 1958), 20–21.
75 Heidegger, *Introduction to Philosophy — Thinking and Poetizing*, trans. Phillip Jacques Braunstein (Bloomington: Indiana University Press, 2011), 1.
76 Heidegger, *What Is Philosophy*, 20–21.
77 Thus, the *Günther Anders-Journal, Jg. 1. Sonderausgabe zur Tagung "Schreiben für übermorgen." Forschungen zu Werk und Nachlass von Günther Anders*, Online. http://www.guenther-anders-gesellschaft.org/wp-content/uploads/2017/12/m%C3%BCller-2017.pdf, was founded as a conference publication of a 2014 conference on his work as the journal of the Austrian Internationale Günther Anders-Gesellschaft.

Chapter 1

1 Sven-Olov Wallenstein counts four lectures in "The Historicity of the Work of Art in Heidegger." In Marcia Sá Cavalcante Schuback, Hans Ruin (eds.), *The Past's Presence: Essays on the Historicity of Philosophical Thinking* (Huddinge: Södertörns högskola, 2005), here: 144. It is useful to look at Jacques Taminiaux, "On Heidegger's Interpretation of the Will to Power as Art." *New Nietzsche Studies*, 2, nos. 1 & 2 (Winter 1999): 1–22. See too David Espinet and Tobias Keiling (eds.), *Heideggers Ursprung des Kunstwerks. Ein Kooperativer Kommentar* (Frankfurt am Main: Klostermann, 2011). See too, as this is also related in this context, Jacques Taminiaux, "Arendt, disciple de Heidegger?" *Études Phénoménologiques*, 1, no. 2

(1985): 111–36. Heidegger's theme is the origin of the artwork as such. Focusing on Walter Benjamin's (different) study of the work of art in an age of "technological" reproducibility. the technological era, Christopher P. Long recounts the seeming convergence of themes in his reflection, "Art's Fateful Hour: Benjamin, Heidegger, Art and Politics." *New German Critique*, No. 83, *Special Issue on Walter Benjamin* (Spring–Summer, 2001): 89–115.

2   See Heidegger, "Einblick in das was ist, Bremer Vorträge 1949." In *Bremer und Freiburger Vorträge*, GA 79 (Frankfurt am Main: Klostermann, 1994), 5–80.

3   The accounts of the history of the Frankfurt School tend to be enthusiastically partisan (think Habermas loyalists), counting in some names and excluding others. Thus, Anders does not appear in Emil Walter-Busch, *Geschichte der Frankfurter Schule. Kritische Theorie und Politik* (Munich: Fink, 2010), although Hannah Arendt is duly mentioned as Heidegger's 'Geliebte'.

4   See Babich, "Constellating Technology: Heidegger's *Die Gefahr*/The Danger." In Babich and Dimitri Ginev (eds.), *The Multidimensionality of Hermeneutic Phenomenology* (Frankfurt am Main: Springer, 2014), 163.

5   Thus, Christian Dries makes this claim as point of departure for his "Technischer Totalitarismus: Macht, Herrschaft und Gewalt bei Günther Anders," *Etica & Politica/Ethics & Politics: Rivista di filosofia*, 15, no. 2 (2013): 175–98; but other specialists have also found it difficult to read Anders on the issue of power. Liessmann, for example, reads Anders in correspondence with Arendt. See Liessmann's "Thought after Auschwitz and Hiroshima: Günther Anders and Hannah Arendt." *Enrahonar*, 46 (2011): 123–35; but cf., Babette Babich, "Martin Heidegger on Günther Anders and Technology: On Ray Kurzweil, Fritz Lang, and Transhumanism." *Journal of the Hannah Arendt Center for Politics and Humanities at Bard College*, 2 (2012): 122–44, as well as Édouard Jolly, "Entre légitime défense et état d'urgence. La pensée andersienne de l'agir politique contre la puissance nucléaire." *Etica & Politica/Ethics & Politics: Rivista di filosofia*, 15, no. 2 (2013); in addition, in the same online locus: Vallori Rasini, "Il potere della violenza. Su alcune riflessioni di Günther Anders." "*Etica & Politica/Ethics & Politics: Rivista di filosofia*," 15, no. 2 (2013): 258–70. See too James Dawsey, "Ontology and Ideology: Günther Anders's Philosophical and Political Confrontation with Heidegger." *Critical Historical Studies*, 4, no. 1 (Spring 2017): 1–37.

6   There are exceptions. For some of these, in media theory, see Timo Kaerlein, "Playing with Personal Media: On an Epistemology of Ignorance." *Culture Unbound*, 5 (2013): 651–70, as well as the reference to Fuchs earlier and Müller's wide-ranging analysis on media and digital culture in *Prometheanism*. See again Konrad Paul Liessmann (ed.), "Thought after Auschwitz and Hiroshima: Günther Anders and Hannah Arendt." *Enrahonar*, 46 (2011): 123–35, in addition, again, to Schraube, "Torturing Things Until They Confess" and, for a commentary that was, alas, for a long time itself unreceived, Van Dijk's *Anthropology in the Age of Technology*.

7   Dawsey uses this term to describe Anders in "Ontology and Ideology: Günther Anders's Philosophical and Political Confrontation with Heidegger." *Critical Historical Studies*, 4, no. 1 (Spring 2017): 1–37, concluding by observing that "Günther Anders role in the 'Heidegger wars' has still not received due attention in the Anglophone world" (37) but limiting himself exclusively to Alan Milchman's and Alan Rosenberg's 2003 essay, "Martin Heidegger and the Political: New Fronts in the Heidegger Wars." See *Review of Politics*, 65, no. 3 (Summer 2003): 439–49, to make his case.

8 When we make decisions, we can more easily project or predict possible or desired positive consequences than possible or even likely negative outcomes which are then discovered in practice. See, for a systematic discussion in a related context, Katinka Waelbers, "Technological Delegation: Responsibility for the Unintended." *Science and Engineering Ethics*, 15, no. 1 (March 2009): 51–68. I explore this further with regard to automation and invisibility in Babich, "Necropolitics and Techno-Scotosis." *Philosophy Today*, 65, 2 (2021).

9 Sometimes this is argued to be a result of what is called the "empirical turn," a focus on technology rather than the sociological or the political, a turn that has the convenience of assuming as point of departure what might otherwise need to be argued. See, for "empirically" inclined special overview, Paul Brey, "Philosophy of Technology after the Empirical Turn." *Techné: Research in Philosophy and Technology*, 14, no. 1 (2010): 1–11, as well as Peter-Paul Verbeek, *What Things Do: Philosophical Reflections on Technology, Agency, and Design* (College Station: Pennsylvania State University Press, 2005).

10 See in the first instance the many studies of Jacques Ellul, for one example, Ellul's *The Technological Society*, trans. John Wilkinson (New York: Knopf/Vintage, 1967 [1964]), as well as the many studies of Paul Virilio and Jean Baudrillard. Less critically but no less usefully, see Michel de Certeau's reflections on the "arts of doing," *The Practice of Everyday Life*, trans. Steven Rendall (Berkeley: University of California Press, 1984). See too, although more directly concerned with labour as indeed education, Stanley Aronowitz et al.

11 Bruno Latour might be noted in the foreground, though there are others, but see for example, and with reference to anthropological reflection on contemporary culture perhaps first and foremost authored with Steve Woolgar, *Laboratory Life: The Construction of Scientific Facts* (Beverly Hills: Sage, 1979), as well as Latour's obviously titled, *Aramis or the Love of Technology*, trans. Catherine Porter (Cambridge: Harvard University Press, 1996), and perhaps most instructively, if for Anglophone readers, elusively argued, *The Pasteurization of France*, trans. Alan Sheridan and John Law (Cambridge: Harvard University Press 1988 [1984]), in addition to his *We Have Never Been Modern*, trans. Catherine Porter (New York: Harvester Books, 1993). See for discussion and further references, Babich, "Hermeneutics and Its Discontents in Philosophy of Science: On Bruno Latour, the 'Science Wars,' Mockery, and Immortal Models." In Babich (ed.), *Hermeneutic Philosophies of Social Science* (Berlin: de Gruyter, 2017), 163–88.

12 I cite exemplary discussions throughout and in the notes to follow, but see, in particular, Langdon Winner, John Street, Gilbert Germaine, all cited later in the text. Thus it should not be assumed that no one had yet considered the problem from the point of view of political theory; hence, see also Michael Allen Gillespie, "The Future of Political Theory: Using the Canon to Prepare for Tomorrow." February 23, 2017. Online. https://humanitiesfutures.org/papers/future-political-theory-using-canon-prepare-tomorrow/. I note that Tracy B. Strong includes reflection on some of these questions in the final chapter, "At Home Alone: The Problems of Citizenship in Our Age" of Strong, *Learning One's Native Tongue: Citizenship, Contestation, and Conflict in America* (Chicago: University of Chicago Press, 2019), 287–314.

13 George Kateb, "Technology and Philosophy." *Social Research*, 64, no. 3 (1997): 1225–46.

14 See Kateb further but also see too Mark Blitz, "Understanding Heidegger on Technology." *The New Atlantis*, 41 (Winter 2014): 63–80.

15 John McCormick, *Carl Schmitt's Critique of Liberalism: Against Politics as Technology* (Cambridge: Cambridge University Press, 1997). See too the recent contributions

to David Pan's edited issue of *Telos* on *Carl Schmitt and the Critique of Technical Rationality*, No. 187 (2019).
16 See Langdon Winner, *Autonomous Technology. Technics-out-of-Control as a Theme in Political Thought* (Cambridge: MIT Press, 1977) and Winner, *The Whale and the Reactor*. Also useful for further if similarly mainstream minded, if not (to date) received as such: although very insightfully articulated: Germain, *A Discourse on Disenchantment: Reflections on Politics and Technology* (Albany: State University of New York Press, 1993), in addition to John Street, *Politics and Technology* (New York: Guilford Press, 1992), among others such as (in many respects well received) Dominique Janicaud or Silvio Vietta or, indeed and to be sure, Hans-Peter Hempel.
17 This is a personal communication, expressed on more than one occasion.
18 See in particular, although he wrote many books, Ellul, *The Technological Society* (New York: Knopf, 1964 [1954]), including an introduction by Robert K. Merton, who also ensured the translation and publication, with an introduction (also ensured from Thomas Kuhn) to Ludwik Fleck, *The Genesis and Development of a Scientific Fact* (Chicago: University of Chicago Press, 1981 [1935]).
19 Here it is common to refer to Marshall McLuhan and, see too, from another but related vantage point, Neil Postman, especially his *Technopoly*. One may argue that today's popular contenders on the level of serious critical reception include names like the late Bernard Stiegler and efforts to retrieve, not unrelated to Stiegler, Gilbert Simondon, as well as Arthur Kroker. One would wish here to be able to add Stanley Aronowitz but given an array of different concerns his engagement with technology remained philosophically desultory (not to speak of engaging Anders), which may also underscore the disciplinary difference between philosophy and sociology, but which difference is not in evidence to the same extent in thinkers like Virilio or de Certeau or Baudrillard.
20 Cf. too Hans Ruin, "The Inversion of Mysticism: *Gelassenheit* and the Secret of the Open in Heidegger." *Religions*, 10, no. 1 (2019). https://www.mdpi.com/2077-1444/10/1/15/htm.
21 Byung-Chul Han, *In the Swarm: Digital Prospects*, trans. Erik Butler (Cambridge, MA: The MIT Press, 2017), and Han's *The Burnout Society*, trans. Erik Butler (Stanford, CA: Stanford University Press, 2015).
22 For a constitutive discussion of this conventional problem, see Verbeek, *What Things Do*.
23 Heidegger, *Überlegungen XII-XV (Schwarze Hefte 1939-1944)*, GA 96 (Frankfurt am Main: Klosterman, 2014), 198. For reflections on this constellation and the deliberately prepared dimensionality of the Black Notebooks, see Babich, "Heidegger Hermeneutik." In Alfred Denker und Holger Zaborowski (eds.), *Heidegger Jahrbuch 12. Zur Hermeneutik der Schwarzen Hefte* (Freiburg im Breisgau: Alber, 2019), as well as Babich, "Heidegger on Nietzsche's 'Rediscovery' of the Greeks: *Machenschaft* and *Seynsgeschichte* in the Black Notebooks." *Journal of the British Society for Phenomenology*, 51, no. 2 (April 2020): 110–23.
24 See for a broad discussion, especially including José Ardillo, *Ellul, La Liberté dans un monde fragile: Écologie et pensée libertaire* (Paris: L'Échappée, 2018) and see, too, the contributions to Jerónimo et. al. (eds.), *Jacques Ellul and the Technological Society in the 21st Century* (Frankfurt am Main: Springer, 2013).
25 This was often well intentioned as prejudice tends to suppose itself. Thus, Carl Mitcham did seek to engage Ellul and, later Illich, but not Anders. For his part, Paul Durbin

likewise failed to engage Anders. Peter-Paul Verbeek, owing to be sure to van Dijk, does engage Anders, but Verbeek's focus excludes the concerns Anders raises, which must be read, as I shall argue, through Virilio and Baudrillard, and of course Heidegger.

26 Thus, one can read Gottlieb's edited collection compiling Arendt's (and not coincidentally also Anders' in the case of the first chapter) *Reflections on Literature and Culture*, ed. Susannah Young-Ah Gottlieb (Stanford: Stanford University Press, 2007), and so too generic discussions of philosophy and literary authors. Traditionally, these have tended to be oblique readings.

27 See Anders, "Kafka: Ritual Without Religion: The Modern Intellectual's Shamefaced Atheism." *Commentary* (December 1949): 560-9, and Anders, *Kafka, Pro und Contra. Die Proceß-Unterlagen* (Munich: Beck, 1953 [1951]): in English as Anders, *Franz Kafka*, trans. A. Steer and A. K. Thorlby (London: Bowes & Bowes, 1960). See too Gellen's "*Kafka, Pro and Contra*: Günther Anders's Holocaust Book." In Arthur Cools and Vivian Liska (eds.), *Kafka and the Universal* (Berlin: de Gruyter, 2016), 283–304, as well as, to be sure, the article by Bernhard Fetz, "Zwischen Heidegger, Kafka, und der Atombombe: Zur veröffentlichten und unveröffentlichten Essayistik des Schriftstellers und Philosophen Günther Anders." In Michael Ansel, Jürgen Egyptien, and Hans-Edwin Friedrich (eds.), *Der Essay als Universalgattung des Zeitalters: Diskurse, Themen und Positionen zwischen Jahrhundertwende und Nachkriegszeit* (Amsterdam: Brill, 2016, 283–97), as well as the comparative discussion included in a collection on Bourdieu and the sociology of art, once again, Sonolet's "Literature and Modernity: Günther Anders, Hannah Arendt, and Theodor W. Adorno – Interpreters of Kafka." In Jeffrey A. Halley and Sonolet (eds.), *Bourdieu in Question* (Amsterdam: Brill, 2018), 426–41.

28 Goethe is a cliché of course. See, if hardly directly focused on Anders, Bettina Meier, *Goethe in Trümmern: Zur Rezeption eines Klassikers in der Nachkriegszeit* (Frankfurt am Main: Springer, 2019) and Goethe appears as well, although less on Anders than the title might suggest as it focuses on Benjamin's essay on Goethe's *Wahlverwandschaften*, in Elio Matassi, "Die Musik Philosophie bei Walter Benjamin und Günther Anders." In Bernd Witte and Mauro Ponzi (eds.), *Theologie und Politik. Walter Benjamin und ein Paradigma der Moderne* (Berlin: Erich Schmidt, 2005), 212–22.

29 Anders, *Die molussische Katakombe: Roman* (Munich: Beck, 1992), would, two decades later be expanded in a new 2012 edition *mit Apokryphen und Dokumenten aus dem Nachlass*.

30 Astrid Nettling, "Halsstarriger Streiter in Sachen Vernunft," *Deutschlandfunk* 11.07.2012. https://www.deutschlandfunk.de/halsstarriger-streiter-in-sachen-vernunft.700.de.html?dram:article_id=214615.

31 See my own political reflection on reception in philosophy with respect to hermeneutic approaches to the philosophy of science, that is, both the natural and the social science, Babich, "Are They Good? Are They Bad? Double Hermeneutics and Citation in Philosophy, Asphodel and Alan Rickman, Bruno Latour and the 'Science Wars.'" In Paula Angelova, Andreev Jaassen and Emil Lessky (eds.), *Das Interpretative Universum* (Würzburg: Königshausen & Neumann, 2017), 259–90.

32 See, for a discussion, the clear account offered in a chapter of the same title (though, manifestly, one should read the entirety of): Robert Bernasconi's *How to Read Sartre* (London: Granta, 2007), here: 70–81.

33 Fuchs, "Günther Anders' Undiscovered Critical Theory of Technology."

34 See, most recently, self-help style authoritarian optimism of Steven Pinker's *Enlightenment Now*. The fact that one feels compelled to engage Pinker is an

epiphenomenon of the circumstance that is the received view. One debates the Harvard intellectual. See for a helpful example, however, Riskin's exercise in refuting (this is to state the obvious) of Pinker's argumentation, "Pinker's Pollyannish Philosophy and Its Perfidious Politics."

35 I refer here to Steve Fuller and Veronika Lipinska, *The Proactionary Imperative: A Foundation for Transhumanism* (London: Palgrave Macmillan, 2014), and, on Ihde, for example, see the contributions to Evan Selinger (ed.), *Postphenomenology: A Critical Companion to Ihde* (Albany: State University of New York Press, 2006).

36 Illich's readings of technology can be difficult to grasp inasmuch as he articulates these from and including a spiritual dimension that is difficult for current readers. But see Ivan Illich, *Selbstbegrenzung: eine politische Kritik der Technik* (Munich: C.H. Beck, 1998 [1973]) as well as, very importantly, inasmuch as this is increasingly relevant for discussion today: Ivan Illich, *H2O and the Waters of Forgetfulness: Reflections on the Historicity of 'Stuff'* (Dallas: Dallas Inst Humanities & Culture, 1985). Note here that the broader sense of this historicity also maps onto what I call Heelan's "material hermeneutics." See for a discussion, Babette Babich, "Material Hermeneutics and Heelan's Philosophy of Technoscience." *AI & Society*, Vol. 35 (2020). April 14, 2020. Online first. https://link.springer.com/article/10.1007/s00146-020-00963-7.

37 See the discussion of Kittler here in Chapter 8, "Radio Ghosts."

38 See again Timo Kaerlein, "Playing with Personal Media: On an Epistemology of Ignorance." *Culture Unbound*, 5 (2013): 651–70. In sociology proper, other names would include Bourdieu, and, cited earlier, de Certeau and Aronowitz, balancing his reading of technology between the radical and the conventional, if excludes mention of Anders (among others) and (also not engaging Anders) C. Fred Alford.

39 See, however, for an invaluable discussion, Arne Johan Vetlesen, *The Denial of Nature: Environmental Philosophy in the Era of Global Capitalism* (London: Routledge, 2015), which should be read contra Habermas more generally and others such as, more particularly, Steve Vogel as this has led to dangerous tendencies in environmental thinking such that Heisenberg's *bon mot*, as Heidegger cites this in *The Question Concerning Technology*, trans. William Lovitt (San Francisco: Harper Torchbooks, 1977 [1962]), whereby the human encounters increasingly only the human, augmented as this is in a media-focused world acquires theoretical respectability, the respectability of blinders but a mindset nonetheless. For Vetlesen's most recent discussion key with regard to debates concerning post- and transhumanism, as these concerns also bear on the current reading of Anders, see his *Cosmologies of the Anthropocene: Panpsychism, Animism, and the Limits of Posthumanism* (London: Routledge, 2019).

40 Exceptions include Hartmut Böhme, *Fetischismus und Kultur: Eine andere Theorie der Moderne* (Reinbek bei Hamburg: Rowohlt, 2006). For an effort to integrate some of the more mainstream of these concerns, see Konrad Paul Liessmann (ed.), *Günther Anders: philosophieren im Zeitalter der technologischen Revolutionen* (Munich: C.H. Beck, 2002). See, for an overview in Dutch, where technology studies have since become more conventional, Achterhuis, van Dijk & Tijmes (eds.), *De maat van de techniek. Zes filosofen over techniek: Günther Anders, Jacques Ellul, Arnold Gehlen, Martin Heidegger, Hans Jonas en Lewis Mumford* (Baarn: Ambo, 1992).

41 John Gray, "Why the Humanities Can't Be Saved." Online. https://unherd.com/2019/08/why-the-humanities-cant-be-saved/. Gray goes back to Roger Kimball's claim in 1990 of a kind of biting the hand that feeds one savagery, whereby "an academic *nomenklatura* controlled sectors of higher education and used its position to attack

the values of the societies that funded it." Both draw their own analysis a little too limitedly, as if the problem were a matter of academia and more specifically of tenure, to wit Kimball's rather inaccurate title, *Tenured Radicals*. Per contra, the radicals in question hadn't been radical since the late 1960s, and by the time they became professors, much less tenured professors, had long since given up any activity that might have justified the title. But the claim continues apace, and because it is a complicated point, I review it in "Philosophy Bakes No Bread." *Philosophy of the Social Sciences*, 48, no. 1 (2018): 47–55, along with a follow-up essay, "Good for Nothing: On Philosophy and Its Discontents." In Diego Bubbio and Jeff Malpas (ed.), *Why Philosophy?* (Berlin: de Gruyter, 2019), 123–50.

42 John Gray, "Why the Humanities Can't Be Saved."
43 See further, usefully, Rens van Munster and Casper Sylvest, "Appetite for Destruction: Günther Anders and the Metabolism of Nuclear Techno-Politics." *Journal of International Political Theory*, 15, no. 3 (2019): 332–48, and Christophe David and Dirk Röpcke. "Günther Anders, Hans Jonas et les antinomies de l'écologie politique." *Ecologie & politique*, 2, no. 29 (2004): 193–213. See too, again, Christian Dries, *Die Welt als Vernichtungslager. Eine kritische Theorie der Moderne im Anschluss an Günther Anders, Hannah Arendt und Hans Jonas* (Bielefeld: Transcript, 2012).
44 Cited in Paul van Dijk, *Anthropology in the Age of Technology*, 6. Cf. Greffrath's interview with Anders, published posthumously in 2002 in honour of what would have been his 100th birthday, "Lob der Sturheit." *Die Zeit*, "Zeitläufte," 28/2002.
45 "Aber wenn's uns doch gelänge, / abzuwerfen unsre Last, / und wir stunden, als Gestänge / in Gestänge eingepaßt, // als Prothesen mit Prothesen / in vertrautestem Verband, / und der Makel war gewesen, / und die Scham schon unbekannt—" (AM I, 39) Compare this to Kurt Vonnegut's 1961 sci-fi story, with all-too-patent parallels to the current mask and vaccine mandates, precisely *qua* 'handicaps,' but also computer and cell phone chirp notifications: "Harrison Vergeron."
46 But see for a discussion of sex robots quite in terms of this gendered differential, Babich, "On Passing as Human and Robot Love." In Carlos Prado (ed.), *How Technology is Changing Human Behavior* (Santa Barbara: Praeger, 2019), 17–26, as well as my "Robot Sex, Roombas, and Alan Rickman." *de Gruyter Conversations: Philosophy & History*, August 17, 2017. Online. https://blog.degruyter.com/robot-sex-roombas-alan-rickman/.
47 See Babich, "The Essence of Questioning after Technology: Techne as Constraint and Saving Power." *British Journal of Phenomenology*, 30, no. 1 (January 1999): 106–24.
48 Heidegger, *Die Frage nach der Technik*, 1.
49 For one discussion of Heidegger's understanding of freedom informed by the political reception of Heidegger's thinking, see Mark Basil Tanzer, *Heidegger: Decisionism and Quietism* (Amherst, NY: Humanity Books, 2002). See, in addition, Peter Trawny, *Irrnisfuge. Heideggers An-Archie* (Berlin: Matthes & Seitz Verlag, 2014), and indeed: Reiner Schürmann, *Heidegger on Being and Acting: From Principles to Anarchy* (Bloomington: Indiana University Press, 1987).
50 Herbert Marcuse, *One Dimensional Man* (Boston: Beacon Press, 1964). Once again, I note that this book, although it ought not to perhaps continue to be as useful as it is, given the changes on the tech horizon, remains the best book on technology and society and the political simply because it has not been bettered. And part of the

reason it has not been outclassed is that, like Erwin Chargaff's uncomprehending colleagues, political theorists seem to constitute a set dedicated to excluding not only Anders, as mentioned earlier, but critical theorists in general, to the mutual detriment of both groups.

51 See, again, for another take on this same question of names and naming, Adi Armon, "The Parochialism of Intellectual History." I earlier speak to a similar set of questions as these were originally inspired by Heidegger's French translator, François Vézin, in his "L'étendue du désastre" (8 août 2014), in the aftermath of the Black Notebooks scandal and given the constellation of names posed as excluded or silence with respect to Heidegger, Babich, "Heidegger et ses Juifs." In Joseph Cohen and Raphael Zagury-Orly (eds.) *Heidegger et les Juifs* (Paris: Grasset, 2015), 411–54 (a related version appears in English as "Heidegger's *Judenfrage*").

52 See here Reinhard Ellensohn, *Der andere Anders: Günther Anders als Musikphilosoph* (Bern: Peter Lang, 2008). Elsewhere I note that the musicologist F. Joseph Smith had for years sought to draw attention to Anders' work only to be rebuffed (and ultimately silenced to inattention by those interested in acoustic phenomenology) by the same Don Ihde, who also silently borrowed his research.

53 Note here the title of Steve Fuller's *Humanity 2.0. What It Means to be Human Past, Present and Future* (London: Palgrave Macmillan, 2011).

54 I discuss Baudrillard along with Virilio in the latter sections of Babette Babich, "The Question of the Contemporary in Agamben, Nancy, Danto" as well as: "Musical 'Covers' and the Culture Industry: From Antiquity to the Age of Digital Reproducibility." *Research in Phenomenology*, 48, no. 3 (2018): 385–407.

55 I note an email communication from David Michaelis, Anders' nephew, who wrote to me at the end of 2013 in response to my review of Maier-Katin's 2010 *Stranger from Abroad: Hannah Arendt, Martin Heidegger, Friendship and Forgiveness* (New York: Norton, 2010). Michaelis wrote, here cited *verbatim*: "When Gunther asked me—or gave me the book 'burning conscience,' I always had a hard time grappling with it. I was 14 years old. . . . Your review of the book *Strangers from Abroad* is a fair assessment of Gunter and Hanna relationship. It is really important that Anders philosophical work will be introduced to the English-speaking world. His point was always that no one can translate his German special language into English. I think if he would have met you today he would reconsider."

56 See for one assessment of one of the reasons for what is clearly an overdetermined question, somehow managing to avoid the most obvious benefit of the playing of overloud music, even without jukeboxes and thus utterly owner-driven "noise," as the tactic insures a relatively speedier turnover than a quieter ambiance might invite: lingering over conversation, with dinner. See Kate Wagner, "How Restaurants Got So Loud. Fashionable Minimalism Replaced Plush Opulence. That's a Recipe for Commotion." *The Atlantic*, November 27, 2018. Online. https://www.theatlantic.com/technology/archive/2018/11/how-restaurants-got-so-loud/576715/, and see, too, the *Vox* essay by Julia Belluz. "Why Restaurants Became So Loud—and How to Fight Back. "I can't hear you." *Vox*, July 27, 2018. Online. https://www.vox.com/2018/4/18/17168504/restaurants-noise-levels-loud-decibels.

57 If only for a discussion of GPS, see Greg Milner, *Pinpoint: How GPS Is Changing Technology, Culture, and Our Minds* (New York: Norton, 2017). I refer to Milner's work on music/audio technology in *The Hallelujah Effect: Music, Performance Practice and Technology* (London Routledge, 2016 [2013]).

58 See Babich, *The Hallelujah Effect* as well as, for a more focused reflection overall, including still further references, Babich, "On *The Hallelujah Effect*."
59 Marcuse, *One Dimensional Man*. It is not irrelevant that, since 2000, one of the main research resources for scholarship, including essential links to further research on Anders, was hosted by Harold Marcuse, Herbert Marcuse's grandson and professor of history at the University of California at Santa Barbara. Marcuse gives his own account of their relationship, and for another discussion, see Christian Fuchs, "Zu einigen Parallelen und Differenzen im Denken von Günther Anders und Herbert Marcuse." In Dirk Röpcke and Raimund Bahr (eds.), *Geheimagent der Masseneremiten* (St. Wolfgang: Art & Science, 2003), 113–27, and cf., for a discussion in connection with Heidegger, Martin Woessner, "Hermeneutic Communism: Left Heideggerian's Last Hope." In Silvia Mazzini and Owen Glyn-Williams (eds.), *Making Communism Hermeneutical* (Frankfurt am Main: Springer, 2017), esp. 40–3.
60 Max Horkheimer and Theodor Adorno, *Dialectic of Enlightenment* (London: Verso, 2016 [1976]). For a general discussion not specifically centred on Anders (and not particularly "Nietzschean"), see Rolf Wiggershaus, "The Frankfurt School's 'Nietzschean Moment.'" Gerd Appelhans, trans. *Constellations*, 8, no 1 (2001): 144–7.
61 See again, the chapters on desire, male and female, in Babich, *The Hallelujah Effect* and, again, and importantly, Manne's *Down Girl*.

Chapter 2

1 Anders, "Wesen und Eigentlichkeit nämlich bei Heidegger (1936)." In *Über Heidegger* (Munich: Beck, 2001).
2 Anders, "On the Pseudo-Concreteness of Heidegger's Philosophy." *Philosophy and Phenomenological Research*, 8, no. 3 (March, 1948): 337–71. I discuss this striking image as it also haunts Nietzsche's inaugural lecture on Homer in Basel, which I read in connection with Hans Ruin's 2019 study of physical anthropology, articulated in dialogue with Michel de Certeau (on Michelet and history): Hans Ruin, *Being with the Dead: Burial, Ancestral Politics, and the Roots of Historical Consciousness* (Stanford: Stanford University Press, 2019) in: Babich, "Blood for the Ghosts: Reading Ruin's *Being With The Dead* With Nietzsche." *History and Theory*, 59, no. 2 (June 2020): 255–69.
3 Theodor Adorno, *Negative Dialectics*, trans. E. B. Ashton (London: Routledge, 1973 [1966]), 75.
4 To wit, Anders, *Die Weltfremdheit des Menschen. Schriften zur philosophischen Anthropologie*.
5 Adorno, *Negative Dialectics*, 124. Others in addition to Adorno and Anders shared this view; see Gail Soffer, "Heidegger, Humanism, and the Destruction of History" as well as, and beyond the scope of the present chapter, a range of discussions of Heidegger and ecology and ecofeminism, and so on.
6 Some readers can take this ambiguity to write Anders in the posthuman tradition, beyond the "anthropocene." See Christopher John Müller, "Desert Ethics: Technology and the Question of Evil in Günther Anders and Jacques Derrida." *Parallax*, 21, no. 1 (2015): 87–102 and Andreas Beinsteiner, "Cyborg Agency: The Technological Self-Production

of the (Post-) Human and the Anti-Hermeneutic Trajectory," *Thesis Eleven*, 153, no. 1 (2019): 113–33 as well as for a statement of the optimistic sentiment of humanist transhumanism: Mark O'Connell's 2018 "hackeristically" styled manifesto: O'Connell, *To Be a Machine: Adventures Among Cyborgs, Utopians, Hackers, and the Futurists Solving the Modest Problem of Death* (London: Granta, 2018). Cf. further, note 10.

7 See here the contributions to Enders' and Zaborowski's curated issue of the 2014 *Jahrbuch für Religionsphilosophie*, Vol. 13 (Freiburg: Alber, 2014), on Nietzsche, along with Eugen Biser, *Nietzsche—Zerstörer oder Erneuerer des Christentums?* (Darmstadt: Wissenschaftliche Buchgesellschaft, 2002); it is worth reviewing Ernest Fortin's writings on Nietzsche, and, for a focus on recent conventional authors, see Hans Ruin, "Saying *Amen* to the Light of Dawn: Nietzsche on Praise, Prayer, and Affirmation." *Nietzsche-Studien,* 48 (2019): 99–116.

8 Friedrich Nietzsche, *Also Sprach Zarathustra. Ein Buch für Alle und Keinen* (Chemnitz: Verlag von Ernst Schmeitzner, 1883). Online. http://www.zeno.org/Philosophie/M/Nietzsche,+Friedrich/Also+sprach+Zarathustra/Zarathustras+Vorrede, "Vorrede," 3.

9 Ibid., 4.

10 See for example, Stefan Lorenz Sorgner, "From Nietzsche's Overhuman to the Posthuman of Transhumanism," *English Language and Literature*, 62, no. 2 (2016): 163–76, in addition to Tuncel (ed.), *Nietzsche and Transhumanism* as well as Nick Bostrom, "Transhumanist Values." *Journal of Philosophical Research*, 30 (2005): 3–14, along with Steve Fuller, *Nietzschean Meditations: Untimely Thoughts at the Dawn of the Transhuman Era* (Basel: Schwabe, 2019) and Jeffrey Bishop, "Nietzsche's Power Ontology and Transhumanism: Or Why Christians Cannot Be Transhumanists." In Steve Donaldson and Ron Cole-Turner (eds.), *Christian Perspectives on Transhumanism and the Church: Chips in the Brain, Immortality, and the World of Tomorrow* (Frankfurt am Main: Springer, 2018), 117–35.

11 "Yet if I derived my existence from myself, then I would neither doubt nor want, nor lack anything at all; for I should have given myself all perfections of which I have any idea." René Descartes, *Meditations on First Philosophy*, trans. John Cottingham (Cambridge: Cambridge University Press, 1996), 33.

12 Ibid.

13 "Wesen und Eigentlichkeit nämlich bei Heidegger (1936)." In *Über Heidegger* (Munich: Beck, 2001), 33.

14 "d. h. das Sein des Menschen ist in der vulgären ebenso wie in der philosophischen 'Definition' umgrenzt als ζῷον λόγον ἔχον, das Lebende, dessen Sein wesenhaft durch das Redenkönnen bestimmt ist." Heidegger, *Sein und Zeit* (Tübingen: Max Niemeyer, 1984), 48. Hereafter: SZ.

15 Anders, "Wesen und Eigentlichkeit," 37.

16 Ibid.

17 I foreground this stylistic tactic in Babich, "A Musical Retrieve of Heidegger, Nietzsche, and Technology: Cadence, Concinnity, and Playing Brass." *Man and World*, 26 (1993): 239–60.

18 "Die Uneigentlichkeit kann vielmehr das Dasein nach seiner vollsten Konkretion bestimmen in seiner Geschaftigkeit, Angeregtheit, Interessiertheit, Genußfähigkeit" (SZ, 43).

19 "*Ego certe, domine, laboro hic et laboro in me ipso: factus sum mihi terra difficultatis et sudoris nimii.*" Augustine, *Confessions*, trans. Henry Chadwick (Oxford: Oxford University Press, 1991), X, 25, 193.

20 This has been done, to be sure, especially with reference to time. See, for a start, Jean Grondin, "Heidegger und Augustine." In: Ewald Richter, ed., *Die Frage nach der Wahrheit* (Frankfurt am Main: Klostermann, 1997), 161–73 in addition to the late Friedrich-Wilhelm von Hermann, "Die 'Confessiones' des Heiligen Augustinus im Denken Heideggers," *Questio*, 1 (2001): 113–46.
21 See Ethan Kleinberg, *Generation Existential: Heidegger's Philosophy in France 1927-1961* (Ithaca: Cornell University Press, 2005).
22 This is the point of departure for Pierre Hadot as he reflects on the path of historical philology, citing Pierre Courcelle's conventional-literary as opposed to *biographical* account, scandalous then, as Hadot emphasized that the scandal would endure. To quote Hadot, "Alerted by his profound knowledge of Augustine's literary procedures and the traditions of Christian allegory, Courcelle dared to write that the fig tree could well have a purely symbolic value, representing the 'mortal shadow of sin,' and that the child's voice could also have been introduced in a purely literary way indicate allegorically the divine response to Augustine's questioning." Pierre Hadot, *Philosophy as a Way of Life* (Oxford: Blackwell, 1994).
23 See the first chapter of David Blair Allison, *Reading the New Nietzsche* (Lanham: Rowman and Littlefield, 2001).
24 See Matthias Rath, *Der Psychologismusstreit in der deutschen Philosophie* (Freiburg i. Briesgau: Karl Alber Verlag, 1994).
25 Gottlob Frege, *Die Grundlagen der Arithmetik. Eine logisch mathematische Untersuchung über den Begriff der Zahl* (Breslau: Wilhelm Koebner, 1884), x.
26 See Jean-Claude Monod, "'L'interdit anthropologique' chez Husserl et Heidegger et sa transgression par Blumenberg." *Revue germanique internationale*, 10 (2009): 221–36, as well as Françoise Dastur, "La critique heideggérienne de l'anthropologisme," In: *Heidegger et la pensée à venir* (Paris: Vrin, 2011). And see more broadly, Kai Haucke, "Anthropologie bei Heidegger. Über das Verhältnis seines Denkenszur philosophischen Tradition," *Phil. Jahrbuch*, 105. Jahrgang / II (1998): 321–45. For a discussion (in two parts) of Heidegger and (Bourdieu's) anthropology, see James F. Weiner, "Anthropology contra Heidegger Part I: Anthropology's Nihilism." *Critique of Anthropology*, 12, no. 1 (1992): 75–90 and "Anthropology contra Heidegger Part II: The Limit of Relationship." *Critique of Anthropology*, 13, no. 3 (1993): 285–301.
27 Kant, *Metaphysical Foundations of Natural Science*, trans. Michael Friedmann (Stanford: Stanford University, 2004), viii. Kant includes chemistry in his roster of non-sciences, and it can seem that this goes too far but philosophy of chemistry has very only recently habilitated itself, and it is indisputable that physics remains the queen of the sciences. See Michael Bennett McNulty, "What is Chemistry, for Kant?" *Kant Yearbook*, 9 (2017): 85–112, as well as Baird, Scerri, and McIntyre's introduction to their collection on the topic, "Introduction: The Invisibility of Chemistry." In Davis Baird, Eric Scerri, and L. McIntyre (eds.), *Philosophy of Chemistry: Synthesis of a New Discipline* (Dordrecht: Springer, 2006), 3–18, as well as Jaap Van Brakel. "Kant's Legacy for the Philosophy of Chemistry." In Davis Baird, Eric Scerri, and L. McIntyre (eds.), *Philosophy of Chemistry: Synthesis of a New Discipline* (Dordrecht: Springer, 2006), 69–91.
28 Baird, Scerri, and McIntyre, "Introduction: The Invisibility of Chemistry," 3.
29 Adorno, *Negative Dialectics*, 124.
30 Ibid.
31 See Babich, "Nietzsche's Aesthetic Tension and Hume's Standard of Taste." In: *Reading David Hume's "Of the Standard of Taste"* (Berlin: de Gruyter, 2019), 236. See on

Kant and the extraterrestrials, Peter Szendy, *Kant in the Land of Extraterrestrials: Cosmopolitical Philosofictions*, trans. Will Bishop (New York: Fordham University Press, 2013), as well as, earlier, David L. Clark, "Kant's Aliens: The Anthropology and Its Others." *The New Centennial Review*, 1, no. 2 (Fall 2001): 201–89, in addition to Holger Schmid's afterword to a French translation of Kant's text: *Sur les extraterrestres: Théorie du ciel* (Paris: Editions Manucius, 2019). But see too, Tyke Nunez, "Logical Mistakes, Logical Aliens, and the Laws of Kant's Pure General Logic." *Mind*, 128, no. 512 (October 2019): 1149–80.

32 Adorno, *Aesthetic Theory*, trans. Robert Hullot-Kentnor (London: Bloomsbury, 1997), 62.
33 Ibid.
34 Anders, *Die atomare Drohung. Radikale Überlegungen zum atomaren Zeitalter* (Munich: Beck. 1993) , 57, cf. 106f.
35 See, my retrospective essay, Babette Babich, "Ivan Illich's *Medical Nemesis* and the 'Age of the Show': On the Expropriation of Death," *Nursing Philosophy*, 19, no. 1 (2018): 1–13.
36 "The events of Auschwitz and Hiroshima can indeed be repressed in memory (insofar as they ever penetrated into it)—and that is in fact what has happened. But what cannot be repressed by contrast is their repeatability. Since these two events—i.e., now for more than twenty years—so-called 'natural death' has become an obsolete special privilege and the possibility of humanity's violent self-annihilation continuously virulent. And since then we have been continuously defined by this non-stop possibility." Anders, *Die atomare Drohung*, 57. Cf. van Munster and Sulvest, "Appetite for Destruction."
37 See Fuller as cited above in addition to Vincent Blok's insightful: "Denken als Handlung. Heideggers Besinnung auf das Wesen des Menschen im Zeitalter des human enhancement." In Holger Zaborowsky and Alfred Denker (eds.), *Heidegger Jahrbuch 10* (Freiburg im Breisgau: Alber, 2017), 265–79.
38 See for a discussion of the broader European tradition in ethnography, Dennis Johannßen, "Mensch und Dasein in Heideggers' Sein und Zeit." In Thomas Ebke and Caterina Zanfi (eds.), *Das Leben im Menschen oder der Mensch im Leben? Deutsch-Französische Genealogien zwischen Anthropologie und Anti-Humanismus* (Potsdam: Universitätsverlag Potsdam, 2017), 91–104. For another discussion, more precise, terminologically regarded, see Annette Sell, "Leben führen—Dasein entwurfen: Zur systematischen und gesellschaftspolitischen Bedeutung von Plessners anthropologischem und Heideggers fundamental-ontologischem Konzept des Menschen." In Kevin Liggieri and Julia Gruevska (eds.), *Vom Wissen um den Menschen: Philosophie, Geschichte, Materialität* (Freiburg im Breisgau: Alber, 2018), 46–61. On anthropology, see Sato, "The Way of the Reduction via Anthropology: Husserl and Lévy-Bruhl, Merleau-Ponty and Lévi-Strauss." *Bulletin d'analyse phénoménologique*, X, no. 1 (2014): 1–18. See, too, Alfred Schütz, "Phenomenology and the Social Sciences," In Maurice Natanson (ed.), *Collected Papers I. The Problem of Social Reality* (Dordrecht: Kluwer (1972)), 118–39. Cf. Philippe Cabestan, "Phénoménologie, anthropologie: Husserl, Heidegger, Sartre." *Alter*, 23 (2015): 226–42, and, again, as cited earlier, Dastur. Anders had been both Husserl's and Max Scheler's student, among others. See Christophe David, "Fidélité de Günther Anders à l'anthropologie philosophique : de l'anthropologie négative de la fin des années 1920 à *L'obsolescence de l'homme*." *L'Homme et la société. La Question anthropologique*, no. 181 (2011/3): 165–80. See too Karl-Siegbert Rehberg, "Philosophical Anthropology from the End of World War I to the 1940s and in a Current Perspective." *Iris*, 1, no. 1 (2009): 131–52.

39 Adorno, *Aesthetic Theory*, 24. For Adorno, "The painfulness of experimentation finds response in the animosity toward the so called isms: programmatic, self-conscious, and often collective art movements. This rancor is shared by the likes of Hitler, who loved to rail against 'these im- and expressionists,' and by writers who out of a politically avant-garde zealousness are wary of the idea of an aesthetic avant-garde." Ibid.

40 Martin Kusch, "The Sociology of Philosophical Knowledge: A Case Study and a Defense." In Kusch (ed.), *The Sociology of Philosophical Knowledge* (Dordrecht: Springer, 2000), 15–38. Kusch gives the locus here as "1910, p. 516." See too Kusch's valuable discussion: *Psychologism* as well as the sections on Heidegger in Kusch, *Language as Calculus vs. Language as Universal Medium*, 135ff.

41 Reiner Schürmann, *On Heidegger's Being and Time* (London. Routledge, 2008), 56.

42 Ibid.

43 Françoise Dastur, "The Critique of Anthropologism in Heidegger's Thought." In James E. Faulconer and Mark A. Wrathall (eds.), *Appropriating Heidegger* (Cambridge: Cambridge University Press, 2000), 119–34.

44 Daniel Dahlstrom, *The Heidegger Dictionary* (London: Bloomsbury, 2013).

45 I make the connection between Heidegger and Nietzsche on questioning and science in Babich, "On Heidegger on Education and Questioning." In Michael A. Peters (ed.), *Encyclopedia of Educational Philosophy and Theory* (Singapore: Springer, 2017), 1641–52, and "The Essence of Questioning after Technology: *Techne* as Constraint and Saving Power," *British Journal of Phenomenology*, 30, no. 1 (January 1999): 106–24.

46 Cf. here Marco Cavallaro, "Der Beitrag der Phänomenologie Edmund Husserls zur Debatte über die Fundierung der Geisteswissenschaften." *Phänomenologische Forschungen* (2013): 77–93, as well as Dieter Lohmar, "On Some Motives for Husserl's Genetic Turn in his Research on a Foundation of the Geisteswissenschaften." *Studia Phaenomenologica*, 18 (2018): 31–48, and for a different logical take on a theory of everything social, assuming as the author says, that "social theory is not just 'social philosophy for failed philosophers.'" Frédéric Vandenberghe, "Empathy as the Foundation of the Social Sciences and of Social Life: A Reading of Husserl's Phenomenology of Transcendental Intersubjectivity." *Sociedade e Estado, Brasília*, 17, no. 2 (2002): 563–85.

47 See George A. Schrader Jr., "Heidegger's Ontology of Human Existence." *The Review of Metaphysics*, 10, no. 1 (1956): 35–56. See too and in general the contributions to Zaborowsky and Denker, eds., *Heidegger Jahrbuch 10* (Freiburg im Briesgau: Alber, 2017), particularly Raimon Paez Blanch, "Dasein und Mensch bei Heidegger. Eine Überlegung anlässlich des 'Humanismusbriefes.'" In Holger Zaborowsky and Alfred Denker (eds.), *Heidegger Jahrbuch 10* (Freiburg im Briesgau: Alber, 2017), 165–77, and especially Zaborowski, "Bedingungen und Möglichkeiten des Humanismus – heute. Jaspers, Heidegger und Levinas zur Frage nach dem Menschen." In Zaborowsky and Denker (eds.), *Heidegger Jahrbuch 10*, 251–64.

48 This fluidity opens a set of possibilities. See my reflections on the question of the human in Heidegger's *Letter on Humanism*, "Heideggers 'Brief über ‚Humanismus'. Über die Technik, das Bösartige des Grimmes—und das Heilen." In Alfred Denker and Holger Zaborowski (eds.), *Heidegger und der Humanismus. Heidegger-Jahrbuch, Bd. 10* (Freiburg: Verlag Karl Alber, 2017), 237–50, as well as Matthew Calarco, "'Another Insistence of Man': Prolegomena to the Question of the Animal in Derrida's Reading of Heidegger." *Human Studies*, 28, no. 3 (November, 2005): 317–34, and,

more conventionally, Simon James, "Phenomenology and the Problem of Animal Minds." *Environmental Values*, 18 (2009): 33–49.
49 It is for this reason that Perlman can point to the distinction between speaking of *Mensch* (Perlman writes "man") and *Dasein* between Heidegger and Heschel. See Lawrence Perlman, *The Eclipse of Humanity: Heschel's Critique of Heidegger* (Berlin: Walter de Gruyter, 2016), see esp. 29–31.
50 Schürmann, *On Heidegger's Being and Time*, 57.
51 Ibid.
52 Ibid.
53 Ibid.
54 See Babich, "Being on Television: Wisser—Heidegger—Adorno." In John Rose (ed.), *52nd Annual Heidegger Conference* (Baltimore: Goucher College, 2018), 81–95. A sharper version of this was presented in German at a May, 2018 meeting at the Heidegger Archiv in Messkirch, and see too, with reference to geoengineering, "Between Heidegger and Adorno: Airplanes, Radios, and Sloterdijk's Atmoterrorism." *Kronos Philosophical Journal*, VI (2017 [2018]): 133–58.
55 Heidegger, *Überlegungen XII-XV*, 198.
56 Anders, *Die Kirschenschlacht. Dialoge mit Hannah Arendt*, ed. Gerhard Oberschlick (Munich: Beck, 2011).
57 Anders, "On the Pseudo-Concreteness of Heidegger's Philosophy," 346, note 11.
58 There are fellow travellers, with variously different approaches, such as Calarco, *Zoographies* or Francione and Charlton, *Eat Like You Care* or Donaldson and Kymlicka, *Zoopolis*, or indeed, and I tend to note such contributions for personal reasons, as I began as a biologist before turning to philosophy, Marc Bekoff, *Wild Justice*, as well as, in sociology, Jeffrey Bussolini, who works on important themes of recognition and respect between disciplinary fields and has a critical and valuable reflection, "Felidae and Extinction: 'Victim' and 'Cause.'"
59 "*Ackerbau ist jetzt motorisierte Ernährungsindustrie, im Wesen das Selbe wie die Fabrikation fon Leichen in Gaskammern und Vernichtungslagern, das Selbe wie die Blockade und Aushungern on Landern, das Selbe wie die Fabrikation von Wasserstoffbomben.*" Heidegger, *Bremer und Freiburger Vorträge*, 27. I have been writing about this since my 1992 essay, "Heidegger's Silence."
60 Adorno himself takes the title from a popular overview: Paul Eiper's *Tiere siehen dich an* (Berlin: Dietrich Reimer / Ernst Vohsen, 1928), and which also becomes a politicized vehicle as a 1944 documentary, a fact after the war to justify a (humanistic) concern/unconcern with animals.
61 Theodor Adorno, *Minima Moralia*, trans. E. F. N. Jephcott (London: New Left, 1997 [1974]), §68, 105.
62 Ibid.
63 Anders (Stern) (1934/35) "Une interprétation de l'*a posteriori*." *Recherches philosophiques*, IV (1934–1935): 65–80, and (1936–7) "Pathologie de la liberté. Essai sur la non-identification." *Recherches Philosophiques*, VI (1936–1937): 22–54.
64 Hannes Bajohr, in his review of the Anders' posthumous *Die Weltfremdheit des Menschen*, reminds us that drafts of what appears in Anders' *Die Antiquiertheit des Menschen* had already appeared in French in 1937-8, an esoteric reference made even more salient as the, "Pathologie de la liberté [The Pathology of Freedom]," is missing the German original, such that our *only* English access to Anders' text is via the French. Cf. Bajohr, 'World-Estrangement as Negative Anthropology: Günther Anders's Early Essays." *Thesis Eleven*, 153 (2019): 141–53.

65 Cf., Babich, "Nietzsche's *Antichrist*: The Birth of Modern Science out of the Spirit of Religion." In Markus Enders and Holger Zaborowski (eds.), *Jahrbuch für Religionsphilosophie* (Freiburg i. Briesgau: Alber, 2014), 134–54, together with Babich, "Adorno on Science and Nihilism, Animals, and Jews," *Symposium: Canadian Journal of Continental Philosophy/Revue canadienne de philosophie continentale*, 14, no. 1, (2011): 110–45.

66 Daniel Dahlstrom, "Heidegger's Transcendentalism." *Research in Phenomenology*, 35, no. 1 (2005): 35.

67 See Kostas Axelos, *Einführung in ein künftiges Denken. Über Marx und Heidegger* (Berlin: Walter de Gruyter, 1966 [1961]).

68 Dominique Janicaud, *Powers of Rationality: Science, Technology, and the Future of Thought* (Bloomington: Indiana University Press, 1994 [1985]).

69 See for this articulation, Patrick Aidan Heelan, *Quantum Mechanics and Objectivity* (The Hague: Nijhoff, 1965).

70 Anders, "Wesen und Eigentlichkeit," 38.

71 Denis de Rougement, "On the Devil and Politics." *Christianity and Crisis*, 1 (June 2, 1941): 2, and cf. *The Devil's Share*.

72 For a discussion from the point of view of Ignatius of Loyola of some of the good faith complexities of this Dominican sense of evil and of heresy, see Antonio de Nicolas' importantly historical and geographically contextualized discussion of the Spanish Inquisition in *Powers of Imagining: Ignatius de Loyola: A Philosophical Hermeneutic of Imagining through the Collected Works of Ignatius de Loyola* (Albany: State University of New York Press, 1986), beginning with 9–10.

73 See Hannah Arendt, *Eichmann in Jerusalem: A Report on the Banality of Evil* (New York: Viking, 1963), Jacques Maritain, *Thomas and the Problem of Evil* (Milwaukee: University of Marquette Press, 1942), and Susan Neiman, *Evil in Modern Thought: An Alternative History of Philosophy* (Princeton: Princeton University Press, 2002). See also, although more elliptically, Jacob Taubes, *Occidental Eschatology*, trans. David Ratmoko (Stanford: Stanford University Press, 2009 [1947]), as well as, bringing in Löwith's *Meaning in History*, notable in the current context given Löwith's subtitle, *The Theological Implications of History* (Chicago: University of Chicago Press, 1949), Willem Styfhals, "Evil in History: Karl Löwith and Jacob Taubes on Modern Eschatology." *Journal of the History of Ideas*, 76, no. 2 (April 2015): 191–213, and Marin Terpstra, "The Management of Distinctions: Jacob Taubes on Paul's Political Theology." In Gert Jan van der Heiden, George Henry van Kooten, and Antonio Cimino (eds.), *Saint Paul and Philosophy* (Berlin: de Gruyter, 2017), 251–68. And see, particularly useful in connection with Anders, Manfred Frings, "Max Scheler: A Novel Look at the Origin of Evil." *Philosophy and Theology*, 6, no. 3 (Spring 1992): 201–11.

74 Wolfgang Palaver, "The Respite: Günther Anders' Apocalyptic Vision in Light of the Christian Virtue of Hope." In Bischof, Dawsey and Fetz (eds.), *The Life and Work of Günther Anders,* 82. Palaver's reference, thinned as this is via the language of rational choice, is to Dupuy, Jean-Pierre. *Pour un catastrophisme éclairé. Quand l'impossible est certain* (Paris: Seuil, 2002).

75 Hannah Arendt, in a tiny solecism, asserts that de Rougement establishes his reading of the devil via G. K. Chesterton rather than Baudelaire. For a political, historical discussion of the furor de Rougement raised in general, see Jeffrey Mehlman, *Émigré New York: French Intellectuals in Wartime Manhattan, 1940–1944* (Plunkett Lake Press, 2019).

76 For political reasons of exile, one always has to read Anders with (and through) the French, as he himself would have read his Baudelaire and his Gide and his de

Rougement: "Mes chers frères, n'oubliez jamais, quand vous entendrez vanter le progrès des lumières, que la plus belle des ruses du Diable est de vous persuader qu'il n'existe pas!" Charles Baudelaire. "Le Joueur Généreux." *Le Figaro*: February 7, 1864.
77 Martin Buber, *Gottesfinsternis* (Zürich: Manesse Verlag, 1953).
78 Ernst Bloch, "Die riesige Gebietskategorie des Bösen ist eine der am wenigsten durchdachten, sie kommt fast nur adjektivisch vor und dann matt, so etwa in der Phrase vom blutbesudelten Hitlerregime." *Experimentum Mundi* (Frankfurt am Main: Suhrkamp, 1976), 231.
79 See, Giorgio Agamben, *Una domanda*. April 13, 2020. https://www.quodlibet.it/giorgio-agamben-una-domanda. I discuss this originally in an online lecture on the 20th of May 2020 and now in print in Babich, "Pseudo-Science and 'Fake' News: 'Inventing' Epidemics and the Police State." In Irene Strasser (ed.), *The Psychology of Global Crises and Crisis Politics - Intervention, Resistance* (Singapore: Springer, 2021).
80 See on teaching, especially today in the wake of the zoomification of the university, Babich, "Reading Nietzsche's 'Educational Institutions' with Jaspers & MacIntyre on 'The Idea of the University'—and Severus Snape." *Existenz*, Vol. 15, no. 2 (Fall 2020).
81 Anders, *Gewalt, Ja oder Nein? Eine notwendige Diskussion* (Munich: Knaur, 1987).
82 Jacob Taubes (ed.), *Der Fürst dieser Welt. Carl Schmitt und die Folgen* (Munich: Fink, 1983).

Chapter 3

1 In English reception, this goes way back; thus, one may note Marshall Montgomery, "Hölderlin and 'Diotima'." *The Modern Language Review*, 7, no. 2 (April, 1912): 193–207, as well as the recent ventriloquism, mixing as advertised, "fiction and non-fiction," David Farrell-Krell, *The Recalcitrant Art: Diotima's Letters to Holderlin and Related Missives*, trans. Douglas F. Kenney and Sabine Menner-Bettscheid (Albany: SUNY Press, 2000), in which Krell, via Gontard, writes himself into Hölderlin's place. See also the first section of my essay on Wallace Stevens focusing on several recent and technically "romantic" readings of "Ganymede" variations in Hölderlin, Goethe, Schiller (and Schumann) "Wallace Stevens, Heidegger, and the 'Virile Hölderlin': Poetry and Philosophy and The Travelogue of the Mind." *Borderless Philosophy*, 3 (June 2020): 1–31.
2 I write about some of the tricky details that can be involved in the masculinist world of philosophy when it comes to such pairing and our assessments, enthusiastic and not, pro and not, of the same in my "Great Men, Little Black Dresses, & the Virtues of Keeping One's Feet on the Ground." *MP: An Online Feminist Journal*, 3, no.1 (August 2010): 57–78.
3 See for references to the literature and discussion, my essay, "Reading Lou von Salomé's Triangles." *New Nietzsche Studies*, 8, nos 3 & 4 (2011/2012): 82–114.
4 This begins, to be sure, with Lou herself: see Lou Andreas Salomé, *You Alone Are Real to Me: Remembering Rainer Maria Rilke*, trans. Angela von der Lippe (Manchester: Carcanet, 2006 [1928]) and Rolf S. Günther, *Rainer Maria Rilke und Lou Andreas Salome: auf welches Instrument sind wir gespannt: Traumerzählung* (Würzburg: Königshausen & Neumann, 2005).

5 Usually a matter of the salacious, even when treated by academic authors, the relationship between Sartre and de Beauvoir was of a piece with their philosophy, most notably Sartre's own. See on this, patently, Edward Fullbrook and Kate Fullbrook, *Sex and Philosophy: Rethinking de Beauvoir and Sartre* (London: Continuum, 2008), and see too the contributions, including a text by Edward Fullbrook, to Christine Daigle and Jacob Golomb, eds., *Beauvoir and Sartre: The Riddle of Influence* (Bloomington: Indiana University Press, 2009).

6 See my aforementioned essay for a discussion of the politics of who we name by name, that is, on a first-name basis, even when we do not happen to know them personally. See further, my Babich, "Jaspers, Heidegger, and Arendt: On Politics, Science, and Communication." *Existenz. An International Journal in Philosophy, Religion, Politics, and the Arts*, 4, no. 1 (Spring 2009). Online. https://existenz.us/volumes/Vol.4-1Babich.html. There are many discussions of Arendt and Heidegger, to be sure, but see Lütkehaus's *Hannah Arendt–Martin Heidegger: Eine Liebe in Deutschland* (Marburg: Basilisken-Presse, 1999). But see, too, antagonistically minded, Emmanuel Faye, *Arendt et Heidegger: Extermination nazie et destruction de la pensée* (Paris: Albin Michel, 2016).

7 I will be focusing on the triangle; most discussions of Heidegger and Arendt detail the dyad, and there is also Carl Djerassi's play on the quadrate, *Foreplay: Hannah Arendt, The Two Adorno, and Walter Benjamin* (Madison: University of Wisconsin Press, 2011). In addition, there is also the focus on the better-known name in all these discussions; this is one critical meaning of the currently popular foregrounding of "public philosophy," such as we may see in the focus of Margarethe von Trotta (Dir.) 2013, *Hannah Arendt*. See, for a discussion, also foregrounding Jaspers by way of yet another triangulation with Heidegger, Babich, "Thinking on Film: Jaspers, Scholem, and Thinking in Margarethe von Trotta's *Hannah Arendt*." *German Politics and Society*, Issue 118, 34, no. 1 (Spring 2016): 77–92.

8 Lüdger Lütkehaus, "'In der Mitte sitzt das Dasein'. Die Philosophen Günther Anders und Peter Sloterdijk lesen zweierlei Heidegger." *Die Zeit*, January 24, 2002.

9 Anne Carson, *Eros, The Bittersweet* (Princeton: Princeton University Press, 1986), 12–13.

10 Ibid., 13.

11 Andler hypothesizes the pathos (and staging) of famous triangular studio photograph of Lou von Salomé and Paul Rée and Nietzsche, describing the puzzle of the 1882 Lucerne photograph including a crouching Lou Salomé in a garden cart—which Lou recalls as "little (far too little!)" in her posthumously published (1951) memoire. Charles Andler, *Nietzsche sa vie et sa pensée II. Le pessimisme esthétique de Nietzsche. La maturité de Nietzsche* (Paris: Gallimard, 1958 [1920–1931]), 440–1. Anders' footnote references medieval woodcuts and sculptures depicting Aristotle on all fours and Phyllis on his back. Cf. Carl Albrecht Bernoulli, *Franz Overbeck und Friedrich Nietzsche: eine Freundschaft*, 2 Vols (Jena: E. Diedrichs, 1908). See, again, for further references, Babich, "Reading Lou von Salomé's Triangles."

12 See Kerstin Putz, "Nachwort: Korrespondenzen Hannah Arendt and Günther Anders." In: Anders/Arendt, *Schreib doch mal hard facts über Dich*, Kertin Putz (ed.), (Munich: Piper, 2016), 229. See too, quite as Putz herself draws on, Iven's 2013 "Spurensuche: Hannah Arendt und Günther Anders in Nowawes." *Mitteilungen der Studiengemeinschaft Sanssouci e.V.*, 18 Jg. (2013): 122–34. All of this should be set, this adds complexities, in the context of Anders, *Kirschenschlacht*, 5.

13 Although Heidrun Friese does not take account of the complexities I engage in my discussion of Lou and Nietzsche's encounters, she benefits from Günter Wohlfart's

discussion of the "blink of an eye," "*Augenblick*," see Friese, "Leseszenen: Gelehrte lessen vor." In Alf Lüdtke and Reiner Prass (eds.), *Gelehrtenleben: Wissenschaftspraxis in der Neuzeit* (Cologne: Böhlau, 2008), 252ff. Friese emphasizes the tendency that captures women who seek as Arendt sought, in her aspiration to talk openly with Heidegger, which is to be sure an aspiration for or towards intellectual conversation, to have "effect," or academic importance, in their nature as women. Thus, even as Jonas focuses on what he assesses to have been an intellectual lack of parity in her relationship to Anders, Arendt is still assessed in and via her position as a woman, as Anders' wife.
14 Alexander Nehamas, *Only a Promise of Happiness: The Place of Beauty in a World of Art* (Princeton: Princeton University Press, 2007), 53.
15 Arendt/Anders, *Schreib doch mal hard facts über Dich. Briefe 1939 bis 1975, Texte und Dokumente*. ed. Kerstin Futz (Munich: Piper, 2016), 33/35.
16 In our era, 'texts' are ubiquitous and only make matters worse, more insidious, as one cannot "miss" a text even if one has no time to respond properly, as one has seen (even if one has not read) the text and, worse yet, the other will also know, be that other the sender himself or herself or government or corporate surveillance, that one has "seen" the text. See Babich, "Texts and Tweets: On The Rules of the Game." *The Philosophical Salon: Los Angeles Review of Books*, May 30, 2016. Online. https://thephilosophicalsalon.com/texts-and-tweets-on-the-rules-of-the-game/; but see also Chapter 9 on media control and "überveillance," and Anders' idea of radio range and connectivity as tether or leash.'
17 Maurice Merleau-Ponty, "Indirect Language and the Voices of Silence." In *Signs*, trans. Richard McCleary (Chicago: Northwestern University Press, 1964).
18 Giuseppina Moneta, "Profile." In Babich (ed.), *From Phenomenology to Thought, Errancy, and Desire: Essays in Honor of William J. Richardson, S.J.* (Dordrecht: Kluwer, 1995), 206.
19 Ibid., 205.
20 Martin Heidegger, *On the Way to Language*, trans. Peter Hertz (New York: Harper & Row, 1971), 58.
21 Moneta, "Profile," 207.
22 Arendt/Anders, *Schreib doch mal hard facts über Dich*, 86.
23 Ibid., 87.
24 Moneta, "Profile," 206.
25 On Maier-Katkin's account, Heinrich Blücher almost necessarily becomes the loyal husband, a counterpart to Heidegger's wife Elfriede. We learn that Heidegger found Blücher congenial because of his insights into Nietzsche. Regrettably the point is not developed, but given Arendt's own interest in Nietzsche (it can be argued that she studs her footnotes with references to Nietzsche, as she does in the latter pages of *The Human Condition*, to appeal to Heidegger) one wonders if these references also spoke to Blücher.
26 See, just for one example, Christian Dries's afterword sketch of the relationship between Anders and Arendt in Anders, *Die Kirschenschlacht*.
27 Gillian Rose, *Love's Work: A Reckoning with Life* (New York: Schocken Books, 1998).
28 I quote here the first three verses to give a sense of Schiller's poem: "In einem Tal bei armen Hirten /.
    Erschien mit jedem jungen Jahr, / Sobald die ersten Lerchen schwirrten, / Ein Mädchen, schön und wunderbar. // Sie war nicht in dem Tal geboren, / Man wußte nicht, woher sie kam, / Doch schnell war ihre Spur verloren, /.

Sobald das Mädchen Abschied nahm. // Beseligend war ihre Nähe / Und alle Herzen wurden weit; / Doch eine Würde, eine Höhe / Entfernte die Vertraulichkeit."

29 Daniel Maier-Katkin, *Stranger from Abroad: Hannah Arendt, Martin Heidegger, Friendship and Forgiveness* (New York: Norton, 2010), 22.
30 For a direct account of Taubes' cavalier attitudes in this regard, see my "Ad Jacob Taubes." *New Nietzsche Studies*, 7, nos. 3 & 4 (2007): v–x, an editorial titled to echo Taubes' own small book on Carl Schmitt, cited note 42 below.
31 Arendt/Anders, *Schreib doch mal hard facts über Dich*. Cf., too Neumann, "Noch Einmal: Hannah Arendt, Günther Stern/Anders mit bezug auf den jüngst komplettierten Briefwechsel zwischen Arendt und Stern und unter Rekurs auf Hannah Arendts unveröffentlichte Fabelerzählung Die weisen Tiere." In Bernd Neumann, Helgard Mahrdt, and Martin Frank (eds.), *"The angel of history is looking back": Hannah Arendts Werk* (Würzbach: Königshausen u. Neumann, 2001), 107–26.
32 See on Arendt and Blücher, Neumann, *Hannah Arendt und Heinrich Blücher*.
33 One titular exception is Jenny Lyn Bader's play, recently performed in both New Jersey and Berlin: *Mrs. Stern Wanders the Prussian State Library*.
34 The Fluxus artist and philosopher, Bazon Brock tells the current author that Heidegger's wife, Elfriede, had confided to him that Heidegger, dating back to an illness early in their marriage, was constitutionally impotent, entailing that the majority of his affairs were things of the spirit rather than the body, as such. This is neither unheard of nor implausible; it may, as I have argued elsewhere, be the quasi-rule. In literal terms, however, see Paul Feyerabend's *Killing Time: The Autobiography of Paul Feyerabend* (Chicago: University of Chicago Press, 1995) on the matter of marriage and impotence and its complications. Nothing is simple.
35 Cited in Maier-Katkin, *Stranger from Abroad*, 75.
36 Ibid., 52.
37 This would be the *Leitmotif* and complement to Anders's reflection on Prometheanism in Anders, *The Antiquatedness of Humanity*. Emphasis added.
38 Maier-Katkin, *Stranger from Abroad*, 222.
39 Ibid.,137ff.
40 Ibid., 147.
41 Ibid., 149–50.
42 Jacob Taubes, *To Carl Schmitt* (New York: Columbia University, 2013 [1987]), 9–10. See for a discussion, Christoph Schmidt, "The Leviathan Crucified. A Critical Introduction to Jacob Taubes' 'The Leviathan as Mortal God'." *Political Theology*, 19 (2018): 172–92.
43 "Na, der guckt mich an, der Beamte, mit Genuß und Sadismus, ha, das dauer' drei Monate bis so'n Zettel bearbeitet wird." Jacob Taubes, *Die politische Theologie des Paulus: Vorträge gehalten an der Forschungsstätte der evangelischen Studiengemeinschaft in Heidelberg, 23-27 February 1987* (Munich: Wilhelm Fink 1995 [1993]), 134.
44 As Taubes explains, having gone to the head librarian to inquire further, "Soldaten fahren da in die Enklave, holen die Bücher, stecken die in die Hosen, bringen sie runter und so weiter." Ibid.
45 Taubes, *To Carl Schmitt*, 11.
46 Ibid.
47 This is drawn, as I am reminded by Tracy Strong, from Arendt's July 20, 1963, letter written from her 370 Riverside Drive address (in NY) to Gershom Scholem, addressed as Gerhardt: "Ich liebe in der Tat nur meine Freunde und bin zu aller anderen Liebe völlig unfähig." *Hannah Arendt and Gershom Scholem, Der*

Briefwechsel, Marie Luise Knott (ed.) (Frankfurt am Main: Suhrkamp Verlag, 2010), 439. See Lüdger Lütkehaus, "Unversöhnte Dissonanz. Der Briefwechsel zwischen Hannah Arendt und Gershom Scholem." *Neue Zürcher Zeitung*, October 4, 2010. See also Annette Vowinckel, *Geschichtsbegriff und Historisches Denken bei Hannah Arendt* (Cologne: Böhlau, 2001), 183.

48 David Michaelis. *Personal communication*, December 12, 2019, 20:07, via Facebook.
49 This is noted in its own range of complexity by Kerstin Putz herself, who edited the correspondence between Arendt and Anders, *Schreib doch mal hard facts über Dich. Briefe 1939 bis 1975, Texte und Dokumente*. ed. Kerstin Putz (Munich: Piper, 2016), 252. Putz underlines this complexity in her own "Afterword," citing Eva Michaelis-Stern's contribution "Trägt ihn mit stolz, den gelben Fleck!" In F. A. Krummacher (ed.), *Die Kontroverse, Hannah Arendt: Eichmann - und die Juden* (Munich: Nymphenburger Verlagshandlung, 1964), 152–60.

Chapter 4

1 Translation by Kata Gellen in her analysis *"Kafka, Pro and Contra,"* 284.
2 There are a number of discussions of Kojève. See for a recent discussion with a title reminiscent of what would have been (but for publisher's preferences) the title of the current volume—*Black Star*—Jeff Love, *The Black Circle: A Life of Alexander Kojève* (New York: Columbia University Press, 2018) and for a measured overview, see Ethan Kleinberg, *Generation Existential: Heidegger's Philosophy in France 1927-1961* (Ithaca: Cornell University Press, 2005).
3 Gellen, *"Kafka, Pro and Contra,"* 283.
4 Günther Anders and Hannah Arendt, *Schreib doch mal hard facts über Dich. Briefe 1939 bis 1975, Texte und Dokumente*, Kerstin Putz (ed.) (Munich: Piper, 2016), 179.
5 Virilio, *Art and Fear*, trans. Julie Rose (London: Continuum, 2003 [2000]).
6 Anders, *Franz Kafka*, 32.
7 Ibid.
8 Ibid., 9.
9 See Ernst Schraube, "'Torturing Things Until They Confess': Günther Anders' Critique of Technology." *Science as Culture*, 14, no. 1 (March 2005): 77–85 and on trial metaphorics in Kant, see Howard Caygill's "Taste and Civil Society." In Babich (ed.), *Reading David Hume's ›Of the Standard of Taste‹* (Berlin: de Gruyter, 2019) , 177–212.
10 Ibid., *Franz Kafka*, 42.
11 See my "Nietzsche and Eros between the Devil and God's Deep Blue Sea: The Erotic Valence of Art and the Artist as Actor—Jew—Woman." *Continental Philosophy Review*, 33 (2000): 159–88.
12 Anders, *Franz Kafka*, 51.
13 Ibid., 51–2.
14 Certainly Arendt writes on Valentine's Day in 1956 to Anders, requesting his Kafka book.
15 Rilke, *Duino Elegies*, cited in Anders, *Franz Kafka*, 60.
16 Anders, *Franz Kafka*, 60.
17 Ibid., 61.
18 Arendt and Stern, "Rilke's *Duino Elegies*," In Arendt, *Reflections on Literature and Culture*, 1.

19 Hans-Georg Gadamer, *The Relevance of the Beautiful*, trans. Nicholas Walker (Cambridge: Cambridge University Press, 1987), 37. Not dissimilar to Anders (or indeed Adorno), Gadamer tells us that his reading of the beautiful in the artwork articulates a specifically philosophical anthropology, recounting the key turns of his essay, "First, we looked for the anthropological foundations of art in the phenomenon of play as an excess. For it is constitutive of our humanity that our instincts are under-determined and we therefore have to conceive of ourselves as free and live with the dangers that this freedom implies. This unique characteristic determines all human existence in the most profound fashion. And here I am following the insights of philosophical anthropology developed by Scheler, Plessner, and Gehlen under Nietzsche's inspiration." Ibid. 46.
20 Gadamer, *The Relevance of the Beautiful*, 34, citing Rilke, *Duino Elegies* VII.
21 Gadamer, *The Relevance of the Beautiful*, 34.
22 The posthumously collaged version of Adorno's *Aesthetic Theory* appeared in 1970, before Gadamer's original lectures given in 1974. However, and to be sure, and this is in the spirit of Gadamerian history of effects, *Wirkungsgeschichte*, Adorno had given lecture courses in Frankfurt on Aesthetic Theory beginning in 1950 and continuing for almost two decades. It is worth reading Max Paddison's review of the original translation 1984 by Christian Lenhardt. Here too it can be useful to read Geoff Waite, "Radio Nietzsche, or, How to Fall Short of Philosophy." In Bruce Krajewski (ed.), *Gadamer's Repercussions: Reconsidering Philosophical Hermeneutics* (Berkeley: University of California Press, 2004), 169–211.
23 Gadamer, *The Relevance of the Beautiful*, 53. For Arendt and Anders, and this will be their difference with and as opposed to Gadamer's echo of Hölderlin, it will not be "enough for the transformation simply to say the sayable to the angel; it endures only in repeated retelling (7th Elegy). The human being undertakes this rescue because he therein finds access to the 'other relation.'" Arendt and Stern, "Rilke's *Duino Elegies*," 7.
24 "Aber das Saitenspiel tönt fern aus Gärten: villeicht, daß / Dort ein Liebendes spielt oder ein einsamer Mann / Ferner Freunde gedenkt und der Jugendzeit." Hölderlin, *Brot und Wein*. My translation. I discuss (some) of the challenges of thinking the relation between the inner city and the market and the garden as this goes back, as Nietzsche notes, to Epicurus. See Babich, "Epicurean Gardens and Nietzsche's White Seas." In Vinod Acharya and Ryan Johnson (eds.), *Epicurus and Nietzsche* (London: Bloomsbury, 2020), 52–67.
25 Arendt and Stern, "Rilke's Duino Elegies." In Arendt, *Reflections on Literature and Culture*, 1.
26 Ibid., 2.
27 Here, to be specific, and this is, for Anders, always also a reference to Busoni as much as to Schopenhauer, where the "word music does not signify any worldly thing encounterable in the world." Anders, "Musikalischen Situationen," 16.
28 Arendt and Stern, "Rilke's Duino Elegies," 4.
29 Ibid., 23.
30 Ibid.
31 See Chapter 5.
32 Rilke, *Das Stundenbuch* (Leipzig: Insel Verlag, 1905), 9.
33 Deutsch, "Preface to Rilke, *The Book of Hours*," 6.
34 Rilke, *Das Stundenbuch*, 9.
35 Günther Stern, mit Hannah Arendt, "Rilkes Duineser Elegien."
36 Kohn, "Introduction" to: Hannah Arendt, *Understanding 1930–1954*, xv.
37 Arendt and Stern, "Rilke's *Duino Elegies*."
38 I am indebted to Holger Schmid for this.

39 Arendt, "Introduction," 32.
40 Ibid.
41 Ibid., 41.
42 Arendt and Stern, "Rilke's *Duino Elegies*," 1.
43 *Shema Yisrael, Adonai eloheinu Adonai ehad* (Deuteronomy 6:4)
44 Arendt and Stern, "Rilke's *Duino Elegies*," 1.
45 Ibid., 2.
46 Ibid., 3–4.
47 Ibid., 4.
48 Ibid., 4.
49 Ibid., 5.
50 Ibid., 6.
51 Ibid., 9.
52 Ibid.
53 Ibid., 10.
54 Rilke, *Duineser Elegien*, 7. My translation.
55 Taubes, *Die Politische Theologie des Paulus*, 77.
56 Rilke, *Duineser Elegien*, 7. My translation.
57 Rilke, *Duineser Elegien*, 7–8. Author's translation.
58 Arendt and Stern, "Rilke's *Duino Elegies*," 12–13.
59 Arendt and Stern, "Rilke's *Duino Elegies*," 12.
60 Heidegger, "Wozu Dichter?," 252. See for discussion, Babich, "Heidegger and Leonard Cohen: 'You Want it Darker," *Religions*, 12, 488 (July 2021).
61 Ibid., 253.
62 Heidegger, "Wozu Dichter?," 269.
63 Rilke, from *The Book of Pilgrimage, I, 24*, as cited in Heidegger, "Wozu Dichter?," 269. My translation.
64 Heidegger, "Wozu Dichter?," 290. My translation.
65 Thus, it is to be hoped as I have already expressed, I repeat, my hope that someone finds his or her way to the critical sensitivity, the light humour, and above all, perhaps, to the Schelerian but also Beethovenian depth of heart needed to translate *Die Kirschenschlacht. Dialogue mit Hannah Arendt*. Anders, *Die Kirschenschlacht*.
66 Anders, *La bataille des cerises*, trans. Philippe Ivernel (Paris: Rivages, 2013).
67 Dücker, "Blühende Kirschbäume" in *Machen–Erhalten–Verwalten*, here: 184.
68 Anders, *Besuch im Hades*, 191.
69 Cf. Liessmann, *Günther Anders: philosophieren im Zeitalter der technologischen Revolutionen*, 115.
70 Raulff, "Asche und Ambivalenz," 83. Citing Anders, *Besuch im Hades*, 203.
71 Ibid. Citing Reemtsma, *Vertrauen und Gewalt*, 339.
72 Gerhard Oberschlick, "Editorische Notiz." In Anders, *Die Kirschenschlacht*, 61–72.
73 Anders, *Die Kirschenschlacht. Dialoge mit Hannah Arendt*, ed. Gerhard Oberschlick, (Munich: Beck, 2011). To this must be added, reflecting at the age of eighty-six, the same title, featuring four interviews and an unwritten letter, circa 1988, on lice and such as was his material wont, crippled as he was by arthritis, Anders, *Schriften zur philosophischen Anthropologie*, 331.
74 Allen and Axiotis, "Heidegger on the Art of Teaching," 32. Five years later, this essay now clearly indicated as an imaginary exercise, avowing the parallel with Socrates' apology, itself becomes the basis for a genuine translation, Allen and Axiotis, *L'Art d'enseigner de Martin Heidegger*.

75 Allen and Axiotis, "Heidegger on the Art of Teaching," 29. See too for useful, if approximately analytic and non-Heideggerian, discussion of the *logos spermatikos*, albeit *qua* glossed, whereby Heidegger is represented as characterizing rhetoric not in Platonic terms of "wind-eggs" but as the third in the series of the trivium, "the bastard son of academe." Ibid., 35.
76 See Jacques Derrida, "Des Tours de Babel" in and "Appendix" (the original French text), following thereupon, 208-248, and Walter Benjamin, "The Task of the Translator."
77 We will come back to this in the last chapter.

Chapter 5

1 Walter Benjamin, "Theses on the Philosophy of History." In Hannah Arendt (ed.), *Illuminations: Essay and Reflections* (New York: Schocken, 1969), 257.
2 Ibid., 257-58.
3 See further, already cited earlier: Daniel Ogden, *In Search of the Sorcerer's Apprentice: The Traditional Tales of Lucian's Lover of Lies* (Swansea: Classical Press of Wales), 2007.
4 Jürgen Moltmann, *The Coming of God: Christian Eschatology* (Fortress Press, 2004), 217, here Moltmann cites Anders, *Endzeit und Zeitenende. Gedanken über die atomare Situation* (Munich: Beck, 1972 [1959]).
5 Latour, *The Pasteurization of France*.
6 For example, the contributions in Liessmann, ed., *Günther Anders kontrovers* or else Margret Lohmann's dissertation, *Philosophieren in der Endzeit* or, again, Lütkehaus, *Philosophieren nach Hiroshima* as well as Lütkehaus' *Schwarze Ontologie*. I have cited English language and other studies in the earlier chapters.
7 "What then is time'? If no one asks me, I know what it is. If I want to explain it to someone who asks, I don't know." Augustine, *Confessions*, trans. Henry Chadwick (Oxford: Oxford University Press, 1991), XI, 14.
8 See my discussion, "Reading Nietzsche's 'Educational Institutions' with Jaspers & MacIntyre on 'The Idea of the University' — and Severus Snape."
9 See Peter Sloterdijk, *Terror from the Air*, trans. Amy Patton and Steve Corcoran (Los Angeles: Semiotext(e), 2009 [In German in 2002]).
10 Anders and Eatherly, *Burning Conscience*, 13. See too the more mainstream title, Huie, *The Hiroshima Pilot*.
11 Anders and Eatherly, *Burning Conscience*, 13.
12 Ibid.
13 Anders and Eatherly, *Burning Conscience*, 1.
14 Ellul, *The Technological Society*. The original, 1954, title was in advance of Anders' work: *La technique ou l'enjeu du siècle*. See Christophe David, "Günther Anders et la question de l'autonomie de la technique," which also includes a brief note on Taubes on Anders which to be sure Taubes regarded in terms of his own eschatological perspective, as he discusses this in his posthumously gathered reflections on Paul—David cites the French translation (and locus): Taubes, *La théologie politique de Paul* (Paris: Seuil, 1999), 164—and articulated between the right (Carl Schmitt) and the left.
15 Anders and Eatherly, *Burning Conscience*, 11.
16 Ibid.

17 Sophocles, *Oedipus at Colonus*, "μὴ φῦναι τὸν ἅπαντα νικᾷ λόγον τὸ δ', ἐπεὶ φανῇ . . ." (1224f).
18 This common quotation from Yeats is cited here following John Holcombe's online critical notice: "Translating Sophocles 3," a notice that draws for its own part on Michael Gilleland's scholarly, critical discussion, "Yeats and Sophocles."
19 A "straight flush" is jargon for a poker hand of five cards in sequence and of which there are better and worse kinds. In Eatherly's case, the name of his B-29 Superfortress was illustrated on the nose of the plane with a depiction of a toilet bowl with a downed Japanese pilot in the toilet and using the toilet seat as a flotation device with a disembodied hand on the right-hand side poised to pull the chain, for a "straight flush."
20 Giorgio Agamben, *What Is an Apparatus and Other Essays* (Stanford: Stanford University Press, 2009), 49.
21 Ibid., 12.
22 Ibid., 11.
23 Anders and Eatherly, *Burning Conscience*, 20.
24 Stern, *Über das Haben*.
25 Anders and Eatherly, *Burning Conscience*, 12.
26 Ibid., 13.
27 In this same context, Anders claims Jules Verne as the patron saint of modern technology: "*The Prophet of the Technological Revolution*." AM II, 428.
28 Berthold Hoeckner, *Programming the Absolute*.
29 I discuss Anders in connection with Adorno on the space of sound and in connection with Nietzsche on time and rhythm in *The Hallelujah Effect*.
30 Hoeckner, to be sure, as a historian of music, does not attend to the breadth of this array as I am discussing Anders here, and prefers the more common constellation, as most scholars do, of the names he cares to name in his own study.
31 Hoeckner's reflections are broad ones but I argue here that to have the measure he wishes and needs, even more damned names (in the Fortean sense) are required than Anders' own. I am speaking of the now-nearly forgotten Siegmund Levarie, and I discuss this (in another context) in *The Hallelujan Effect*; see 7 as well as 196ff.
32 Hoeckner, *Programming the Absolute*, 16.
33 Ibid.
34 Ibid., 17.
35 Ibid. See for the same citation, Benjamin, GS V: 1, 578, Günther Figal's chapter: "Aesthetic Experience of Time" in Figal, *For a Philosophy of Freedom and Strife*, 121.
36 J. Robert Oppenheimer, on the thoughts and reactions on July 16, 1945, at the Trinity atomic bomb test site. "We knew the world would not be the same. A few people laughed. . . . A few people cried. . . . Most people were silent. I remembered the line from the Hindu scripture the Bhagavad Gita; Vishnu is trying to persuade the prince that he should do his duty, and to impress him takes on his multi-armed form, and says, 'Now I am become death, the destroyer of worlds.' I suppose we all thought that, one way or another." In *The Decision to Drop the Bomb*, NBC documentary, 1965.
37 Peter Sloterdijk, *Critique of Cynical Reason* (Minneapolis: University of Minnesota Press, 1987), 1312.
38 Günther Anders, *Endzeit und Zeitenende. Gedanken über die atomare Situation* (Munich: Beck, 1972 [1959]).
39 The historian Charles Patterson's *Eternal Treblinka: Our Treatment of Animals and the Holocaust* thus borrows from Isaac Bashevis Singer. For the quote here, see Patterson, *Eternal Treblinka*, 183.

40   Robert Jungk, *Strahlen aus der Asche. Geschichte einer Wiedergeburt* (Munich: Scherz, 1991 [1959]), 317; the English edition was published only two years later in 1961 as *Children of the Ashes. The Story of a Rebirth* (London: Heinemann, 1961).
41   Holger Nehring, "Cold War, Apocalypse and Peaceful Atoms. Interpretations of Nuclear Energy in the British and West German Anti-Nuclear Weapons Movements, 1955–1964." *Historical Social Research*, 29, no. 3 (2004): 150–70. Nehring also refers to Anders in the same time era.
42   See David S. Bertolotti, "The Atomic Bombing of Hiroshima." In Bertolotti, *Culture and Technology* (Bowling Green: Bowling Green State University Popular Press, 1984), 81–112.
43   W. G. Sebald, *On the Natural History of Destruction* (New York: Modern Library, 2004).
44   Sebald, "Air War and Literature." In *On the Natural History of Destruction*, 30.
45   Karl Löwith, "European Nihilism: Reflections on the Intellectual and Historical Background of the European War." In Löwith, *Martin Heidegger, and European Nihilism*. Gary Steiner, trans. (New York: Columbia University Press, 1995), 173–284, as well as Karl Jaspers, *The Atom Bomb and the Future of Man*, trans. E. B. Ashton (Chicago: University of Chicago Press, 1961). For the specifically American context here, see Herman Kahn, *On Thermonuclear War* (Princeton: Princeton University Press, 1962), and for a discussion of Kahn from a present-day context, Sharon Ghamari-Tabrizi, *The Worlds of Herman Kahn: The Intuitive Science of Thermonuclear War* (Cambridge: Harvard University Press, 2005).
46   To this end it can be useful to read Jeffrey Bussolini's "Los Alamos as Laboratory for Domestic Security Measures: Nuclear Age Battlefield Transformations and the Ongoing Permutations of Security," both in the current context and with respect to the closing theme, if this is not his question, of the geopolitics that is the geoengineering that Sloterdijk for his part traces and that for Anders requires the consideration of the atrocities that took place in the Pacific at the hands of the good guys, on the right side, the American soldiers quite beyond the atrocities we are still attempting to think with respect to Europe.
47   See Babich, "Constellating Technology: Heidegger's *Die Gefahr* / The Danger."
48   Eatherly and Anders, *Burning Conscience*.
49   See again, Sloterdijk, *Critique of Cynical Reason*. Cf. Babich, "Sloterdijk's Cynicism."
50   Adorno, *Minima Moralia*, §148, 233.

Chapter 6

1   The formula is striking and, again, I recommend as already cited at the outset, Schraube's study, "Torturing Things Until They Confess."
2   See, again, Gabriel Marcel, *Being and Having*, trans. Katharine Farrer, (Glasgow: University Press, 1949).
3   Marcel: "Herr Gunter Stern's book *Ueber das Haben* (published at Bonn by Cohen, 1928)" in Marcel, *Being and Having*, 157.
4   Ibid., 158.
5   Marcel on Anders, in ibid., 157.
6   Johann Wolfgang von Goethe, *Faust – Eine Tragödie. Beide Theile in Einem Bande* (Stuttgart und Augsburg, J. G. Cotta'schen Buchhandlung, 1858), Kapitel 6.

7   I refer to Adorno's 1937 *Zur Metakritik der Erkenntnistheorie* (the dating of which Adorno recalls in the foreword to his 1966 *Negative Dialectics* and which he subsequently revised in the 1950s). In his preface to the 1970 edition, Adorno writes that he began work on his "Three Studies of Hegel" as early as 1934, making 1934-7 the years of composition. See *Zur Metakritik der Erkenntnistheorie. Drei Studien zu Hegel*.
8   As cited in Mark Levene, *The Meaning of Genocide: Genocide in the Age of the Nation State* (London: Tauris, 2005), 35; see also Scott Straus, "Contested Meanings and Conflicting Imperatives: A Conceptual Analysis of Genocide." *Journal of Genocide Research*, 3, no. 3 (2011): 366.
9   Levene, *The Meaning of Genocide*, 35.
10 Günther Anders, *Visit Beautiful Vietnam* (Cologne: Pahl-Rugenstein, 1968).
11 There is, it should be noted, a relative dearth of research on this question with the received view mustered, as might be expected, on the side of denial. But see Fenn, "Biological Warfare in Eighteenth-Century North America," and tellingly, given the locus of publication *The Journal of American Folklore*, Adrienne Mayor, "The Nessux Shirt in the New World: Smallpox Blankets in History and Legend." *The Journal of American Folklore*, 108, no. 427 (1995): 54–77 But see: Kristine B. Patterson and Thomas Runge, "Smallpox and the Native American." *American Journal of Medical Science*, 342, no. 4 (April 2002): 216–22.
12 Rolf Tiedemann, "'Do you know What it Will Look Like?' On the Relevancy of Adorno's Theory of Society." Sean Nye, trans. *Cultural Critique*, 70 (Fall 2008): 123–36, here 126.
13 Jacques Derrida, *The Animal That Therefore I Am*, trans. David Wills (New York: Fordham University Press, 2008), 25–6.
14 See here, critically and in particular: Debra Bergoffen, *Contesting the Politics of Genocidal Rape: Affirming the Dignity of the Vulnerable Body* (New York: Routledge, 2011) *Rape* as well as Allison Ruby Reid-Cunningham, "Rape as a Weapon of Genocide." *Genocide Studies and Prevention*, 3, no. 3 (December 2008): 279–96; Sherrie L. Russell-Brown, "Rape as an Act of Genocide." *Berkeley Journal of Law*, 20 (2003): 350–74.
15 See Regina M. Schwartz, *The Curse of Cain: The Violent Legacy of Monotheism* (Chicago: University of Chicago Press, 1997).
16 Cited in Schwartz, *The Curse of Cain*. I thank Holger Schmid for calling Schwartz's book to my attention.
17 The economic geographer, Donald Worster, highlights just this in his book, *Nature's Economy: A History of Ecological Ideas* (Cambridge: Cambridge University Press, 1977), although this is hardly his central point.
18 Patterson, *Eternal Treblinka*.
19 Briton Cooper Busch, *The War Against the Seals: A History of the North American Seal Fishery* (Montreal: McGill-Queen's Press, 1987).
20 Greg Grandin, *The Empire of Necessity*, 1 and Grandin, "The Other 'Moby Dick': Melville's 'Benito Cereno' is an Analogy for American Empire." Here it is useful to cite the extended title: "Captain Ahab isn't the only Melville character that stands to teach us something about unhinged American power."
21 See further, among others in Jason Frank's edited collection, *A Political Companion to Herman Melville* (Lexington: University Press of Kentucky, 2013), Strong, "Follow Your Leader," here 281–309 as well as Balfour, "What Babo Saw," 259–80.
22 Anders AMI II, 432, cites Kant's *Critique of Practical Reason*, I, 8.
23 Anders, *Die Antiquirtheit des Menschen II*. Note to be sure that this text exists in French translation under the name: "Monsieur autrement." See a discussion, largely

in response to the first volume, by Thierry Simonelli, "Le monde, 'Vu à la télé'" in the online journal founded by Angèle Kremer-Marietti, *Dogma—Revue électronique de Sciences Humaines*.

24 See Bernasconi & Mann, "The Contradictions of Racism." See too Vall's edited collection, *Race and Racism in Modern Philosophy* and cf. Farr, "Locke, Natural Law, and New World Slavery."
25 Walter Benjamin, "The Work of Art in the Age of Mechanical [Technical] Reproduction." *Illuminations* (New York: Schocken, 1968), 241.
26 Paul Virilio, *Art & Fear*, trans. Julie Rose (London: Continuum, 2003 [2000]), 16.
27 Ibid., 17.
28 Bruno Latour, "Biography of an Investigation: On a Book about Modes of Existence." Cathy Porter, trans. *Social Studies of Science*, 43, no. 2 (2013): 292.
29 Lewis Carroll, *Jabberwocky and Other Poems* (Mineola: Dover Publications, Inc., 2001), 20.

Chapter 7

1 For a discussion, specifically directed to Francophone reception, Christian Sommer, "«Ni homme, ni capucin, c'est un Dasein». Remarques sur *Über Heidegger*," *Tumultes*, n° 28-29, 1/2 (2007): 51-68. In the context not of a reading of Anders but Heidegger and a reflection on reproductive technology, Jill Drouillard used Anders' phrase as subtitle for her lecture, "Heidegger's Sexless Community," published in the 2018 *Proceedings of the Heidegger Circle*.
2 Adorno, *Aesthetic Theory*, 13-14.
3 See, for discussion, the last section, "Afterword/Afterworld: On Embryonic Mosaics and Chimeras, Animal Farm for the 21st Century" of Babich, "Ivan Illich's Medical Nemesis and the 'Age of the Show,'" 11-13.
4 Anders, *Besuch im Hades*, 195.
5 Ibid., 432.
6 Jean Baudrillard, *Intelligence of Evil or, The Lucidity Pact* (London: Bloomsbury, 2013 [2004]) in addition to his *La Guerre du Golfe n'a pas eu lieu* (Paris: Editions Galilée, 1991).
7 Anders, *Commandments in the Atomic Age* in: Anders and Eatherly, *Burning Conscience*, 11.
8 Nietzsche, *Dionysos-Dithramben*: "Zwischen Raubvögeln," *Nietzsche's Werke. 1. Abtheilung. Band VIII*, 424.
9 Nietzsche, 1888, 16 W II 7a, NF 1887-1889, 488.
10 Virilio, *La procédure silence*, in English as *Art & Fear*. See, especially the section entitled: "A Pitiless Art."
11 Virilio, "A Pitiless Art," in: Virilio, *Art & Fear*, 28-9.
12 Benjamin, "The Work of Art in the Age of Mechanical [Technical] Reproduction," 241.
13 Thus, the editor of Virilio's *Art & Fear*, John Armitage, explains this focus as a result of Virilio's Catholicism.
14 Modern physiology has hardly abandoned pain experiments to this day, all designed "to determine how all the different body parts react to pain." Virilio, *Art & Fear*, 32. Thus, one scholar rightly compares "Haller's achievement . . . as accomplishing for physiology what Copernicus, Newton, and Huygens accomplished for their fields."

15. Here Virilio seems to anticipate the age of Trump and "the desolation of modern times with their cardboard cut-out dictator that keeps popping up, whether it be Hitler or the 'Futurist', Mussolini, Stalin or Mao Zedong." Virilio, *Art & Fear*, 33.
16. Pierre Duhem, "German Science and German Virtues." See further, Babich, "Heidegger's Jews," here 144.
17. Patterson, *Eternal Treblinka*.
18. See, again, Babich, "Afterword/Afterworld" in "Ivan Illich's Medical Nemesis and the 'Age of the Show.'"
19. Martine Rothblatt, "Biology is Technology." Youtube Lecture. Online. https://www.youtube.com/watch?v=wSZgrEtakz8. See Jason Koebler, "Martine Rothblatt Wants to Grow Human Organs in Pigs at This Farm." Motherboard. *Vice*. June 24, 2015.
20. Nick Thieme, "The Gruesome Truth About Lab-Grown Meat: It's Made by Using Fetal Cow Blood." *Slate*, July 11, 2017."
21. See again, Virilio 'A Pitiless Art" as well as Joachim Müller-Jung, "Das Schwein, dein Spender. Vermenschlicht: gentechnisch veränderte Ferkel aus München" in addition to further references in the notes to Babich, "Ivan Illich's Medical Nemesis and the 'Age of the Show.'" And see the contributions to Taupitz, Jochen and Marion Weschka, eds., *Chimbrids—Chimeras and Hybrids in Comparative European and International Research: Scientific, Ethical, Philosophical and Legal Aspects* (Frankfurt: Springer Science & Business Media, 2009).
22. See Michael's "Preface," *Überveillance and the Social Implications of Microchip Implants*, xxiv. The collection includes Albrecht's "Microchip Induced Tumors in Laboratory Rodents and Doges," 281–318.
23. See here, for a very tame example of cyborg—elsewhere I note that the wearing of contact lenses is adequate to qualify one as cyborg, however anticlimactic, Trafton's 2019, "Storing Medical Information Below the Skin's Surface." Cf. Donnelly et al., *Microneedles for Drug and Vaccine Delivery and Patient Monitoring* (Oxford: John Wiley & Sons, 2018).

Chapter 8

1. See here Carl Størmer's 1928 report on a letter received from the Oslo radio engineer, Jørgen Hals, in Størmer's "Shortwave Echoes and the Aurora Borealis." In Paris, in the same year, Størmer also published a longer, more explicative account: "Sur un écho d'ondes électromagnétiques courtes arrivant plusieurs secondes apres le signal émis, et son explication d'après la théorie des aurores boréales." In a more "nominal" echo, more recently, a Norwegian documentary directed by yet another Carl Størmer, *The Ghost Radio Hunter*, follows the electronic artist-pair, Per Martinsen and singer Aggie Peterson, tracking similar phenomena in recording the 2013 Frost album, *Radiomagnetic*. Anders' "Spuk und Radio" concerns rather lower atmospheric heights, radio echoes of the same music heard in passing along a corridor or a street.
2. Johann Wolfgang von Goethe, *Faust. Der Tragödie zweiter Teil* (Stuttgart und Tübingen: J. G. Cotta'schen Buchhandlung, 1832 [1808]), 314.
3. Tracy B. Strong, *Politics Without Vision: Thinking Without a Bannister in the Twentieth Century* (Chicago: University of Chicago Press, 2012), 160. I am grateful to Tracy Strong for reminding me of Freud's quotation of Goethe in response to my constant observation of Goethe's importance for Anders.

4 Friedrich Kittler, *Gramophone, Film, Typewriter*, trans. Geoffrey Winthrop-Young and Michael Wutz (Stanford: Stanford University Press, 1999), 13.
5 Stern (Anders), "Spuk im Radio" (on the title page of the journal) "Spuk und Radio" on the essay itself.
6 See for a discussion, including further reference to the literature on this study, Babich, "Adorno's Radio Phenomenology," in addition to Babich, *The Hallelujah Effect*.
7 See Adorno, *Current of Music. Elements of a Radio Theory*.
8 Kittler, *Gramophone, Film, Typewriter*, 97.
9 See on, for example, the contributions to Janz, ed., *Hermeneutics: Place and Space*, as well as, more broadly, the contributions to Pavlos Kontos and Veronique Fóti, eds., *Political and Philosophical Essays for Jacques Taminiaux* and for a more Ricoeur-minded approach, Paul Downes, *Concentric Space as a Life Principle Beyond Schopenhauer, Nietzsche and Ricoeur*.
10 See, including illustrations, Babich, *The Hallelujah Effect*, 121ff.
11 Anders, "Zur Phänomenologie des Zuhörens (Erläutert am Hören impressionistischer Musik)," in *Musikphilosophische Schriften*, 211–225.
12 Adorno's text was translated, as Ernest McClain (1918–2014) recounts this, by George Simpson.
13 Anders, "*Philosophische Untersuchungen über musikalische Situationen*," in *Musikphilosophische Schriften*, 15–140, but see too the correspondence concerning, including an application to the Guggenheim Foundation to find support for the preparation of an English version of his original 1937 manuscript, here: 145–73.
14 Adorno writes on this in *Current of Music*. For a specific discussion, see Babich, "On The Hallelujah Effect."
15 See further, again, Babich, "Adorno's Radio Phenomenology."
16 See, for a discussion, Cornelia Epping-Jäger, "Lautsprecher." In Daniel Morat and Hans Jakob Ziemer (eds.), *Handbuch Sound: Geschichte-Begriffe-Ansätze* (Frankfurt: Springer, 2017), esp. 396–7. And see too Thadeusz, "Nazi-Labor in Oberfranken. Geheimwaffen aus dem Burgverlies."
17 Heidegger, *Überlegungen XII–XV*, GA 96, 92.
18 Martin Heidegger, *Beiträge zur Philosophie (Vom Ereignis)*. Gesamtausgabe 65, F.W. von Hermann, ed., (Frankfurt am Main: Klostermann, 1989), GA 65, 131.
19 Theodor Lessing, *Der Lärm. Eine Kampfschrift gegen die Geräusche unseres Lebens* (Wiesbaden: Bergmann, 1908).
20 Cited in John L. Snelly, *The Nazi Revolution* (Boston: B.C. Heath, 1959), 7.
21 See, in addition to the third section of Babich, *The Hallelujah Effect*, very specifically and in physical detail Thanos Vovolis, *Prosopon: The Acustical Mask in Greek Tragedy and in Contemporary Theatre* (Stockholm University of the Arts, 2009).
22 Babich, "Adorno's Radio Phenomenology," and, again, *The Hallelujah Effect*.
23 Reinhard Ellensohn attends to Husserl in his book, citing the authority of Malwida Husserl to testify to Anders' phenomenological facility. See Ellensohn, *Der andere Anders*.
24 Stern and Anders, "Spuk und Radio," 66.
25 Kittler, *Gramophone, Film, Typewriter*, 99. Kittler references the Jürgenson wave in his introduction, ibid., 13, via Walter Rathenau as well as Cocteau's "radio theory," including a reference to "*doppelgänger*," ibid., 192.
26 Stern and Anders, "Spuk und Radio," 66.
27 To this extent, this differs from Edvin Østergaard's thoughtful analysis of shadow (and echo). See Østergaard, "Echoes and Shadows: A Phenomenological Reconsideration of Plato's Cave Allegory," *Phenomenology & Practice*, 13, no. 1 (2019): 20–33.

28 Heidegger, *Überlegungen XII-XV*, GA 96, 265. I discuss this instantiation of radio and its phenomenological significance in Babich, "Heidegger's Black Night." In Ingo Farin and Jeff Malpas (eds.), *Reading Heidegger's Black Notebooks 1931–1941* (Cambridge, MA: MIT Press, 2016), 59–86.
29 Jean Baudrillard, "Requiem for the Media." In *For a Critique of the Political Economy of the Sign*, here 169.
30 Baudrillard, "Requiem for the Media," 169. Cf. further Jean Baudrillard, *Intelligence of Evil or, The Lucidity Pact* (London: Bloomsbury, 2013 [2004]).
31 Jean Baudrillard, "Requiem for the Media." In *For a Critique of the Political Economy of the Sign*, 170.
32 Ibid., 169.
33 Babich, "Screen Autism, Cellphone Zombies, and GPS Mutes."
34 Knelangen contradicts Anders' assertion that "only Adorno would have understood" the tonal reference of rising and ascending tones (cf. Knelangen, "Günther Anders und die Musik oder 'Der Klavier-spieler mit dem Zeichenstift,'" here: 79), per contra, see Babich, "*O, Superman!* Or Being Towards Transhumanism."
35 Kittler, *Gramophone, Film, Typewriter*, 37.
36 Ibid., 100.
37 Ibid., 103.
38 Ibid., 107–8.
39 Ibid., 111.
40 See for one author who draws a connection here Timo Kaerlein, "Playing with Personal Media" and see too, albeit largely with approbative reference to Simondon, Giovanni Carrozzini, "Technique et humanisme."
41 This preconditioning is the condition for what has been called the current age of omnipresent surveillance, well beyond any Benthamite or Foucauldian panopticon. There are many discussions, but see, in addition Galič et al., "Bentham, Deleuze and Beyond" as well as, including further references, Hauptman, "Surveillance" and, for a recent popular account, Louise Matsakis, "How the Government Hides Secret Surveillance Programs." *Wired*, January 9, 2018.
42 Adorno, *Current of Music*, 73ff. See for discussion and further references, Babich, *The Hallelujah Effect*, 144f.
43 This is the reason for my title, and I do this from the start, writing on "Effects, Mediations, and Primes." See Babich, *The Hallelujah Effect*, 1.
44 See further, Illich's "The Age of the Show" and Debord, *The Society of the Spectacle*.
45 Once again, I recall, as cited earlier that the philosopher and gaming designer names our smartphones "pocket robots." See Babich and Bateman, "Your Plastic Pal Who's Fun to Be With," and Bateman's *The Virtuous Cyborg*. Thus, I reflect not only on the gender presumptions (and deficiencies) of sex robots Babich, "Robot Sex, Roombas, and Alan Rickman," as well as the relation between twitter and erotic communication. Cf. "Texts and Tweets."
46 Hofheinz, "Arab Internet Use" but see too Morozov, *The Net Delusion* as well as Frischmann and Selinger, *Re-Engineering Humanity* and, very conventionally, David Patrikarakos, *War in 140 Characters: How Social Media Is Reshaping Conflict in the Twenty-First Century* (New York: Basic Books, 2017).
47 See: https://www.youtube.com/watch?time_continue¼3&v¼vwSRqaZGsPw. See, for a discussion, Lester, "Gil Scott-Heron: The Revolution Lives On." In addition to an array of cultural influences and echoes, including remix, this is also the self-reflexive title of a documentary that should have been a point-counterpoint of the Venezuelan coup

contra Hugo Chavez (see here: the philosopher and film theorist and producer, Rod Stoneman, *Chavez: The Revolution Will Not Be Televised*. Cf. Babich, "The Revolution Will be Televised").
48 This is thematized as our consumption of mass media in Anders' interview with Manfred Bissinger.

Chapter 9

1 For example, the aptly titled "Pariser Musikbriefe."
2 Anders, "Philosophische Untersuchungen über musikalische Situationen," including grant applications and other documents: MS, 145–173.
3 Theodor Adorno, "What National Socialism Has Done to the Arts." In Richard Leppert (ed.), *Essays on Music*, trans. Susan Gillespie (Berkeley: University of California Press, 2002), 376.
4 Ibid.
5 Lydia Goehr's *The Imaginary Museum of Musical Works: An Essay in the Philosophy of Music* (Oxford: Oxford University Press, 1992).
6 Adorno, "Analytical Study of the *NBC Music Appreciation Hour*," and "'What a Music Appreciation Hour Should Be': Exposé," in *Current of Music*.
7 I am indebted for the reference to Adorno's smile to Holger Schmid.
8 See Roholt, "On the Divide: Analytic and Continental Philosophy of Music," but, as if to confirm the aforementioned distinction, Roholt discusses mainstream *analytic* academics, with the exception of Bourdieu inasmuch as Roholt follows Simon Critchley's claim that "there is no" such contingent, contra Reiner Schürmann, who wrote on the distinction. See Schürmann, "Concerning Philosophy in the United States." Today, analytic philosophy increasingly occupies *both* sides such that few representatives of Schürmann's style of philosophising remain.
9 Roger Scruton, "Is Adorno a Dead Duck?" Lecture: *Royal Musical Association Music and Philosophy Study Group 2nd Annual Conference*, King's College London, 20 J.
10 Max Paddison, *Adorno's Aesthetics of Music* (Cambridge: Cambridge University Press, 1997).
11 See here: Lütkehaus, "Der 'Atomphilosoph' scheitert an der Musik."
12 Stern, *Die Rolle der Situationskategorie bei den logische Sätzen*.
13 See, on guilt, Carl Schmitt, *Über Schuld und Schuldarten. Eine terminologische Untersuchung* (Berlin: Duncker & Humblott, 1977 [1910]) and further, with respect to Schmitt and music, Angela Reinthal: "*Mich hält ein reines Intervall*". *Carl Schmitt und die Musik* (Berlin: Duncker & Humblott, 2019).
14 Lütkehaus, "Der 'Atomphilosoph' scheitert an der Musik."
15 Maurice Halbwachs, "La mémoire collective chez les musiciens." *Revue philosophique* (March–April 1939): 136–65.
16 Dilthey, "Other Persons and their Expressions of Life," In *Descriptive Psychology and Historical Understanding* (The Hague: Martinus Nijhoff, 1977), 222/140.
17 Ibid.
18 Ibid., cited in Anders, "Philosophische Untersuchungen über musikalische Situationen," 39.
19 Dilthey, "Other Persons and their Expressions of Life," 140.
20 Ibid., 138.

21 Ibid., 138.
22 Ibid., 139.
23 Ibid., 141.
24 See Eugen Fink, *Oase des Glücks. Gedanken zu einer Ontologie des Spiels* (Freiburg: Alber, 1957), and cf. my, "Artisten Metaphysik und Welt-Spiel in Fink and Nietzsche."
25 Cf. Adorno's 1955 "The Aging of the New Music," an "aging" which proceed as it might had done little to comesticate the new music for the concert programme, as Gadamer also attests in his 1974 *The Relevance of the Beautiful*, and which still seems to hold to the current day.
26 Gadamer, *The Relevance of the Beautiful*.
27 See further for a mainstream overview, Gratzer's encyclopedia entry, "Listening to Music: An Art in It Self or Not," citing the variations on the art of listening per se, as well as the generic concept of "musical appreciation," 465–6, including reference to Schmid's broader study, *Kunst des Hörens. Orte und Grenzen philosophischer Spracherfahrung*. Gratzer, who also refers to Konrad Liessman, is writing in the context of a handbook that grew out of an Austrian conference on the theme of music, foregrounding listening; thus, it is also worth noting the articles by Maus, "Listening and Possessing" and not at all least: Christiane, "Everybody in the concert hall should be devoted entirely to the music," 441–60 and 477–99, respectively.
28 See, in addition to the Heideggerian and musicologist, F. Joseph Smith, Erik Wallrup's *Being Musically Attuned*. Wallrup attends to the tuning or attunement, *Stimmung* of music, not unlike Anders' own reflections on *Zuhören*, which in turn derive from Heidegger. See too Wallrup's earlier, explicitly Heideggerian, but with the focus on belonging, still reminiscent of Anders, "Music, Truth and Belonging," and again sidestepping Anders, Thwaites, "Heidegger and Jazz." See too in passing, as Fanon notes one Günther Stern's importance for his own reflections, Garcia's instructively, phenomenologically, and sociologically titled *Listening for Africa*, 11.
29 I have been reflecting on this question for some time. See for a general discussion of masculine and feminine aesthetics in Nietzsche's use of the contrast, complete with an illustration, relevant to Anders, of Nietzsche's commissioned woodcut of *Prometheus Unbound*, in connection with Beethoven, Babich, *The Hallelujah Effect*, 255.
30 Arendt and Stern, "Rilke's Duino Elegies," 4.
31 Many popular authors on media have drawn on and reference these studies and, in philosophy Jean Baudrillard dedicated, as did Paul Virilio, considerable energy to reflecting on the consequences of what Baudrillard called the "Sociological System of Objects and their Consumption." See Baudrillard, *The System of Objects* and Virilio's *Desert Screen*.
32 Anders, "'What use is the Moon?' (Molussian saying). At no time does the doubt ever arise that something might not have a possible use." AM II, 32.
33 This is also the connection that McClain seeks to make: see Anders, "Rückfrage nach dem Element des Tönes," MS, 98ff. And see my discussion of the acroamatic in Nietzsche in Babich, "Reading Nietzsche's 'Educational Institutions.'"
34 Adorno, "The Social Situation of Music." In Adorno, *Essays on Music, Selected, with Introduction, Commentary, and Notes by Richard Leppert* (Berkeley: University of California Press, 2002), 391–36.
35 Adorno's "On the Fetish Character of Music," 289ff.
36 Ibid., 289.
37 Ibid.
38 Ibid., 290.

39 Julian Johnson, "'The Elliptical Geometry of Utopia': New Music Since Adorno." In Berthold Hoeckner (ed.), *Apparitions: Essays on Adorno and Twentieth-Century Music* (London: Routledge, 2013), 69. See too, Daniel Chua, "Drifting: The Dialectics of Adorno's Philosophy of New Music." In Hoeckner (ed.), *Apparitions*, 1–18.

40 Sonolet, already noted earlier, draws a parallel of this kind by comparing Anders with Bourdieu at the outset of "Literature and Modernity," specifically citing Bourdieu's *habitus* in order to articulate "the nature of power-dominated cultural consumption in France, as is demonstrated in his extended study on taste." Nevertheless, Anders' focus is both more critical and more fundamental.

41 Heidegger, "In der besorgenden Umsicht gibt es dergleichen Aussagen "zunächst" nicht." (SZ 157).

42 To be sure this too is a reference offered as a negative distinction: "Auf dem Grunde dieses existenzial primären Hörenkönnens ist so etwas möglich wie *Horchen*, das selbst phänomenal noch ursprünglicher ist als das, was man in der Psychologie 'zunächst' als Hören bestimmt, das Empfinden von Tönen und das Vernehmen von Lauten" (SZ, 163).

43 [*Das Horen konstituiert sogar die primare und eigentliche Offenheit des Daseins für sein eigenstes Seinkonnen*]. Heidegger here adds the important reference to the voice of the friend, the reference is disputed, moving between Augustinian conscience and the Aristotelian commonplace of another self: "*als Horen der Stimme des Freundes, den jedes Dasein bei sich tragt.*" To be sure this too is a reference offered as a negative distinction: "*Auf dem Grunde dieses existenzial primären Hörenkönnens ist so etwas möglich wie Horchen, das selbst phänomenal noch ursprünglicher ist als das, was man in der Psychologie "zunächst" als Hören bestimmt, das Empfinden von Tönen und das Vernehmen von Lauten*" (SZ, 163).

44 [*"Zunachst" hören wir nie und nimmer Geräusche und Lautkomplexe, sondern den knarrenden Wagen, das Motorrad. Man hört die Kolonne auf dem Marsch, den Nordwind, den klopfenden Specht, das knisternde Feuer.*] Ibid.

45 Note that Adorno engages Anders (as "Stern") in Adorno's *Current of Music* in "Space Ubiquity," 80f. Cf., Babich, "Adorno's Radio Phenomenology."

46 On the Dionysiac in Nietzsche, see the final chapter of Babich, *Nietzsches Antike*, 299–344.

47 This is different from the sheer description of the phenomenon and assessment of different responsive profiles—not everyone, famously enough, experiences the phenomenon, thus leading to the straightforward research assessment in terms of personality types and issues of suggestibility/susceptibility. See Fredborg, Clark, and Smith, "An Examination of Personality Traits Associated with Autonomous Sensory Meridian response (ASMR)" and Ahuja, "It feels good to be measured."

48 Stern/Anders, "The Acoustic Stereoscope." *Philosophy and Phenomenological Research*, 10, no. 2 (December 1949): 238.

49 Ibid.

50 Ibid., 239.

51 Ibid.

52 The View-Master, which had existed for some time, first introduced as a gadget for the 1939 World's Fair, was redesigned in 1958 by the Afro-American designer Charles Harrison for Sears, Roebuck & Company, to be rereleased to great success.

53 Anders, "The Acoustic Stereoscope," 239.

Chapter 10

1. See, again, Greg Milner, *Pinpoint* and cf., Jordan Frith, *Smartphones as Locative Media* (London: Polity, 2015). And see too for reflections on the whereness of "whereness," Robin Mannings, *Ubiquitous Positioning* (Norwood: Artech House, 2008), here 19f. And cf. Alan Oxley, *Uncertainties in GPS Positioning: A Mathematical Discourse* (London: Elsevier, 2017). See also Feldstein, "The Global Expansion of AI Surveillance," Bartlett, *The Dark Net* and my "Screen Autism, Cellphone Zombies, and GPS Mutes."
2. Adorno, "Musical Situation," *Current of Music*, 83.
3. Ibid.
4. See for a discussion the author's "Looking Right, Reading Left."
5. Thus, in addition to Laurence Paul Hemming's foreword to Ernst Jünger, *The Worker: Dominion and Form*, trans. Bogdan Costea (Evanston: Northwestern University Press, 2017), see Bogdan Costea and Kostas Amiridis, "Ernst Jünger, Total Mobilisation and the Work of War." *Organisation*, 24, no. 4 (2017): 1–23, and Antoine Bousquet, "Ernst Jünger and the Problem of Nihilism in the Age of Total War." *Thesis Eleven*, 132, no. 1 (2016): 17–38. And see, too, very insightfully, Vincent Blok, *Ernst Jünger's Philosophy of Technology: Heidegger and the Poetics of the Anthropocene* (New York: Routledge, 2017).
6. Pietsch, et al., eds., *Berechenbarkeit der Welt?: Philosophie und Wissenschaft im Zeitalter von Big Data* (Frankfurt: Springer, 2017), and see Lin's "Ethics of Hacking Back."
7. I cite Joseph Nye Welsh, counsel for the US Army under investigation for communist activities, delivered June 9, 1954, during the Army-McCarthy Hearings in Washington, DC.
8. There are thousands of engineers, architects, chemists, and so on who have pointed to this particularly naked emperor, but, for example, note Harry G. Robinson, III: "The collapse was too symmetrical to have been eccentrically generated. The destruction was symmetrically initiated to cause the buildings to implode as they did."
9. Rosa Brooks, *How Everything Became War and the Military Became Everything: Tales from the Pentagon* (New York: Simon & Schuster, 2016).
10. See here Françoise Levie, *L'homme qui voulait classer le monde. Paul Otlet et le Mundaneum* (Brussels: Les Impressions Nouvelles, 2006).
11. Cf. in this constellation the contributions to the late Frank Hartmann's collective volume, *Vom Buch zur Datenbank*. And cf., too, my "Necropolitics and Techno-Scotosis," *Philosophy Today*, 65 (2021): 305–324.
12. Vannevar Bush, "As We May Think." *Atlantic Monthly*, 176 (1945): 101–8.
13. Kittler, *Gramophone, Film, Typewriter*, 13.
14. Don Gordon, *Electronic Warfare: Element of Strategy and Multiplier of Combat Power* (New York: 1981).
15. Kittler, *Gramophone, Film, Typewriter*, 13.
16. Goodman, *Sonic Warfare: Sonic Warfare: Sound, Affect, and the Ecology of Fear* (Cambridge: MIT Press, 2010).
17. Edward Bernays, *Crystallizing Opinion* (New York: Liveright, 1961).
18. Paul Forman, "Kausalität, Anschaulichkeit, and Individualität, or How Cultural Values Prescribed the Character and the Lessons Ascribed to Quantum Mechanics." In N.

Stehr and V. Meja (eds.), *Society and Knowledge: Contemporary Perspectives in the Sociology of Knowledge* (New Brunswick: Transaction Books, 1984), 333–47.

19 Paul Forman, "Behind Quantum Electronics: National Security as Basis for Physical Research in the United States, 1940–1960." *Historical Studies in the Physical and Biological Sciences*, 18 (1987): 149–229, and see too Forman's earlier essay: "Weimar Culture, Causality, and Quantum Theory, 1918–1927: Adaptation by German Physicists and Mathematicians to a Hostile Intellectual Environment."

20 Street, *Politics and Technology*, 44.

21 See, for one example, Steve Fuller, *Humanity 2.0: What It Means to Be Human Past, Present, and Future* (London: Palgrave Macmillan, 2011) and see too Robert Frodeman, *Transhumanism, Nature, and the Ends of Science* (New York: Routledge, 2019).

22 I owe this account to Arun Tripathi, via email, personal communication.

23 Ihde, *Listening and Voice*. It should also be said that Ihde drew for that study both on the insights of his colleague Patrick Heelan, the hermeneutic phenomenologist of perception and philosopher of science, author of *Space-Perception and the Philosophy of Science*, and Smith, author of *The Experiencing of Musical Sound*. Even more than Anders, Smith was ignored by his contemporaries. Smith, editor of *Understanding the Musical Experience*, was also author of *Jacobi Leodiensis Speculum musicae*.

24 This makes an ongoing difference for philosophical reflection, and I have addressed it on numerous occasions. See perhaps most accessibly the first of the dialogue initiated by Chris Bateman—"The Last of the Continental Philosophers," in addition to a book looking at the implications of the analytic turn for Francophone philosophy, the frankly titled: *La fin de la pensée?* as well as "Politik und die analytische-kontinentale Trennung in der Philosophie. Zu Heideggers sprechender Sprache, Nietzsches lügender Wahrheit und der akademischen Philosophie" in: Babich, "Eines Gottes Glück voller Macht und Liebe" and "An Impoverishment of Philosophy," and, most recently, "Good for Nothing: On Philosophy and Its Discontents."

25 This connection is evident on Harold Marcuse's web page subpage dedicated to Anders, featured for many years on the Marcuse website, http://marcuse.faculty.history.ucsb.edu/anders.

26 Quite as David Gill writes, and although Anders is not his reference, there is a clear parallel as "Ellul is often dismissed as a backward-looking, world-fleeing pessimist, and a superficial reading of his work sometimes invites this response." See Gill, "Jacques Ellul and Technology's Trade-Off," as well as Rose, "Errors of Thamus" and Matlack, "Confronting the Technological Society."

27 See Ogden, "The Apprentice's Sorceror." I am grateful to Joel Relihan for discussion.

28 For a discussion of Goethe's indebtedness to Wieland's translation of Lucian and the subsequent history of effects or influence, see Ernst Ribbat's wonderfully titled (from the perspective of a text on Günther Anders), "'Die ich rief, die Geister . . .' Zur späten Wirkung einer Zaubergeschichte Lukians," especially "Mysteriose Poesie," 290ff.

29 See for an analysis, Merton, "The Unanticipated Consequences of Purposive Social Action," here 903. As cited above, Goethe writes his own paradox into the mouth of Mephisto: "*Teil von jener Kraft Die stets das Böse will und stets das Gute schafft*"/"Part of that Power which would/The Evil ever do, and ever does the Good"]. Goethe, *Faust*, Ch. 6.

30 See Anders, *Gewalt, Ja oder Nein?* See too, systematically on this very topic, Orrin H. Pilkey and Linda Pilkey-Jarvis, *Useless Arithmetic: Why Environmental Scientists Can't Predict the Future* (New York: Columbia University Press, 2007). As well as, as already cited and as the phenomenon is not new, David S. Bertolloti, Jr., *Culture and Technology*.

31 See Alexander Stingl's *The Digital Coloniality of Power: Disobedience in the Social Sciences and the Legitimacy of the Digital Age* (Lanham, MD: Lexington Books, 2015).
32 Here Anders refers the reader to his effort in his 1956 volume, to analyse television quite and already in this same direction: AM I, 99.
33 See Yasha Levine, *Surveillance Valley: The Secret Military History of the Internet* (New York: Public Affairs, 2018), as well as the contributions to the collection already cited above, M. Michael and K. Michael (eds.), *Uberveillance and the Social Implications of Microchip Implants: Emerging Technologies* (Hersey: IGI Glbal, 2013) as well as, modulated for today's mainstream, Zuboff, *The Age of Surveillance Capitalism*.
34 Julie Accardo and M. Ahmad Chaudhry, "Radiation Exposure and Privacy Concerns Surrounding Full-Body Scanners in Airports." *Journal of Radiation Research and Applied Sciences*, 7, no. 2 (April 2014): 198–200. See also Rebekka Murphy, "Note, Routine Body Scanning in Airports: A Fourth Amendment Analysis Focused on Health Effects," *Hastings Const. L.Q.*, 39 (2012): 915.
35 Anders, "Being Without Time," 145.
36 Ibid., 146.
37 Ibid., 148.
38 Ibid.
39 Ibid., 149.
40 Ibid., 142.
41 Ibid., 148–9.
42 Adorno, *Current of Music*, 128ff.
43 Ibid., 137.
44 Ibid.
45 Ibid., Emphasis added.
46 See Kittler's *Aufschreibsysteme* and his *Gramophone, Film, Typewriter*.
47 There is an abundance of literature on this. See for pop summaries focused on the United States, Gerald Carson, "The Piano in the Parlor," *American Heritage*, 17, no. 1 (December 1965): 54–9 and Lynn Spigel, *Make Room for TV: Television and the Family Ideal in Postwar America* (Chicago: University of Chicago Press, 1992). Baudrillard's *System of Objects* offers a discussion from a French perspective, as does, in a different sense Bourdieu's *Distinction*, and in Germany the theme is a focus for Heidegger fairly early on.
48 See, for example, again, Spigel, *Make Room for TV*.
49 See Anders, "The Outdatedness of Privacy."
50 Thus, several authors characterize this using the language of "Effect," drawing out the seemingly programmed efficacy of the gambling trick, that is, the use of a like button.
51 Anders, "Being Without Time," 148.

Chapter 11

1 As Gadamer writes, to cite the quote in context: "it is the hermeneutic identity that establishes the unity of the work. To understand something, I must be able to identify it. For there was something there that I passed judgment upon and understood. I

identify something as it was or as it is, and this identity alone constitutes the meaning of the work." *The Relevance of the Beautiful*, 25.

2 Adorno, "Situation" in: *Aesthetic Theory*, 38. Emphasis added.
3 In "Totalitarianism without Terror," which we might today rename, as Anders does himself, as "soft totalitarianism" and should be read in correspondence with Hannah Arendt (and Herbert Marcuse), in: Anders, AM II, 241.
4 It is this that Adorno overlooks in his discussion of Anders in *Current of Music*, 136, 142, etc.
5 Hawaii in general was influential in pop music in the 1950s and earlier. Indeed, one unfortunate title going back to Tin Pan Alley in the mid-1910s, "Oh How She Could Yacki Hacki Wicki Wacki Woo"—performed by Eddie Cantor no less. Cf., Tyler, *Hit Songs, 1900-1955*, 87.
6 Babich, "Getting to Hogwarts."
7 See here Babette Babich, "Hallelujah and Atonement." In Jason Holt (ed.), *Leonard Cohen and Philosophy* (Chicago: Open Court, 2014), 123–34.
8 Cf. Adorno, "On the Fetish-Character in Music and the Regression of Listening" (1938).
9 See too Veit Erlmann, *Reason and Resonance* (New York: Zone Books, 2010).
10 Thus, asked by a friend, Bill Richardson, for a song at the age of ninety-six, *I am Gonna Live Until I Die*, it turned out to be Frankie Laine's recording (rather than Frank Sinatra's). Cf. Finch, "Twelve Songs That Everyone Thinks are by Other Artists" as well as, technico-legalistically, Catherine L. Fisk, "The Modern Author at Work on Madison Avenue." In Paul K. Saint-Amour (ed.), *Modernism and Copyright* (Oxford: Oxford University Press, 2011), 173–94.
11 To look at this I draw on phenomenology to look at recording under all mediatic forms, radio and social distribution (this includes television and cable and YouTube references) as it also involves the invisibility of the effects of the same (this is the culture industry) on our consciousness, individual and collective, the "space" and "time" as Adorno would have it, of musical sound as it was and could/might be as it is and is not.
12 Anders, "On the Pseudo-Concreteness of Heidegger's Philosophy," 370.
13 See again, *The Hallelujah Effect* as well as Babich, "Hallelujah and Atonement."
14 See Nietzsche, On the Theory of Quantitifying Rhythm." James W. Halporn, trans. *New Nietzsche Studies*, 10, nos 1 & 2 (2016): 69–78.
15 Nietzsche's original notes to his *Zur Quantierenden Rhythmik* suggest a link to Aristoxenus. Cf., albeit without reference to Nietzsche, Williams, *The Aristoxenian Theory of Musical Rhythm*.
16 Ullrich von Wilamowitz-Möllendorf, "Future Philology." Gertrude Postl, Babette Babich, and Holger Schmid, trans. *New Nietzsche Studies*, 4, nos 1 and 2 (2000): 1–32.
17 The theologian, David Strauss, as writer and confessor.
18 That is, Nietzsche, *Vom Nutzen und Nachteil der Historie für das Leben*.
19 In addition to Benjamin's famous essay, see Adorno's *Current of Music* as well as "On the Fetish Character of Music and the Regression of Music" and "The Schema of Mass Culture," the first two essays in *The Culture Industry*, 29–60 and 61–97. See too the chapter "Sociological Aspects" in Adorno and Eisler, *Composing for the Films*, 30–41.
20 See the author's "The Revolution will be Televised."
21 Mark Zwonitzer and Charles Hirshberg, *Will You Miss me When I'm Gone? The Carter Family & Their Legacy in American Music* (New York: Simon & Schuster, 2004).
22 Bennett, *Becoming a Rock Musician*.
23 See for context and (some) discussion, Leppert's "Commentary," esp. 219f.

24 Von Bülow to Nietzsche, July 24, 1872. Nietzsche, *Kritische Briefausgabe* [KGB]. Not only does one fail the genius of genius—in Kant's sense—when one's artistry is "natured" rather than "as if" by nature but, as von Bülow noted, one fails far more basically unless learns, *qua* composer, the grammar in which one intends to express oneself. *Sine qua non*. This is the subject of another discussion on Adorno and popular music or jazz and not less the "new music," the rules of composition.

25 Nietzsche's draft to von Bülow contains this more defensive assertion (KGB, 77); the version sent is more polished, invokes a scale of ironies, and is itself somewhat more ironic or distant (ibid., 78–80). Only the (more self-defensive) draft is included in Middleton's *Selected Letters of Friedrich Nietzsche*, 106–7.

26 Michael James Roberts, *Tell Tchaikovsky the News: Rock 'n' Roll, the Labor Question, and the Musicians' Union, 1942–1968* (Durham: Duke University Press, 2014).

27 The concert including this performance was described in a 2000 National Public Radio broadcast by Will Hermes. See transcript here: http://www.npr.org/2000/05/08/1073885/4-33. See for a musicological reading Kyle Gann, *No Such Thing as Silence: John Cage's '4' 33'''* (New Haven: Yale University Press, 2011).

28 See BBC News January 19, 2004. http://news.bbc.co.uk/2/hi/entertainment/3401901.stm

29 See, for example, the German gallery exhibition, SOUNDS LIKE SILENCE *Cage – 4'33'' – Stille 1912 – 1952 – 2012* (Dortmund, HartwareMedienKunstVerein, Dortmunder U, third and sixth floors (Galerie), August 25, 2012 – January 6, 2013) and Ross, "Searching for Silence" discusses one such exhibit, "The Anarchy of Silence," on Cage's career and his myriad connections to other arts.

30 See Agamben's painful, and beautifully, crafted, *Una domanda*, and for discussion, see my "Retrieving Agamber's Questions."

31 See for a discussion the author's "Spirit and Grace, Letters and Voice," and (in passing), "Ivan Illich's *Medical Nemesis* and the 'Age of the Show.'"

32 See Babich, "Heidegger and Leonard Cohen: 'You Want it Darker.'"*Religions* (June 2021).

33 [Wer sich nicht auf der Schwelle des Augenblicks, alle Vergangenheiten vergessend, niederlassen kann, wer nicht auf einem Punkte wie eine Siegesgöttin ohne Schwindel und Furcht zu stehen vermag, der wird nie wissen, was Glück ist, und noch schlimmer: er wird nie etwas tun, was andre glücklich macht.] Nietzsche, *Vom Nutzen und Nachteil der Historie für das Leben* [1874], 1.

34 [*gegen die Zeit und dadurch auf die Zeit und hoffentlich zugunsten einer kommenden Zeit—zu wirke*] Nietzsche [1874] 1980, 247.

Chapter 12

1 Anders, *Hiroshima ist Überall. Tagebuch aus Hiroshima und nagasaki. Der Briefwechsel mit dem Hiroshima-Piloten Claude Eatherly. Rede über die drei Weltkriege* (Munich: Beck, 1982).

2 Anders, *Gewalt, Ja oder Nein? Eine notwendige Diskussion* (Munich: Knaur, 1987), 21–2. See further, Elisabeth Rohrlich, "'To Make the End Time Endless:' The Early Years of Gunther Anders' Fight against Nuclear Weapons." In Günter Bischof, et al. (eds.), *The Life and Work of Günther Anders: émigré, iconoclast, Philosopher, Man of Letters* (Innsbruck: Studienverlag, 2014).

3 Anders, *Gewalt – Ja oder Nein?* 22.

4 There is a lot to read on this complex topic but see just on the issue of theoretical (in-) visibility of atomic waste in particular Pilkey-Jarvis, "Yucca Mountain" and "Giant Cups of Poison" in Pilkey and Jarvis-Pilkey, *Useless Arithmetic*, 45–65/140–63. With explicit mention of Anders, Jogschies thematizes invisibility with respect to atomic catastrophe in his "Zur Chiffrierung von Atomkriegsängsten in Science-Fiction-Filmen und ihrer De-Chiffrierung in der Politik." See too, Babich, "Heidegger and Hölderlin on Aether and Life."
5 We don't, as philosophers cite our colleagues in general, but when we do, we tend not to know the names of outlier thinkers.
6 For a range of further references, see my essay "Adorno and Radio Phenomenology."
7 See Babich, *The Hallelujah Effect*.
8 Just as Marcuse would also work on behalf of the government during the same wartime effort, Adorno was also, in effect, "imported" to the United States. See Levin and von der Linn, "Elements of a Radio Theory" as well as Cavin, "Adorno, Lazarsfeld &The Princeton Radio Project, 1938-1941." On Adorno especially, see Babich, "Adorno's Radio Phenomenology."
9 Babich, "On *The Hallelujah Effect*" as well as Babich, "Texts and Tweets: On The Rules of the Game."
10 See here, again, Frith, *Smartphones as Locative Media*, in addition to the inevitable update that lockdown has dated, as it were, before its time: Frith and Kalin, "Here, I Used to Be" and Adam Alter, *Irresistible: The Rise of Addictive Technology and the Business of Keeping Us Hooked* (New York: Penguin Press, 2017).
11 See again, Pilkey-Jarvis, "Yucca Mountain."
12 Goodman, *Sonic Warfare*, p. xi. Very few people notice the error, though I thank Tracy Strong for calling it to my attention but even the political theorist and former member of SDS, Tracy Strong, got the military rank wrong as "Colonel Kilgore"—rather than *Lieutenant Colonel*. And cf. Kittler "Weltatem."
13 Stern, "Spuk und Radio." This point is already related to Arnheim's *Radio* and Adorno's *Current of Music*.
14 See references and further discussion, Babich, "Adorno's Radio Phenomenology."
15 Smythe, in a footnote to his own text, refers the reader to Ewen's 1976 *Captains of Consciousness* for documentary evidence of what Smythe describes as the "purposiveness with which monopoly capitalism used advertising and the infant mass media for this purpose in the period around and following World War I." Smythe, "Communications," 27.
16 See here, again, Epping-Jäger "Lautsprecher," 396–7.
17 The reference here has been the subject of a 2006 HBO documentary, which is, however strange this may seem, an effective way to drive something from public consciousness, *Hacking Democracy*, and featuring Bev Harris and Kathleen Wynne, director and associate director, respectively, for the non-profit election watchdog group Black Box Voting, not a matter of "Russian hacking" with the Trump election 2016 or indeed with Biden 2020 but, back to the famously disputed election between Gore and Bush, inasmuch as, given digital technology, *any* election, anywhere, anytime, can be hacked without anyone being able to detect the hack *one way or the other*. That is the beauty of the digital; it is the same problem as some librarians have noted this, with the digitization of books and the destruction or prohibition of access, it is the same, to the originals. (See Sare, "A Comparison of HathiTrust and Google Books Using Federal Publications.") The question of digitized books and the dangers (the books are destroyed or access is refused depending on geographic location),

one must be in the United States to use most digitized texts "owned," one off, by the Hathi trust. These questions, like the questions of voting hacks, are complicated. But in the political realm, as the recent US elections demonstrated yet once again, the problems were never resolved: claims of voter fraud, legitimate claims, go back and forth—whatever "legitimate" might mean as the point of digital hacking is precisely that there is no evidence of the hack. More recent discussions of the new kinds of social media "hacking" correspond to ads and the like in those kinds of social digital engineering that we happen to we know of, which does not mean that that we are conscious of such manipulation nor indeed that we know its extent. The manipulation of public opinion is what Adorno and Horkheimer named the "culture industry" and called "programming." Thus, such discussions go back to the beginnings of broadcast technology as this concerns the control of public opinion, what Chomsky discusses in his book *Media Control*.

18 See Milner, *Pinpoint* and see also Sheng-Chih Wang, *Transatlantic Space Politics: Competition and Cooperation Above the Clouds* (London: Routledge, 2013).
19 Anders, *Gewalt, Ja oder Nein?* 31.
20 See Adorno's published research for the Princeton Radio Project, particularly "Music in Radio" and a "Plugging Study," and see further references in Babich, "Adorno's Radio Phenomenology."
21 "If a person gave your body to any stranger he met on his way, you would certainly be angry. And do you feel no shame in handing over your own mind to be confused and mystified by anyone who happens to verbally attack you?" Epictetus: *Enchir.*, §28.
22 Baudrillard, "Requiem for the Media," 170. As Baudrillard, who was the author of the political economy of the sign in which he had analysed this, the media thus make "all processes of exchange impossible (except in the various forms of response simulation, themselves integrated in the transmission process, thus leaving the unilateral nature of the communication intact). This is the real abstraction of the media. And the system of social control and power is rooted in it." Ibid., 169.
23 Baudrillard, "Requiem for the Media," 169.
24 This, of course, drives datification. Indeed, one can even share the post on one's own page—a kind of a super "like" (similar to a retweet).
25 Illich, "The Cultivation of Conspiracy." In Lee Hoinacki and Carl Mitcham (eds.), *The Challenges of Ivan Illich: A Collective Reflection* (Albany: The State University of New York Press, 2002), 233–42, an acceptance speech presented at the Villa Ichon, March 14, 1998, for the Culture and Peace Prize of Bremen.
26 "Viel hat von Morgen an, / Seit ein Gespräch wir sind und hören voneinander, / Erfahren der Mensch; bald sind wir aber Gesang. / Und das Zeitbild, das der große Geist entfaltet, / Ein Zeichen liegts vor uns, daß zwischen ihm und andern / Ein Bündnis zwischen ihm und andern Mächten ist." From: Hölderlin, *Friedensfeier*.
27 In a related hymn, *Versöhnender, der du nimmergeglaubt*, we read: "Ein Chor nun sind wir. Drum soll alles Himmlische was genannt war, / Eine Zahl geschlossen, heilig, ausgehn rein aus unserem Munde. / Denn sieh! es ist der Abend der Zeit, / Die Stunde wo die Wanderer lenken zu der Ruhstatt. / Es kehrt bald Ein Gott um den anderen ein." Hölderlin, *Sämtliche Werke*, Vol. 7, 159.
28 Anders, *Gewalt, Ja oder Nein?* 31.
29 This is the subject of Anders' 1956 *Die Antiquiertheit des Menschen* (Vol. I), but no less thematized as our consumption of mass media in Anders' interview with Manfred Bissinger, already cited earlier: *Gewalt, Ja oder Nein?* [*Violence, Yes or No?*].

30 See my discussion quite on the level of transmission and effect, "Heidegger and Hölderlin on Aether and Life."
31 Baudrillard, "Requiem for the Media," 170.
32 Ibid.
33 Ibid.
34 Ibid.
35 Anders was uncompromising in his conviction that the problem was not the inferiority of Russian versus West-European or American technology (Anders, *Gewalt Ja oder Nein?*, 21), and insisted much rather, just as he earlier declared that "Hiroshima is everywhere" that Chernobyl only attests not to something so adventitious as a possibility but rather a pestilence, an already pernicious "plague," 22.
36 See for a discussion of some of the more everyday challenges of storing nuclear waste precisely in a geological context, Pilkey-Jarvis, cited earlier but to be read in the context of Pilkey's discussion of geological processes.
37 Gehring briefly notes Anders and Chargaff in her discussion of Sloterdijk in "Zwischen Menschenpark and Eugenics."
38 See also, in addition to Fukushima already mentioned, the challenges of nuclear waste and plastic and nanowaste are the newest variations on the same.
39 Anders, *Gewalt, Ja oder Nein?* 23.
40 Ibid.
41 But it is hard to parse this, just as Horst Mahler writes in an immediate more provocative refusal: "Oh mein Gott—was ist das für eine braune Soße, die da aus Deiner Feder geflossen ist!" Horst Mahler: "Ist Dein Mut zu töten wirklich so groß?"
42 Bertolloti, "The Atomic Bombing of Hiroshima," 106.
43 *Newsweek*, August 20, 1945.
44 Bertolloti, "The Atomic Bombing of Hiroshima," 106.
45 The phrase is, in John Ford's 1962 film, *The Man Who Shot Liberty Valance*, based on Dorothy M. Johnson's short story of the same title first printed in the July 1949 issue of *Cosmopolitan*—"When the legend becomes fact, print the legend." As Timothy P. O'Neil points out, "Dorothy M. Johnson, a journalism professor at the University of Montana, had two other stories made into films, *A Man Called Horse* (1970) and *The Hanging Tree* (1959)." See O'Neil, "Two Concepts of Liberty Valance: John Ford, Isaiah Berlin, and Tragic Choice on the Frontier." *Creighton Law Review*, 37 (2004): 471–92, 475.
46 Menachem Begin, *The Revolt: Story of the Irgun* (Jerusalem: Steimatzky, 1977), and Saul Zadka, *Blood in Zion: How the Jewish Guerrillas Drove the British out of Palestine* (London: Brassey's, 1995).
47 Jacob Taubes, *Die Politische Theologie des Paulus*, ed. Aleida and Jan Assmann (Munich: Wilhelm Fink, 1993), 14. Italicized, in English in the original.
48 Anders, *Gewalt, Ja oder Nein?* 24.
49 Cited in van Dijk, *Anthropology in the Age of Technology*, 6. Cf. Greffrath's interview with Anders, «Lob der Sturheit».
50 Anders, "Ten Theses on Chernobyl."

Chapter 13

1 Adorno, *The Jargon of Authenticity*, trans. Knut Tarnowsky and Frederic Will (Evanston: Northwestern University Press, 1973 [1964]), 62.
2 Author's own translation.

3   Paul Ehrlich, *The Population Bomb*. In support of Jeff Gibb's claims in *Planet of Humans* (produced by Michael More and released, *gratis*, Earthday 2020), note that Ehrlich's book was originally published under the imprint of the Sierra Club.
4   Heidegger, "Wozu Dichter?" 269.
5   Ibid., 292.
6   Ibid., 267.
7   Simone de Beauvoir, *The Second Sex*, trans. H. M. Parshley (New York, Vintage. 1989, 1952 [1949]), 267–327.
8   See on the collaboration between Peter Handke and Wim Wenders, Barry, "The Weight of Angels."
9   Cited in Heidegger, "Wozu Dichter?" 269. My translation.
10 Ibid.
11 See for example, Müller, "Desert Ethics," as well as Meyer-Drawe, "Mit ‚eiserner Inkonsequenz' fürs Überleben—Günther Anders." See too, quite focused on the question of globalization, Ehgartner et al., "On the Obsolescence of Human Beings in Sustainable Development," as well as, from a legal perspective, if not quite centred on at least beginning with Anders, Sarat, et al., eds., *The Time of Catastrophe*, here 19–21.
12 See Alvis, "Transcendence of the Negative" in addition to Palaver, "The Respite: Günther Anders' Apocalyptic Vision in Light of the Christian Virtue of Hope" in: Bischof et al. (eds.), *The Life and Work of Günther Anders* as well as Dupuy's *The Mark of the Sacred*.
13 Shelley, *Frankenstein or a Modern Prometheus*, 1818.
14 See here, Müller's "Better than Human" in his *Prometheanism* and Babich, "Geworfenheit und prometheische Scham im Zeitalter der transhumanen Kybernetik."
15 Again, I refer the reader to Bostrom's "In Defense of Posthuman Dignity" as well as, on both sides, the contributions to Tuncel (ed.), *Nietzsche and Transhumanism*.
16 See the contributions to Crawford and Vogt (eds.), *Adorno and the Concept of Genocide*.
17 See Winner, *The Whale and the Reactor* along with Weinberg's "A Wake-Up Call for Technological Somnambulists." For discussion of Anders on violence and nuclear power, see the contributions to Micaela Latini et al., *La grammatica della violenza* and see directly in the wake of Chernobyl and after Anders' death, Bahro, *Avoiding Social and Ecological Disaster* as well as Lütticken, "Shattered Matter, Transformed Forms."
18 Later, Sloterdijk will go on to reflect on "the climate toxins emitted from people. themselves, since, desperately agitated, they stand sealed together under a communication bell-jar: in the pathogenic air conditions of agitated and subjugated publics, inhabitants are constantly re-inhaling their own exhalate." Sloterdijk, *Terror from the Air*, 101. Hereafter cited in the text as TA.
19 Latour, *Facing Gaia: Eight Lectures on the New Climatic Regime* (London: Polity, 2017). Cf. Shiva, *Biopiracy* and Horn, "Air Conditioning."
20 Fuller, "Creating the Covid-19 Story: How to Control a Pandemic's Narrative." Thus, in milder, less world-domination directed stakes, as it were, one can prime reception (this the language of psychology as I explore this in *The Hallelujah Effect* and elsewhere) by speaking, circa January 2015 of a "Bomb Cylone" as the technical term for the assault on the Eastern coast of the United States. See too, as of 2018, Blinder et al., "Bomb Cyclone." For an analysis of the use of Newspaper headlines and cover pages, with respect to the First World War, see again, Bertolotti "The Atomic Bombing of Hiroshima." Today, an ongoing issue, part and parcel of the

Covid-19 story to which Fuller refers, would be bioweapons or the proactionary threat of the same as most of the measures of 2020 were undertaken as precautions in response to a model.
21 See Baudrillard, *Intelligence of Evil or, The Lucidity Pact* in addition to his influential but more cited as a horror notion than actually read 1991 *La Guerre du Golfe n'a pas eu lieu*.
22 Sloterdijk, *Schäume, Luftbeben,* "Airquake" and *Terror from the Air*.
23 See the US Department of Defense 1996 document: "Weather as a Force Multiplier: Owning the Weather in 2025." Sloterdijk dares collegial heresy as he here adverts to public documents available from the US Department of Defense (the United States long ago learned that the best way to conceal its motives was to hide them in plain sight): thus, the organization of what Sloterdijk goes on to describe as "terrorist" means of international aggression is justified, and no one notices any kind of contradiction.
24 See Sharma and Kumar, "Changes in Honeybee Behaviour and Biology under the influence of Cellphone Radiations," as well as the WHO's 2010 Electromagnetic Fields and Public Health: Mobile Phones, Fact sheet No. 193, and note that this is only the latest result of an ongoing project. See *Establishing a Dialogue on Risks from Electromagnetic Fields. A Handbook*. Worth reading here, if only because the research science needed to document the death of the bees was conducted in India, one of the few places as the authors themselves observed, free of cell phone towers (at least in some loci), and which can almost be singularly unnoted, is Visvanathan, "On the Annals of the Laboratory State." Cf. a 2020 European study of the same phenomenon with bees, Thielens et al., "Radio-Frequency Electromagnetic Field Exposure of Western Honey Bees."
25 The lecture title is rather incorrectly rendered as "The Cultivation of Conspiracy" in Hoinacki and Mitcham, eds., *The Challenges of Ivan Illich*.
26 Illich, "The Cultivation of Conspiracy," 237.
27 Ibid., 239.
28 One can instructively read Illich—this has been done for some time after all—as a Catholic theologian. See, for example, the essays in: Illich, *The Powerless Church and Other Selected Writings*.
29 Illich, "The Cultivation of Conspiracy," 239.
30 I discuss Montagnier and Duesberg in Babich, "Calling Science Pseudoscience," quite where mobbing denigrations have so hampered the field of science studies that what was (once) the "strong program" in the sociology of science is so debilitated that it can never rise again. Moewus is a complicated story, vilified by those who research his work as a simple hack (odd as this can seem for research that is meant to explore the history of science but which proceeds), as if Herbert Butterfield had never written his cautionary methodological study, *The Whig Interpretation of History*. See Jan Saap's normativizing account written, as such accounts are written to silence any further questions: *Where the Truth Lies*. Note here and to be sure that if one were to try to raise further questions, no press would publish one's work. See just on this all-too-academic detail, Pilkey (co-authored with his daughter, Pilkey-Jarvis), *Useless Arithmetic*. I discuss some, not all of this, focusing on the targeted calumnies of Alan Sokal (very much contra Latour) in my "Hermeneutics and its Discontents in Philosophy of Science."
31 David Ray Griffin, *The New Pearl Harbor Revisited: 9/11, the Cover-Up, and the Exposé* (Northampton: Olive Branch [Interlink Books], 2008). But although Patrick Aidan

Heelan (1926–2015), philosopher and scientist (and himself a theologian), read Griffin, found his arguments persuasive, Heelan did not write about Griffin.
32 Garrett Stewart, *Closed Circuits: Screening Narrative Surveillance* (Chicago: University of Chicago Press, 2015), ix.
33 Luce Irigaray, *The Forgetting of Air in Martin Heidegger*, trans. Mary Beth Mader (Texas: University of Texas Press, 1999), 166.
34 See for a recent focus on Heidegger and the question as such, Babich, "On Heidegger on Education and Questioning."
35 Shelley, *Frankenstein or a Modern Prometheus*.
36 See for a discussion including xenotransplantation, the closing section of Babich, "Ivan Illich's *Medical Nemesis* and the 'Age of the Show.'"
37 Alphonso Lingis has an important and disquieting reflection on the phenomenology of medical practice as lived for recipients of face (and other) transplants. The outcome of these much-touted media events are less than ideal for recipients.
38 Adorno, *Minima Moralia*, 26.
39 Ibid., 27.
40 Ibid., 28.
41 The term "act of God" has for insurance companies a technical, legal definition. As with the current vaccines, all liability is passed on to the side of the consumer rather than the corporate agent.
42 See further my contribution "Pseudo-Science and 'Fake' News: 'Inventing' Epidemics and the Police State" to Irene Strasser and Martin Dege (eds.) *The Psychology of Global Crises and Crisis Politics – Intervention, Resistance, Decolonization*. (London: Palgrave, 2021).
43 It is not for nothing that de Rougement begins his 1941 essay with the chauvinistic but not inaccurate observation that "what is most lacking in America is belief in the Devil." De Rougement, "On the Devil and Politics," 2.
44 Anders, *Franz Kafka*, 73–4. Note to be sure that Anders is not the only one to read Nietzsche in this light but it has made it very difficult for those who read Kafka to come to terms with Anders.

Bibliography

Günther Anders

Die Antiquiertheit des Menschen 1: Über die Seele im Zeitalter der zweiten industriellen Revolution. Munich: C.H. Beck, 1956.
L'obsolescence de l'homme: Sur l'âme à l'époque de la deuxième révolution industrielle. Paris: EDN/Ivréa, Réed, 2002.
Die Antiquiertheit des Menschen; Zweiter Band. Über die Zerstörung des Lebens im Zeitalter der dritten Industriellen Revolution. Munich: Beck, 1984. [1980]
L'obsolescence de l'homme: Tome 2, Sur la destruction de la vie à l'époque de la troisième révolution industrielle, trans. Christophe David. Paris: Fario, coll. Ivrea, 2011.
"Die Antiquiertheit der Privatheit." In: *Die Antiquiertheit des Menschen 2: Über die Zerstörung des Lebens im Zeitalter der dritten industriellen Revolution*, 210–46. Munich: C.H. Beck, 2002.
Die atomare Drohung. Radikale Überlegungen zum atomaren Zeitalter. Munich: Beck. 1993.
Besuch im Hades. Munich: Beck, 1979.
Endzeit und Zeitenende. Gedanken über die atomare Situation. Munich: Beck, 1972. [1959]
Hiroshima ist Überall. Tagebuch aus Hiroshima und nagasaki. Der Briefwechsel mit dem Hiroshima-Piloten Claude Eatherly. Rede über die drei Weltkriege. Munich: Beck, 1982.
"Kafka: Ritual Without Religion: The Modern Intellectual's Shamefaced Atheism." *Commentary* (December 1949): 560–9.
Kafka, Pro und Contra. Die Proceß-Unterlagen. Munich: Beck, 1953. [1951]
Franz Kafka, trans. A. Steer and A. K. Thorlby. London: Bowes & Bowes, 1960.
Die Kirschenschlacht. Dialoge mit Hannah Arendt, ed. Gerhard Oberschlick. Munich: Beck, 2011.
La bataille des cerises: Dialogues avec Hannah Arendt, suivi d'un essai de Christian Dries, trans. Philippe Ivernel. Paris: Rivages, 2013.
Die molussische Katakombe: Roman. Munich: Beck, 1992.
Die molussische Katakombe: Roman mit Apokryphen und Dokumenten aus dem Nachlass. Munich: Beck, 2012.
Gewalt, Ja oder Nein? Eine notwendige Diskussion. Munich: Knaur, 1987.
Musikphilosophische Schriften. Texte und Dokumente, ed. Reinhard Ellensohn. Munich: Beck, 2017.
"Philosophische Untersuchungen über Musikalischen Situationen." In Anders, *Musikphilosophischen Schriften. Musikphilosophische Schriften. Texte und Dokumente*, 15–176. ed. Reinhard Ellensohn. Munich: Beck, 2017.
"Pathologie de la liberté. Essai sur la non-identification." *Recherches Philosophiques*, VI (1936–1937): 22–54.
"On Beckett's Play *Waiting for Godot*," trans. Martin Esslin and revised by Anders. In Martin Esslin (ed.), *Samuel Beckett: A Collection of Critical Essays*, 140–51. Englewood Cliffs: Prentice Hall, 1965.

"On the Pseudo-Concreteness of Heidegger's Philosophy." *Philosophy and Phenomenological Research*, 8, no. 3 (March 1948): 337–71.
"Reflections on the H Bomb." *Dissent*, 3, no. 2 (Spring 1956): 146–55.
[Stern] *Die Rolle der Situationskategorie bei den "logischen Sätzen". Erster Teil einer Untersuchung über die Rolle der Situationskategorie*. Diss. Freiburg/Brsg. 1924.
[Stern] "Spuk und Radio."*Anbruch*, XII/2 (February 1930): 65–6.
Tagebücher und Gedichte. Munich: Beck, 1985.
"The Obsolescence of Privacy,' trans. Christopher John Müller. *CounterText*, 3, no. 1 (2017): 20–46.
"Ten Theses on Chernobyl." Sixth World Congress of the International Physicians for the Prevention of Nuclear War in 1986. Online. https://libcom.org/library/ten-theses-chernobyl-%E2%80%93-g%C3%BCnther-anders. Accessed 8 March 2018.
[Anders-Stern] "The Acoustic Stereoscope." *Philosophy and Phenomenological Research*, 10, no. 2 (December 1949): 238–43.
"The World as Phantom and Matrix," trans. Norbert Gutterman. *Dissent*, 3, no. 1 (1956): 14–24.
[Stern] *Über das Haben. Sieben Kapitel zur Ontologie der Erkenntnis*. Bonn: Cohen, 1928.
"Une interprétation de l'*a posteriori*." *Recherches philosophiques*, IV (1934–1935): 65–80.
Visit Beautiful Vietnam. Cologne: Pahl-Rugenstein, 1968.
Die Weltfremdheit des Menschen. Schriften zur philosophischen Anthropologie. Munich: Beck, 2018.
"Wesen und Eigentlichkeit nämlich bei Heidegger (1936)." In *Über Heidegger*, 32–38. Munich: Beck, 2001.
"We, Sons of Eichmann: An Open Letter to Klaus Eichmann," trans. Jordan Levinson. 2015. Online. http://anticoncept.phpnet.us/eichmann.htm.
Wir Eichmannsöhne. Offener Brief an Klaus Eichmann. Munich: C.H. Beck, 1964.
Anders, Günther [Stern] with Hannah Arendt. "Rilkes Duineser Elegien." *Neue Schweizer Rundschau*, 23 (1930): 855–871.
Anders, Günther and Hannah Arendt. *Schreib doch mal hard facts über Dich. Briefe 1939 bis 1975, Texte und Dokumente*. Kerstin Putz (ed.) Munich: Piper, 2016.
Anders, Günther and Claude Eatherly, *Burning Conscience: The case of the Hiroshima Pilot, Claude Eatherly, told in his letters to Günther Anders*. Preface by Bertrand Russell; Foreword by Robert Jungk. New York: Paragon, 1961.
Anders, Günther and Mathias Greffrath. "Lob der Sturheit". *Die Zeit*, "Zeitläufte", 28/2002. http://marcuse.faculty.history.ucsb.edu/anders, maintained by the historian, Harold Marcuse.

Secondary

Accardo, Julie and M. Ahmad Chaudhry. "Radiation Exposure and Privacy Concerns Surrounding Full-Body Scanners in Airports." *Journal of Radiation Research and Applied Sciences*, 7, Issue 2 (April 2014): 198–200.
Achterhuis, Hans, Paul van Dijk and Pieter Tijmes, eds., *De maat van de techniek. Zes filosofen over techniek: Günther Anders, Jacques Ellul, Arnold Gehlen, Martin Heidegger, Hans Jonas en Lewis Mumford*. Baarn: Ambo, 1992.
Adams, Tim. "Sherry Turkle: 'I am not anti-technology, I am pro-conversation.'" *The Guardian*, October 18, 2015.

Adorno, Theodor. *Aesthetic Theory*, edited by Gretel Adorno and Rolf Tiedmann, trans. Christian Lenhardt. London: Routledge and Kegan Paul, 1984.
Adorno, Theodor. *Aesthetic Theory*, edited by Gretel Adorno and Rolf Tiedmann, trans. Robert Hullot-Kentor. London: Bloomsbury, 1997. [1970]
Adorno, Theodor. *The Culture Industry*. New York: Routledge, 1991.
Adorno, Theodor. *Essays on Music, Selected, with Introduction, Commentary, and Notes by Richard Leppert*, Berkeley: University of California Press, 2002.
Adorno, Theodor. *Kierkegaard: Construction of the Aesthetic*. Albany: State University of New York Press, 1962.
Adorno, Theodor. *Minima Moralia*, trans. E. F. N. Jephcott. London: New Left, 1997. [1974]
Adorno, Theodor. "Music in Radio." Memorandum, 1938. Archive. Paul Lazarsfeld, ed., Micfrofilm.
Adorno, Theodor. *Nachgelassene Schriften. Abteilung I: Fragment gebliebene Schriften — Band 3: Current of Music. Elements of a Radio Theory*. Frankfurt am Main: Suhrkamp, 2006.
Adorno, Theodor. *Negative Dialectics. Negative Dialectics*, trans. E. B. Ashton. London: Routledge, 1973. [1966]
Adorno, Theodor. "On the Fetish-Character in Music and the Regression of Listening." In *Essays on Music*. Berkeley: University of California Press, 2002.
Adorno, Theodor. "Plugging Study." [and Douglas MacDougald], 1939. Bureau of Applied Social Research records, 1938–1977, Section III, Reports. Columbia University Archive.
Adorno, Theodor. "The Aging of the New Music." *Télos* (77) (1988): 95–116. [1955]
Adorno, Theodor. *The Jargon of Authenticity*, trans. Knut Tarnowsky and Frederic Will. Evanston: Northwestern University Press, 1973. [1964]
Adorno, Theodor. "What National Socialism Has Done to the Arts," 376. In Richard Leppert (ed.), *Essays on Music*, trans. Susan Gillespie, 373–90. Berkeley: University of California Press, 2002.
Adorno, Theodor. *Zur Metakritik der Erkenntnistheorie. Drei Studien zu Hegel*. Frankfurt am Main: Suhrkamp Verlag, 1970. [1937]
Adorno, Theodor and Hanns Eisler. *Composing for the Films*. London: Continuum, 1997 [1947].
Adorno, Theodor with Max Horkheimer. *Dialectic of Enlightenment*, trans. Edmund Jephcott. Stanford: Stanford University Press, 2002. [1944]
Ahuja, N. K. "'It Feels Good to be Measured': Clinical Role-Play, Walker Percy, and the Tingles." *Perspectives in Biology and Medicine* 56 (2013): 442–51.
Agamben, Giorgio. *Una domanda*, 13 April 2020. Online. https://www.quodlibet.it/giorgio-agamben-una-domanda.
Agamben, Giorgio. *What Is an Apparatus and Other Essays*, trans. David Kishik and Stefan Pedatella. Stanford: Stanford University Press, 2009.
Albrecht, Katherine. "Microchip Induced Tumors in Laboratory Rodents and Doges: A Review of the Literature." In M. G. Michaels (ed.), *Uberveillance and the Social Implications of Microchip Implants: Emerging Technologies*, 281–318. Hershey: IGI Global, 2013.
Allen, Valerie and Ares D. Axiotis. "Heidegger on the Art of Teaching." In M. A. Peters and V. Allen (eds.), *Heidegger, Education, and Modernity*, 27–45. Lanham: Rowman & Littlefield, 2002.
Allen, Valerie and Ares D. Axiotis. *L'Art d'enseigner de Martin Heidegger*, trans. Xavier Blandin. Paris: Klincksieck, 2007.

Allison, David. *Reading the New Nietzsche*. Lanham: Rowman & Littlefield, 2001.
Alter, Adam. *Irresistible: The Rise of Addictive Technology and the Business of Keeping Us Hooked*. New York: Penguin Press, 2017.
Alvis, Jason W. "Transcendence of the Negative: Günther Anders' Apocalyptic Phenomenology." *Religions*, 8, no. 59 (2017): 1–16. Online. https://www.mdpi.com/2077-1444/8/4/59
Andler, Charles. *Nietzsche sa vie et sa pensée II. Le pessimisme esthétique de Nietzsche. La maturité de Nietzsche*, 440–1. Paris: Gallimard, 1958 [1920–1931].
Andreas Salomé, Lou. *You Alone Are Real to Me: Remembering Rainer Maria Rilke*, trans. Angela von der Lippe. Manchester: Carcanet, 2006 [1928].
Ardillo, José. *La Liberté dans un monde fragile: Écologie et pensée libertaire*. Paris: L'Échappée, 2018.
Arendt, Hannah. *Eichmann in Jerusalem: A Report on the Banality of Evil*. New York: Viking, 1963.
Arendt, Hannah. "Introduction." In: *Illuminations: Essay and Reflections*, 1, ed. Walter Benjamin, New York: Schocken, 1969.
Arendt, Hannah. *Reflections on Literature and Culture*, ed. Susannah Young-ah Gottlieb. Stanford: Stanford University Press, 2007.
Arendt, Hannah and Günther Stern. "Rilke's *Duino Elegies*." In: Arendt, *Reflections on Literature and Culture*, ed. Susannah Young-Ah Gottlieb, 1–23. Stanford: Stanford University Press, 2007.
Arendt, Hannah and Gershom Scholem. *Hannah Arendt–Gershom Scholem, Der Briefwechsel*. Berlin: Juedischer Verlag, 2010.
Armon, Adi. "The Parochialism of Intellectual History: The Case of Günther Anders." *Leo Baeck Institute Yearbook*, 62 (2017): 225–41.
Attwood, Bain. *The Making of Aborigines*. Sydney: Allen and Unwin, 1989.
Augustine, *Confessions*, trans. Henry Chadwick. Oxford: Oxford University Press, 1991.
Axelos, Kostas. *Einführung in ein künftiges Denken. Über Marx und Heidegger*. Berlin: Walter de Gruyter, 1966. [1961]
Babich, Babette. "Ad Jacob Taubes." *New Nietzsche Studies*, 7, nos. 3 and 4 (2007): v–x
Babich, Babette. "Adorno on Science and Nihilism, Animals, and Jews," *Symposium: Canadian Journal of Continental Philosophy/Revue canadienne de philosophie continentale*, 14, no. 1 (2011): 110–45.
Babich, Babette. "Adorno's Radio Phenomenology: Technical Reproduction, Physiognomy, and Music." *Philosophy & Social Criticism*, 40, no. 10 (2014): 957–96.
Babich, Babette. "Adorno's 'The Answer is False': Archaeologies of Genocide." In Ryan Crawford and Erik M. Vogt. *Adorno and the Concept of Genocide*, 1–17. Leiden: Brill, 2016.
Babich, Babette. "An Impoverishment of Philosophy." Dennis Erwin and Matt Story (eds.), *Purlieu: Philosophy and the University*, 2 (2011): 37–71.
Babich, Babette. "Angels, the Space of Time, and Apocalyptic Blindness: On Günther Anders' *Endzeit—Endtime*." *Etica & Politica/Ethics & Politics: Rivista di filosofia*, 15, no. 2 (2013): 144–74.
Babich, Babette. "Are They Good? Are They Bad? Double Hermeneutics and Citation in Philosophy, Asphodel and Alan Rickman, Bruno Latour and the 'Science Wars.'" In Paula Angelova, Andreev Jaassen, and Emil Lessky (eds.), *Das Interpretative Universum*, 259–90. Würzburg: Königshausen & Neumann, 2017.
Babich, Babette. "Artisten Metaphysik und Welt-Spiel in Fink and Nietzsche." In Cathrin Nielsen and Hans Rainer Sepp (eds.), *Welt denken. Annäherung an die Kosmologie Eugen Finks*, 57–88. Freiburg im Breisgau: Alber, 2011.

Babich, Babette. "Aspect Reflections on R. Scott Bakker's The Three Pound Brain." Online: *The Three Pound Brain*. Online. http://diogenesinthemarketplace.blogspot.com/2015/02/aspect-reflections-on-r-scottbakkers.html

Babich, Babette. "Being on Television: Wisser—Heidegger—Adorno." In John Rose (ed.), *52nd Annual Heidegger Conference*, 81–95. Baltimore: Goucher College, 2018.

Babich, Babette. "Between Heidegger and Adorno: Airplanes, Radios, and Sloterdijk's Atmoterrorism." *Kronos Philosophical Journal*, VI (2017 [2018]): 133–58.

Babich, Babette. "Blood for the Ghosts: Reading Ruin's *Being With The Dead* With Nietzsche." *History and Theory*, 59, no. 2 (June 2020): 255–69.

Babich, Babette. "Calling Science Pseudoscience: Fleck's Archaeologies, Latour's Biography, and Demarcation or AIDS Denialism, Homeopathy, and Syphilis." *International Studies in the Philosophy of Science*, 29, Nr. 1 (2015): 1–39.

Babich, Babette. "Constellating Technology: Heidegger's *Die Gefahr*/The Danger." In Babich and Dimitri Ginev (eds.), *The Multidimensionality of Hermeneutic Phenomenology*, 153–82. Frankfurt am Main: Springer, 2014.

Babich, Babette. "The Essence of Questioning after Technology: *Techne* as Constraint and Saving Power." *British Journal of Phenomenology*, 30, no. 1 (January 1999): 106–24.

Babich, Babette. "Epicurean Gardens and Nietzsche's White Seas." In Vinod Acharya and Ryan Johnson (eds.), *Epicurus and Nietzsche*, 52–67. London: Bloomsbury, 2020.

Babich, Babette. "Friedrich Nietzsche and the Posthuman/Transhuman in Film and Television." In Michael Hauskeller, Thomas D. Philbeck, and Curtis D. Carbonell (eds.), *Palgrave Handbook of Posthumanism in Film and Television*, 45–54. London: Palgrave/Macmillan, 2015.

Babich, Babette. "The Genealogy of Morals and Right Reading: On the Nietzschean Aphorism and the Art of the Polemic." In Christa Davis Acampora (ed.), *Nietzsche's On the Genealogy of Morals*, 171–90. Lanham: Rowman & Littlefield, 2006.

Babich, Babette. "Getting to Hogwarts: Michael Oakeshott, Ivan Illich, and J.K. Rowling on 'School.'" In David Bakhurst and Paul Fairfield (eds.), *Education and Conversation: Exploring Oakeshott's Legacy*, 199–218. London: Bloomsbury, 2016.

Babich, Babette. "Geworfenheit und prometheische Scham im Zeitalter der transhumanen Kybernetik. Technik und Machenschaft bei Martin Heidegger, Fritz Lang und Günther Anders." In Christoph Streckhardt (ed.), *Die Neugier des Glücklichen*, 63–91. Weimar: Bauhaus Universitätsverlag, 2012.

Babich, Babette. "Good for Nothing: On Philosophy and Its Discontents." In Diego Bubbio and Jeff Malpas (eds.), *Why Philosophy?*, 123–50 Berlin: de Gruyter, 2019.

Babich, Babette. "Great Men, Little Black Dresses, & the Virtues of Keeping One's Feet on the Ground." *MP: An Online Feminist Journal*, 3, no. 1 (August 2010): 57–78.

Babich, Babette. "Hallelujah and Atonement." In Jason Holt (ed.), *Leonard Cohen and Philosophy*, 123–34. Chicago: Open Court, 2014.

Babich, Babette. *The Hallelujah Effect: Music, Performance Practice and Technology*. London: Routledge, 2016. [2013]

Babich, Babette. "Heidegger and Hölderlin on Aether and Life." *Études Phénoménologique, Phenomenological Studies*, 2 (2018): 111–33.

Babich, Babette. "Heidegger et ses Juifs." In Joseph Cohen and Raphael Zagury-Orly (eds.), *Heidegger et les Juifs*, 411–54. Paris: Grasset, 2015.

Babich, Babette. "Heidegger Hermeneutik." In Alfred Denker und Holger Zaborowski (eds.), *Heidegger Jahrbuch 12. Zur Hermeneutik der Schwarzen Hefte*. Freiburg im Briesgau: Alber, 2019.

Babich, Babette. "Heidegger on Nietzsche's 'Rediscovery' of the Greeks: *Machenschaft* and *Seynsgeschichte* in the Black Notebooks." *Journal of the British Society for Phenomenology*, 51, no. 2 (April 2020): 110–23.

Babich, Babette, "Heidegger on *Verfallenheit*." *Foundations of Science*, 22, no. 2 (2017): 261–4.

Babich, Babette. "Heidegger's Black Night." In Ingo Farin and Jeff Malpas (eds.), *Reading Heidegger's Black Notebooks 1931–194*, 59–86. Cambridge, MA: MIT Press, 2016.

Babich, Babette. "Heideggers 'Brief über ‚Humanismus'. Über die Technik, das Bösartige des Grimmes – und das Heilen." In Alfred Denker and Holger Zaborowski (eds.), *Heidegger und der Humanismus. Heidegger-Jahrbuch, Bd. 10*, 237–50. Freiburg: Verlag Karl Alber, 2017.

Babich, Babette. "Heidegger's Jews: Inclusion/Exclusion and Heidegger's Anti-Semitism." *Journal of the British Society for Phenomenology*, 47, Nr. 2 (2016): 133–56.

Babich, Babette. "Heidegger's *Judenfrage*." In Micha Brumlik and Elad Lapidot (eds.), *Heidegger and Jewish Thought: Difficult Others*, 135–54. Lanham: Rowman & Littlefield, 2018.

Babich, Babette. "Heidegger's Silence: Towards a Post-Modern Topology." In Charles Scott and Arleen Dallery (eds.), *Ethics and Danger: Currents in Continental Thought*, 83–106. Albany: State University of New York Press, 1992.

Babich, Babette. "Hermeneutics and Its Discontents in Philosophy of Science: On Bruno Latour, the 'Science Wars', Mockery, and Immortal Models." In Babich (ed.), *Hermeneutic Philosophies of Social Science*, 163–88. Berlin: de Gruyter, 2017.

Babich, Babette. "Ivan Illich's *Medical Nemesis* and the 'Age of the Show': On the Expropriation of Death." *Nursing Philosophy*, 19, no. 1 (2018): 1–13.

Babich, Babette. "Jaspers, Heidegger, and Arendt: On Politics, Science, and Communication." *Existenz. An International Journal in Philosophy, Religion, Politics, and the Arts*, 4, no. 1 (Spring 2009). Online. https://existenz.us/volumes/Vol.4-1Babich.html

Babich, Babette. "*Körperoptimierung im digitalen Zeitalter, verwandelte Zauberlehrlinge, und künftige Übermenschsein*." In Andreas Beinsteiner and Tanja Kohn (eds.), *Körperphantasien. Technisierung – Optimierung – Transhumanismus*, 203–26. Innsbruck: Universitätsverlag Innsbruck, 2016.

Babich, Babette. "*La violenza della violenza*." In Michaela Latini, Alessandra Sannella, and Alfredo Morelli (eds.), *La grammatica della violenza Un'indagine a più voci*, 83–98. Milan: Mimesis Edizioni, 2017.

Babich, Babette. "*Love, Actually*: Logic and Misogyny and Analytic Anger." Keynote lecture for SWIP. Spokane Exension. Gonzaga University, May 14, 2020. https://vimeo.com/435110852. Online.

Babich, Babette. "Martin Heidegger on Günther Anders and Technology: On Ray Kurzweil, Fritz Lang, and Transhumanism." *Journal of the Hannah Arendt Center for Politics and Humanities at Bard College*, 2 (2012): 122–44.

Babich, Babette. "Material Hermeneutics and Heelan's Philosophy of Technoscience." *AI & Society*, 35 (2020). April 14, 2020. Online first. https://link.springer.com/article/10.1007/s00146-020-00963-7

Babich, Babette. "Musical 'Covers' and the Culture Industry: From Antiquity to the Age of Digital Reproducibility." *Research in Phenomenology*, 48, no. 3 (2018): 385–407.

Babich, Babette. "A Musical Retrieve of Heidegger, Nietzsche, and Technology: Cadence, Concinnity, and Playing Brass." *Man and World*, 26 (1993): 239–60.

Babich, Babette. "Necropolitics and Techno-Scotosis," *Philosophy Today*, 65, Issue 2 (2021): 305–324.
Babich, Babette. "Nietzsche: Looking Right, Reading Left." *Educational Philosophy and Theory*, November 16, 2020. Online: https://doi.org/10.1080/00131857.2020.1840974
Babich, Babette. "Nietzsche (as) Educator." *Educational Philosophy and Theory. Encounter of East Asian Educational Tradition and Western Modernity*, 51, Issue 9 (2019): 871–885.
Babich, Babette. "Nietzsche and Eros between the Devil and God's Deep Blue Sea: The Erotic Valence of Art and the Artist as Actor — Jew — Woman." *Continental Philosophy Review*, 33 (2000): 159–188.
Babich, Babette. "Nietzsche's Aesthetic Tension and Hume's Standard of Taste." In *Reading David Hume's "Of the Standard of Taste,"* 213–45 Berlin: de Gruyter, 2019.
Babich, Babette. "Nietzsche's *Antichrist*: The Birth of Modern Science out of the Spirit of Religion." In Markus Enders and Holger Zaborowski (eds.), *Jahrbuch für Religionsphilosophie*, 134–54. Freiburg i. Briesgau: Alber, 2014.
Babich, Babette. "O, Superman! Or Being Towards Transhumanism: Martin Heidegger, Günther Anders, and Media Aesthetics." *Divinatio* (January 2013): 83–99.
Babich, Babette. "On Günther Anders, Political Media Theory, and Nuclear Violence." *Philosophy & Social Criticism*, 44 (September 17, 2018): 1–17.
Babich, Babette. "On Heidegger on Education and Questioning." In Michael A. Peters (ed.), *Encyclopedia of Educational Philosophy and Theory*, 1641–52. Singapore: Springer, 2017.
Babich, Babette. "On Necropolitics and Techno-Scotosis." *Philosophy Today*, 65 (2021): 305–24.
Babich, Babette. "On Passing as Human and Robot Love." In Carlos Prado (ed.), *How Technology Is Changing Human Behavior*, 17–26. Santa Barbara: Praeger, 2019.
Babich, Babette. "On *The Hallelujah Effect*: Priming Consumers, Recording Music, and The Spirit of Tragedy." In *Proceedings of the Society for Phenomenology and Media*, 1–12. San Diego: National University Press, 2015.
Babich, Babette. "On the Order of the Real: Nietzsche and Lacan." In David Pettigrew and François Raffoul (eds.), *Disseminating Lacan*, 48–63. Albany: State University of New York Press, 1996.
Babich, Babette. "Pedagogy and Other Defences against the Dark Arts: Professor Severus Snape and Harry Potter." *The Philosophical Salon: Los Angeles Review of Books*, December 28, 2015. Online. https://thephilosophicalsalon.com/pedagogy-and-other-defenses-against-the-dark-arts-professor-severus-snape-and-harry-potter/
Babich, Babette. "Philosophy Bakes No Bread." *Philosophy of the Social Sciences*, 48, Nr. 1 (2018): 47–55.
Babich, Babette. "Pseudo-Science and 'Fake' News: 'Inventing' Epidemics and the Police State." In Irene Strasser and Martin Dege (eds.), *The Psychology of Global Crises and Crisis Politics - Intervention, Resistance*. London: Palgrave, 2021.
Babich, Babette. "The Question of the Contemporary in Agamben, Nancy, Danto: Between Nietzsche's Artist and Nietzsche's Spectator." In Paulo de Assis and Michael Schwab (eds.), *Futures of the Contemporary*, 49–82. Ithaca: Cornell University Press, 2019.
Babich, Babette. "Radio Ghosts: Phenomenology's Phantoms and Digital Autism." *Thesis Eleven. Critical Theory and Historical Sociology*, 153, Issue 1 (2019): 57–74.
Babich, Babette. "Reading Lou von Salomé's Triangles." *New Nietzsche Studies*, 8, nos. 3 and 4 (2011/2012): 82–114.
Babich, Babette. "Reading Nietzsche's 'Educational Institutions' with Jaspers & MacIntyre on 'The Idea of the University' — and Severus Snape." *Existenz*, 15, no. 2 (Fall 2021) In press.
Babich, Babette. "Retrieving Agamben's Questions." *FORhUM Forum za Humanistiko Forum für Humanwissenschaften Forum pour les sciences humaines Forum per gli studi*

umanistici *Forum for the Humanities*, May 28, 2020. Online. http://www.for-hum.com/

Babich, Babette. "Review: Daniel Mayer-Katin, *Stranger from Abroad: Hannah Arendt, Martin Heidegger, Friendship and Forgiveness* (New York: Norton, 2010)." *Shofar: An Interdisciplinary Journal of Jewish Studies*, 29, Nr. 4 (Summer 2011): 189–91.

Babich, Babette. "The Revolution will be Televised," April 26, 2020. Online. https://babettebabich.uk/2020/04/26/the-revolution-will-be-televised/

Babich, Babette. "Robot Sex, Roombas, and Alan Rickman." *de Gruyter Conversations: Philosophy & History*, August 17, 2017. Online. https://blog.degruyter.com/robot-sex-roombas-alan-rickman/

Babich, Babette. "Sloterdijk's Cynicism: Diogenes in the Marketplace." In Stuart Elden (ed.), *Sloterdijk Now*, 17–36 and 186–9. Oxford: Polity, 2011.

Babich, Babette. "Screen Autism, Cellphone Zombies, and GPS Mutes." In C. P. Prado (ed.), *Technology Is Changing Us for Better or Worse*, 65–71. Santa Barbara: Praeger, 2019.

Babich, Babette. "Signatures and Taste: Hume's Mortal Leavings and Lucian." In Babich (ed.), *Reading David Hume's Of the Standard of Taste*, 3–22. Berlin: de Gruyter, 2019.

Babich, Babette. "Spirit and Grace, Letters and Voice." *Journal of the Philosophy of Education*, III (2018): 1–27. Online journal. https://www.academia.edu/36960381/Spirit_and_Grace_Letters_and_Voice

Babich, Babette. "Reading Lou von Salomé's Triangles." *New Nietzsche Studies*, 8, nos. 3 and 4 (2011/2012): 82–114.

Babich, Babette. "Review: Daniel Mayer-Katin, *Stranger from Abroad: Hannah Arendt, Martin Heidegger, Friendship and Forgiveness* (New York: Norton, 2010)." *Shofar: An Interdisciplinary Journal of Jewish Studies*, 29, Nr. 4 (Summer 2011): 189–91.

Babich, Babette. "Texts and Tweets: On The Rules of the Game." *The Philosophical Salon: Los Angeles Review of Books*, May 30, 2016. Online. https://thephilosophicalsalon.com/texts-and-tweets-on-the-rules-of-the-game/

Babich, Babette. "Thinking on Film: Jaspers, Scholem, and Thinking in Margarethe von Trotta's *Hannah Arendt*." *German Politics and Society*, 34, No. 1, Issue 118 (Spring 2016): 77–92.

Babich, Babette. "Tools for Subversion: Illich and Žižek on Changing the World." In Sylvie Mazzinie and Owen Glyn-Williams (eds.), *Making Communism Hermeneutical: Reading Vattimo and Zabala*, 95–111. Frankfurt am Main: Metzler, 2017.

Babich, Babette. "Überlegungen nach Heidegger. 'Reflexionen aus dem beschädigten Leben'". In Michael Medved and Holger Zaborowski (eds.), *Heidegger Jahrbuch 13. Zur Hermeneutik der Schwarzen Hefte*. Freiburg im Breisgau: Alber, 2021. In press.

Babich, Babette. "Wallace Stevens, Heidegger, and the 'Virile Hölderlin': Poetry and Philosophy and The Travelogue of the Mind." *Borderless Philosophy*, 3 (June 2020): 1–31.

Babich, Babette. "Who is Nietzsche's Archilochus? Rhythm and the Problem of the Subject." In Bambach and George (eds.), *Philosophers and their Poets: Reflections on the Poetic Turn in Philosophy since Kant*, 85–114. Albany: State University of New York Press, 2019.

Babich, Babette and Christopher Bateman. "The Last of the Continental Philosophers." *Only a Game*, November 29, 2016. Online. https://onlyagame.typepad.com/only_a_game/2016/11/babich-and-bateman-1.html

Babich, Babette. and Christopher Bateman. "Touching Robots." Online. https://onlyagame.typepad.com/only_a_game/2017/02/babich-and-bateman-touching-robots. Accessed September 1, 2019.

Babich, Babette and Christopher Bateman. "Your Plastic Pal Who's Fun to Be With." Online. https://onlyagame.typepad.com/only_a_game/2017/03/babich-and-bateman-your-plastic-pal-whos-fun-to-be-with.html

Bader, Jenny Lyn. *Mrs. Stern Wanders the Prussian State Library*. Museum of Jewish Heritage.

Bahro, Rudolph. *Avoiding Social and Ecological Disaster: The Politics of World Transformation: An Inquiry into the Foundations of Spiritual and Ecological Politics*. Chicago: Gateway, 1994.

Baird, Davis, Eric Scerri, and L. McIntyre. "Introduction: The Invisibility of Chemistry." In Scerri Baird and L. McIntyre (eds.), *Philosophy of Chemistry: Synthesis of a New Discipline*, 3–18. Dordrecht: Springer, 2006.

Bajohr, Hannes. "World-Estrangement as Negative Anthropology: Günther Anders's Early Essays." *Thesis Eleven*, 153 (2019): 141–53.

Balfour, Lawrie. "What Babo Saw: Benito Cereno and the World We Live In." In Jason Frank (ed.), *A Political Companion to Herman Melville*, 259–80. Lexington: University Press of Kentucky, 2013.

Bambach, Charles and Theodore George (eds.), *Philosophers and their Poets*. Albany: State University of New York Press.

Baram, Marcus. "Fear Pays: Chertoff, Ex-Security Officials Slammed For Cashing In On Government Experience." *Huffington Post*, November 23, 2010, 06:16 pm ET | Updated May 25, 2011. Online.

Barnabas, Renaud. *Desire and Distance*: Introduction to a Phenomenology of Perception. Translated by Paul B. Milan. Stanford: Stanford University Press, 2005. [1999]

Barry, Thomas F. "The Weight of Angels: Peter Handke and 'Der Himmel über Berlin.'" *Modern Austrian Literature*, 23, no. 3/4, Special Issue: *The Current Literary Scene in Austria* (1990): 53–64.

Bartlett, Jamie. *The Dark Net: Unterwegs in den dunklen Kanälen der digitalen Unterwelt*. Kulmbach: Plassen Verlag, 2015.

Bateman, Christopher. *The Virtuous Cyborg*. London: Eyewear Publishing, 2018.

Baudelaire, Charles. "Le Joueur Généreux." *Le Figaro*, February 7, 1864: 5.

Baudrillard, Jean. *For a Critique of the Political Economy of the Sign*. St. Louis: Telos Press, 1981.

Baudrillard, Jean. *Intelligence of Evil or, The Lucidity Pact*. London: Bloomsbury, 2013 [2004]

Baudrillard, Jean. *La Guerre du Golfe n'a pas eu lieu*. Paris: Editions Galilée, 1991.

Baudrillard, Jean. "Requiem for the Media." In *For a Critique of the Political Economy of the Sign*, 164–84. St. Louis: Telos Press, 1981.

Baudrillard, Jean. *The Consumer Society: Myths and Structures*. Paris: Gallimard, 1970.

Baudrillard, Jean. *The System of Objects*, trans. James Benedict. London: Verso, 1996.

Beauvoir, Simone de. *The Second Sex*, trans. H. M. Parshley. New York, Vintage. 1989, 1952 [1949].

Begin, Menachem. *The Revolt: Story of the Irgun*. Jerusalem: Steimatzky, 1977.

Beinsteiner, Andreas. "Cyborg Agency: The Technological Self-Production of the (Post-) Human and the Anti-Hermeneutic Trajectory." *Thesis Eleven*, 153, Issue 1 (2019): 113–33.

Bekoff, Marc. *Wild Justice: The Moral Lives of Animals*. Chicago: University of Chicago Press, 2010.

Belluz, Julia. "Why Restaurants Became So Loud — and How to Fight Back. 'I can't hear you.'" *Vox*, July 27, 2018. Online. https://www.vox.com/2018/4/18/17168504/restaurants-noise-levels-loud-decibels.

Benjamin, Walter. [Detlef Holz]. *Deutsche Menschen. Von Ehre ohne Ruhm. Von Grösse ohne Glanz. Von Würde ohne Sold. Eine Folge von Briefen.* Lucerne: Vita Nova, 1936.
Illuminations: Essay and Reflections, ed. Hannah Arendt. New York: Schocken, 1969.
Benjamin, Walter. "Franz Kafka." In Hannah Arendt (ed.), *Illuminations: Essay and Reflections*, 111–140. New York: Schocken, 1969.
Benjamin, Walter. "Theses on the Philosophy of History." In Hannah Arendt (ed.), *Illuminations: Essay and Reflections*, 253–64. New York: Schocken, 1969.
Benjamin, Walter. "The Task of the Translator." In *Illuminations*, trans. Harry Zohn, edited and introduced by Hannah Arendt, 69–82. New York: Harcourt Brace Jovanovich 1968 [1923].
Benjamin, Walter. "The Work of Art in the Age of Mechanical [Technical] Reproduction." In *Illuminations*, 217–51. New York: Schocken, 1968.
Bennett, H. Stith. *Becoming a Rock Musician.* New York: Columbia University Press, 2018 [1980]. With a new preface by Howard Becker.
Bergoffen, Debra. *Contesting the Politics of Genocidal Rape: Affirming the Dignity of the Vulnerable Body.* New York: Routledge, 2011.
Bernasconi, Robert. *How to Read Sartre.* London: Granta, 2007.
Bernasconi, Robert. "Hegel at the Court of the Ashanti." In Stuart Barnett (ed.), *Hegel after Derrida*, 41–63. London: Routledge, 1998.
Bernasconi, Robert. "With What Must the Philosophy of World History Begin? On the Racial Basis of Hegel's Eurocentrism." *Nineteenth Century Contexts*, 22 (2000): 171–201.
Bernasconi, Robert and Anika Maaza Mann. "The Contradictions of Racism: Locke, Slavery, and the Two Treatises." In Andrew Valls (ed.), *Race and Racism in Modern Philosophy*, 89–107. Ithaca: Cornell University Press, 2005.
Bernays, Edward. *Crystallizing Opinion.* New York: Liveright, 1961.
Bernoulli, Carl Albrecht. *Franz Overbeck und Friedrich Nietzsche: eine Freundschaft.* 2 vols. Jena: E. Diedrichs, 1908.
Bertolotti, David S. *Culture and Technology.* Bowling Green, Ohio: Bowling Green State University Popular Press, 1984.
Bertolotti, David S. "The Atomic Bombing of Hiroshima." In Bertolotti, *Culture and Technology*, 81–112. Bowling Green: Bowling Green State University Popular Press, 1984.
Berry, David. *Critical Theory and the Digital.* London: Bloomsbury, 2014.
Berry, David. "Television and the Art of Symbolic Exchange." In William Merrin (ed.), *Baudrillard and the Media: A Critical Introduction*, 10–27. London: Polity, 2005.
Bischof, Günter, Jason Dawsey, and Bernhard Fetz (eds.), *The Life and Work of Günther Anders: Émigré, Iconoclast, Philosopher, Man of Letters.* London: Taylor & Francis, 2014.
Biser, Eugen. *Nietzsche – Zerstörer oder Erneuerer des Christentums?* Darmstadt: Wissenschaftliche Buchgesellschaft, 2002.
Bishop, Jeffrey. "Nietzsche's Power Ontology and Transhumanism: Or Why Christians Cannot Be Transhumanists." In Steve Donaldson and Ron Cole-Turner (eds.), *Christian Perspectives on Transhumanism and the Church: Chips in the Brain, Immortality, and the World of Tomorrow*, 117–35. Frankfurt am Main: Springer, 2018.
Blanch, Raimon Paez. "Dasein und Mensch bei Heidegger. Eine Überlegung anlässlich des 'Humanismusbriefes.'" In: Holger Zaborowsky and Alfred Denker (eds.) *Heidegger Jahrbuch 10*, 165–77. Freiburg im Breisgau: Alber, 2017.
Blinder, Alan, Patricia Mazzei, and Jess Bidgoodian's headline discussion "'Bomb Cyclone': Snow and Bitter Cold Blast the Northeast." *New York Times*, January 4, 2018.

Blitz, Mark. "Understanding Heidegger on Technology." *The New Atlantis* 41 (Winter 2014): 63–80.

Bloch, Ernst. *Experimentum Mundi*. Frankfurt am Main: Suhrkamp, 1976.

Blok, Vincent. "Denken als Handlung. Heideggers Besinnung auf das Wesen des Menschen im Zeitalter des human enhancement." In Holger Zaborowsky and Alfred Denker (eds.), *Heidegger Jahrbuch 10*, 265–79. Freiburg im Briesgau: Alber, 2017.

Blok, Vincent. *Ernst Jünger's Philosophy of Technology: Heidegger and the Poetics of the Anthropocene*. New York: Routledge, 2017.

Böhme, Hartmut. *Fetischismus und Kultur: Eine andere Theorie der Moderne*. Reinbek bei Hamburg: Rowohlt, 2006.

Boer, Roland. "A Totality of Ruins: Adorno on Kierkegaard." *Cultural Critique*, 83 (Winter 2013): 1–30.

Bostrom, Nick. "In Defense of Posthuman Dignity." *Social Epistemology Review and Reply Collective*, 6, no. 2 (2017): 1–10.

Bostrom, Nick. "Transhumanist Values." *Journal of Philosophical Research*, 30 (2005): 3–14.

Bourdieu, Pierre. *Distinction: A Social Critique of the Judgement of Taste*. London: Routledge, 1986. [1979]

Bousquet, Antoine. "Ernst Jünger and the Problem of Nihilism in the Age of Total War." *Thesis Eleven*, 132, Issue 1 (2016): 17–38.

Brey, Paul. "Philosophy of Technology after the Empirical Turn." *Techné: Research in Philosophy and Technology*, 14, no. 1 (2010): 1–11.

Brooks, Rosa. *How Everything Became War and the Military Became Everything: Tales from the Pentagon*. New York: Simon & Schuster, 2016.

Buber, Martin. *Gottesfinsternis*. Zürich: Manesse Verlag, 1953.

Busch, Briton Cooper. *The War against the Seals: A History of the North American Seal Fishery*. Montreal: McGill-Queen's Press, 1987.

Busch, Wilhelm. *Die fromme Helene*. Munich: Braun & Schneider, 1972.

Bush, Vannevar. "As We May Think." *Atlantic Monthly*, 176 (1945): 101–8.

Bussolini, Jeffrey. "Felidae and Extinction: 'Victim' and 'Cause.'" *Temporal Belongings*, April 20, 2015. Online. www.temporalbelongings.org/presentations2/jeffrey-bussolini-cuny. Accessed August 31, 2019.

Bussolini, Jeffrey. "Los Alamos as Laboratory for Domestic Security Measures: Nuclear Age Battlefield Transformations and the Ongoing Permutations of Security." *Geopolitics*, 16, no. 2 (2011): 329–58.

Butterfield, Herbert. *The Whig Interpretation of History*. London: G. Bell and Sons, 1931.

Cabestan, Philippe. "Phénoménologie, anthropologie: Husserl, Heidegger, Sartre." *Alter*, 23 (2015): 226–42.

Calarco, Matthew. "'Another Insistence of Man': Prolegomena to the Question of the Animal in Derrida's Reading of Heidegger." *Human Studies*, 28, no. 3 (November 2005): 317–34.

Calarco, Matthew. *Zoographies: The Question of the Animal from Heidegger to Derrida*. New York: Columbia University Press, 2008.

Carlyle, Thomas. "The Nigger Question." In *Miscellaneous Essays: Volume 7*, 79–101. London: Chapman and Hall, 1888.

Carr, Nicholas. *The Shallows: What the Internet Is Doing to Our Brains*. New York: W. W. Norton & Co., 2011.

Carson, Anne. *Eros, the Bittersweet*. Princeton: Princeton University Press, 1986.

Carson, Gerald. "The Piano in the Parlor." *American Heritage*, 17, Issue 1 (December 1965): 54–59.
Carroll, Lewis. *Jabberwocky and Other Poems*. Mineola: Dover Publications, Inc., 2001.
Carroll, Richard J., S.J. "Disenchantment, Rationality and the Modernity of Max Weber." *Forum Philosophicum*, 16, no. 1 (2011): 117–37.
Carrozzini, Giovanni. "Technique et humanisme. Günther Anders et Gilbert Simondon." *Appareil* 2 (2008): 1–12.
Cavallaro, Marco. "Der Beitrag der Phänomenologie Edmund Husserls zur Debatte über die Fundierung der Geisteswissenschaften." *Phänomenologische Forschungen* (2013): 77–93.
Cavin, Susan. "Adorno, Lazarsfeld &The Princeton Radio Project, 1938–1941." American Sociological Association Annual Meeting, Sheraton Boston and the Boston Marriott Copley Place, Boston, MA, July 31, 2008. Online. https://www.scribd.com/doc/151660755/Adorno-Lazarsfeld-The-Princeton-Radio-Project-1938-1941. Accessed September 27, 2020.
Caygill, Howard. "The Consolation of Philosophy: 'Neither Dionysus nor the Crucified.'" *Journal of Nietzsche Studies, Futures of Nietzsche: Affirmation and Aporia*, Vol. 7 (Spring 1994): 131–50.
Caygill, Howard, "Taste and Civil Society." In Babich (ed.), *Reading David Hume's ›of the Standard of Taste‹*, 177–212. Berlin: de Gruyter, 2019.
Chomsky, Noam. *Media Control: The Spectacular Achievements of Propaganda*. New York: Seven Stories Press, 2008.
Chua, Daniel. "Drifting: The Dialectics of Adorno's Philosophy of New Music." In Berthold Hoeckner (ed.), *Apparitions: Essays on Adorno and Twentieth-Century Music*, 1–18. London: Routledge, 2013.
Clark, David L. "Kant's Aliens: The Anthropology and Its Others." *The New Centennial Review*, 1, no. 2 (Fall 2001): 201–89.
Clifton, Thomas. "Music as Constituted Object." In F. Joseph Smith (ed.), *In Search of Musical Method*, 73–98. New York: Gordon and Breach Science Publishers, Inc., 1976.
Clifton, Thomas. *Music as Heard: A Study in Applied Phenomenology*. New Haven: Yale University Press, 1983.
Coeckelbergh, Mark. *New Romantic Cyborgs: Romanticism, Information Technology, and the End of the Machine*. Cambridge, MA: MIT Press, 2017.
Costea, Bogdan and Kostas Amiridis. "Ernst Jünger, Total Mobilisation and the Work of War." *Organisation*, 24, Issue 4 (2017): 1–23.
Crawford, Ryan and Erik M. Vogt (eds.), *Adorno and the Concept of Genocide*. Amsterdam: Brill, 2016.
Crowell, Steve. "Does the Husserl/Heidegger Feud Rest on a Mistake? An Essay on Psychological and Transcendental Phenomenology." *Husserl Studies*, 18, no. 2 (2002): 123–40.
Dahlstrom, Dan. "Heidegger's Transcendentalism." *Research in Phenomenology*, 35, 1 (2005): 29–54.
Dahlstrom, Dan. *The Heidegger Dictionary*. London: Bloomsbury, 2013.
Daigle, Christine and Jacob Golomb (eds.), *Beauvoir and Sartre: The Riddle of Influence*. Bloomington: Indiana University Press, 2009.
Darwin, Charles. *The Descent of Man and Selection in Relation to Sex*, 201. New York: D. Appleton, 1874.
Darwin, Charles. *The Voyage of the Beagle*, 228–9. New York: P.F. Collier & Son, 1909.

Darwin, Charles. *Über die Entstehung der Arten um Their- und Pflanzen-Reich durch natürliche Züchtung, oder Erhaltung der vervollkemmneten Rassen im Kampfe um's Daseyn*, trans. Heinrich G. Bronn. Stuttgart: E. Schweizerbart, 1860.

Dastur, Françoise. *"La critique heideggérienne de l'anthropologisme", Heidegger et la pensée à venir*. Paris: Vrin, 2011.

Dastur, Françoise. "The Critique of Anthropologism in Heidegger's Thought." In James E. Faulconer and Mark A. Wrathall (eds.), *Appropriating Heidegger*, 119–34. Cambridge: Cambridge University Press, 2000.

David, Christophe. "Fidélité de Günther Anders à l'anthropologie philosophique: de l'anthropologie négative de la fin des années 1920 à *L'obsolescence de l'homme*." *L'Homme et la société. La Question anthropologique*, no. 181 (2011/3): 165–80.

David, Christophe. "Günther Anders et la question de l'autonomie de la technique." *Ecologie & politique*, 1, no, 32 (2006). Online. http://1libertaire.free.fr/GAnders54.html.

David, Christophe. and Dirk Röpcke. "Günther Anders, Hans Jonas et les antinomies de l'écologie politique." *Ecologie & politique*, 2, no. 29 (2004): 193–213.

Dawsey, James. "The Life of a Rescuer: Eva Michaelis-Stern in Dark Times," June 25, 2019. Online. https://www.google.com/url?sa=t&rct=j&q=&esrc=s&source=web&cd=&cad=rja&uact=8&ved=2ahUKEwiOsZbCutHpAhWzqHEKHeBPC5cQFjAAegQIAhAB&url=https%3A%2F%2Fwww.nationalww2museum.org%2Fwar%2Farticles%2Flife-rescuer-eva-michaelis-stern-dark-times&usg=AOvVaw3AodLtsezuxLID0q3ojPZ0

Dawsey, James. "Ontology and Ideology: Günther Anders's Philosophical and Political Confrontation with Heidegger." *Critical Historical Studies*, 4, no. 1 (Spring 2017): 1–37.

de Certeau, Michel. *The Practice of Everyday Life*, trans. Steven Rendall. Berkeley: University of California Press, 1984.

Debord, Guy. *The Society of the Spectacle: Annotated Edition*, Bureau of Public Secrets [1995], trans. Fredy Perlman and friends. London: Black and Red, 1977. [1967]

Delabar, Walter. "Fabula docet. Zu den erzählenden Texten von Günther Anders und zum Roman 'Die molussische Katakombe.'" *Zeitschrift für Germanistik*. Neue Folge, 2, no. 2 (1992): 300–19.

de Nicolas. *Powers of Imagining: Ignatius de Loyola: A Philosophical Hermeneutic of Imagining through the Collected Works of Ignatius de Loyola*. Albany: State University of New York Press, 1986.

Derrida, Jacques. "Des Tours de Babel," trans. Joseph F. Graham. In Joseph F. Graham (ed.), *Difference in Translation*, 165–207. Ithaca: Cornell University Press, 1985.

Derrida, Jacques.*The Animal That Therefore I Am*, trans. David Wills. New York: Fordham University Press, 2008.

Descartes, René. *Meditations on First Philosophy*, trans. John Cottingham. Cambridge: Cambridge University Press, 1996.

Deutsch, Babette. "Preface." Rilke, *The Book of Hours*, trans. Deutsch, 3–6. New York: New Directions Books, 1941.

Deutsch, Werner and Christliebe El Mogharbel. "Clara and William Stern's Conception of a Developmental Science." *European Journal of Developmental Psychology*, 8, Issue 2 (2011): 135–56.

Djerassi, Carl. *Foreplay: Hannah Arendt, The Two Adorno, and Walter Benjamin*. Madison: University of Wisconsin Press, 2011.

Donaldson, Sue and Will Kymlicka. *Zoopolis: A Political Theory of Animal Rights*. Oxford: Oxford University Press, 2011.

Donnelly, Ryan F., Thakur Raghu Raj Singh, Eneko Larrañeta, and Maeliosa T. C. McCrudden, *Microneedles for Drug and Vaccine Delivery and Patient Monitoring*. Oxford: John Wiley & Sons, 2018.
Downes, Paul. *Concentric Space as a Life Principle Beyond Schopenhauer, Nietzsche and Ricoeur: Inclusion of the Other*. London: Routledge, 2020
Downey, Greg, "Listening to Capoeira: Phenomenology, Embodiment, and the Materiality of Music." *Ethnomusicology*, 46, no. 3 (Autumn 2002): 487–509.
Dries, Christian. "Technischer Totalitarismus: Macht, Herrschaft und Gewalt bei Günther Anders." *Etica & Politica/Ethics & Politics: Rivista di filosofia*, 15, no. 2 (2013): 175–98.
Dries, Christian. *Die Welt als Vernichtungslager. Eine kritische Theorie der Moderne im Anschluss an Günther Anders, Hannah Arendt und Hans Jonas*. Bielefeld: Transcript, 2012.
Drouillard, Jill. "Heidegger's Sexless Community: Ni homme, ni femme — c'est un Dasein." In John Rose (ed.), *Proceedings of the Heidegger Circle*. Goucher College, 2018.
Dücker, Bürckhard. "Blühende Kirschbäume" in: Dücker, *Machen – Erhalten – Verwalten: Aspekte einer performativen Literaturgeschichte*. Göttingen: Wallstein Verlag, 2016.
Duhem, Pierre. "German Science and German Virtues." In *German Science*, trans. John Lyon, 113–26. Chicago: Open Court, 1991. [1916]
Dupuy, Jean-Pierre. *Pour un catastrophisme éclairé. Quand l'impossible est certain*. Paris: Seuil, 2002.
Dupuy, Jean-Pierre. *The Mark of the Sacred*, trans. M. B. De Bevoise. Stanford: Stanford University, 2013.
Ehgartner, Ulrike, Patrick Gould, and Marc Hudson, "On the Obsolescence of Human Beings in Sustainable Development." *Global Discourse. An Interdisciplinary Journal of Current Affairs and Applied Contemporary Thought*, 7, Issue 1(*After Sustainability – What?*) (2017): 66–83.
Ehrlich, Paul. *The Population Bomb*. New York: Ballantine Books, 1968.
Eipper, Paul. *Tiere siehen dich an*. Berlin: Dietrich Reimer / Ernst Vohsen 1928.
Ellensohn, Reinhard. Der *andere Anders: Günther Anders als Musikphilosoph*. Bern: Peter Lang, 2008.
Elmer, Greg, Ganaele Langlois, and Joanna Redden (eds.), *Compromised Data: From Social Media to Big Data*. London: Bloomsbury, 2015.
Elliott, Christopher. "The TSA Has Never Kept You Safe: Here's Why." *Fortune*, June 2, 2015, 12:30 PM. http://fortune.com/2015/06/02/the-teaairport-security-problems/
Ellul, Jacques. The Technological Society, trans. John Wilkinson. New York: Knopf/Vintage, 1967. [1964]. Introductio: Robert K. Merton.
Ellul, Jacques. *La technique ou l'enjeu du siècle*. Paris: Armand Colin, 1954.
Enders, Markus and Holger Zaborowski (eds.), *Jahrbuch für Religionsphilosophie*, Vol. 13. Freiburg: Alber, 2014.
Epictetus, *Enchiridion*, trans. Thomas William Hazen Rolleston. London: Kegan Paul, Trench, & Co, 1881.
Epping-Jäger, Cornelia. "Lautsprecher." In Daniel Morat and Hans Jakob Ziemer (eds.), *Handbuch Sound: Geschichte-Begriffe –Ansätze*, 396–400. Frankfurt: Springer, 2017.
Erlmann, Veit. *Reason and Resonance*. New York: Zone Books, 2010.
Espinet, David and Tobias Keiling (eds.), *Heideggers Ursprung des Kunstwerks. Ein Kooperativer Kommentar*. Frankfurt am Main: Klostermann, 2011.
Fahrenbach, Helmut. "Heidegger und das Problem einer 'philosophischen Anthropologie.'" In Vittorio Klostermann (ed.), *Durchblicke. Martin Heidegger zum 80. Geburtstag*. Frankfurt am Main: Vittorio Klostermann, 1970.

Farr, James. "Locke, Natural Law, and New World Slavery." *Political Theory*, 36, no. 4 (August 2008): 495–522.

Farrell-Krell, David. *The Recalcitrant Art: Diotima's Letters to Holderlin and Related Missives*, trans. Douglas F. Kenney and Sabine Menner-Bettscheid. Albany: SUNY Press, 2000.

Faye, Emmanuel. *Arendt et Heidegger: Extermination nazie et destruction de la pensée*. Paris: Albin Michel, 2016.

Feldstein, Steven. "The Global Expansion of AI Surveillance." September 17, 2019, Carnegie Endowment for International Peace. https://carnegieendowment.org/2019/09/17/global-expansion-of-ai-surveillance-pub-79847

Fenn, Elizabeth A. "Biological Warfare in Eighteenth-Century North America: Beyond Jeffery Amherst." *The Journal of American History*, 86, no. 4 (March 2000): 1552–80.

Fetz, Bernhard. "Zwischen Heidegger, Kafka, und der Atombombe: Zur veröffentlichten und unveröffentlichten Essayistik des Schriftstellers und Philosophen Günther Anders." In Michael Ansel, Jürgen Egyptien, and Hans-Edwin Friedrich (eds.), *Der Essay als Universalgattung des Zeitalters: Diskurse, Themen und Positionen zwischen Jahrhundertwende und Nachkriegszeit*, 283–97. Amsterdam: Brill, 2016.

Feyerabend, Paul. *Killing Time: The Autobiography of Paul Feyerabend*. Chicago: University of Chicago Press, 1995.

Figal, Günter. *For a Philosophy of Freedom and Strife: Politics, Aesthetics, Metaphysics*, trans. Wayne Klein. Albany: State University of New York Press, 1997.

Filk, Christian. "Der Mensch ist größer und kleiner als er selbst Günther Anders' Negative Anthropologie im Zeitalter der '(Medien-)Technokratie.'" *Medienpulse*, 50, no. 2 (2012): 1–19.

Filk, Christian. "'Frei sind die Dinge: Unfrei ist der Mensch' – Zur Entwicklung von Günther Anders Negativer Anthropologie im technisch-medialen Zeitalter." *MEDIENwissenschaft: Rezensionen | Reviews*, 23, no. 3 (2006): 277–91.

Finch Sidd. "Twelve Songs That Everyone Thinks are by Other Artists." *The Brag Media. Tone-Deaf*, February 28, 2019. Online. https://tonedeaf.thebrag.com/12-songs-everyone-thinks-other-artists/

Fink, Eugen. *Oase des Glücks. Gedanken zu einer Ontologie des Spiels*. Freiburg: Alber, 1957.

Fisk, Catherine L. "The Modern Author at Work on Madison Avenue." In Paul K. Saint-Amour (ed.), *Modernism and Copyright*, 173–94. Oxford: Oxford University Press, 2011.

Fleck, Ludwik. *The Genesis and Development of a Scientific Fact*. Chicago: University of Chicago Press, 1981. [1935]

Ford, John. *The Man Who Shot Liberty Valance*. 1962. Film.

Forman, Paul. "Behind Quantum Electronics: National Security as Basis for Physical Research in the United States, 1940–1960." *Historical Studies in the Physical and Biological Sciences*, 18 (1987): 149–229.

Forman, Paul. "Kausalität, Anschaulichkeit, and Individualität, or How Cultural Values Prescribed the Character and the Lessons Ascribed to Quantum Mechanics." In N. Stehr and V. Meja (eds.), *Society and Knowledge: Contemporary Perspectives in the Sociology of Knowledge*, 333–47. New Brunswick: Transaction Books, 1984.

Forman, Paul. "Weimar Culture, Causality, and Quantum Theory, 1918–1927: Adaptation by German Physicists and Mathematicians to a Hostile Intellectual Environment." *Historical Studies in the Physical Sciences*, 3 (1971): 1–115.

Francione, Gary and Anna Charlton. *Eat Like You Care: An Examination of the Morality of Eating Animals*. London: Exempla Press, 2013.

Frank, Jason (ed.). *A Political Companion to Herman Melville*. Lexington: University Press of Kentucky, 2013.
Fredborg, Beverley, Jim Clark and Stephen D. Smith. "An Examination of Personality Traits Associated with Autonomous Sensory Meridian Response (ASMR)." *Frontiers in Psychology*, February 23, 2017.
Frege, Gottlob. *Die Grundlagen der Arithmetik. Eine logisch mathematische Untersuchung über den Begriff der Zahl*. Breslau: Wilhelm Koebner, 1884.
Friese, Heidrun. "Lesezenen: Gelehrte lessen vor." In Alf Lüdtke and Reiner Prass (eds.), *Gelehrtenleben: Wissenschaftspraxis in der Neuzeit*, 252–9. Cologne: Böhlau, 2008.
Frings, Manfred S. "Max Scheler: A Novel Look at the Origin of Evil." *Philosophy and Theology*, 6, Issue 3 (Spring 1992): 201–11.
Frischmann, Brett and Evan Selinger. *Re-Engineering Humanity*. Cambridge: Cambridge University Press, 2018.
Frith, Jordan. *Smartphones as Locative Media*. London: Polity, 2015.
Frith, Jordan and Jason Kalin, "Here, I Used to Be: Mobile Media and Practices of Place-Based Digital Memory." *Space & Culture, Space and Culture*, 19, no. 1 (2016): 43–55.
Frodeman, Robert. *Transhumanism, Nature, and the Ends of Science*. New York: Routledge, 2019.
Fuchs, Christian. "Günther Anders' Undiscovered Critical Theory of Technology in the Age of Big Data Capitalism." *tripleC*, 15, no. 2 (2017): 582–611. Online. https://www.triple-c.at/index.php/tripleC/article/view/898
Fuchs, Christian. "Zu einigen Parallelen und Differenzen im Denken von Günther Anders und Herbert Marcuse." In Dirk Röpcke and Raimund Bahr (eds.), *Geheimagent der Masseneremiten*, 113–27. St. Wolfgang: Art & Science, 2003.
Fuchs, Emil. "Vom Sinn des menschlichen Daseins — Eine Auseinandersetzung mit M. Heidegger." *Deutsche Zeitschrift für Philosophie*, 10, Issue 8 (1962): 982–94.
Fullbrook, Edward and Kate Fullbrook, *Sex and Philosophy: Rethinking de Beauvoir and Sartre*. London: Continuum, 2008.
Fuller, Steve. "Creating the Covid-19 Story: How to Control a Pandemic's Narrative," May 2, 2020. Online. https://iai.tv/articles/creating-the-covid-19-story-auid-1529. Accessed May 2, 2020.
Fuller, Steve. *Humanity 2.0. What It Means to be Human Past, Present and Future*. London: Pallgrave Macmillan, 2011.
Fuller, Steve. *Nietzschean Meditations: Untimely Thoughts at the Dawn of the Transhuman Era*. Basel: Schwabe, 2019.
Fuller, Steve and Veronika Lipinska. *The Proactionary Imperative: A Foundation for Transhumanism*. London: Palgrave Macmillan, 2014.
Gadamer, Hans-Georg. *The Relevance of the Beautiful*, trans. Nicholas Walker. Cambridge: Cambridge University Press, 1987.
Galič, Maša, Tjerk Timan, and Bert-Jaap Koops. "Bentham, Deleuze and Beyond: An Overview of Surveillance Theories From the Panopticon to Participation." *Philosophy & Technology*, 30, no. 1 (2017): 9–37.
Gann, Kyle. *No Such Thing as Silence: John Cage's '4' 33'''*. New Haven: Yale University Press, 2011.
Garcia, David F. *Listening for Africa: Freedom, Modernity, and the Logic of Black Music's African Origins*. Durham: Duke University Press, 2017.
Gehring, Petra. "Zwischen Menschenpark and Eugenics." In Günter Figal (ed.), *Humanismus*, 81–112. Mohr Siebeck, 2003.

Gellen, Kata. "Kafka, Pro and Contra: Günther Anders's Holocaust Book." In Arthur Cools and Vivian Liska (eds.), *Kafka and the Universal*, 283–304. Berlin: de Gruyter, 2016.
Germain, Gilbert G. *A Discourse on Disenchantment: Reflections on Politics and Technology*. Albany: State University of New York Press, 1993.
Ghamari-Tabrizi, Sharon. *The Worlds of Herman Kahn: The Intuitive Science of Thermonuclear War*. Cambridge, MA: Harvard University Press, 2005.
Gibbs, Jeff. *Planet of Humans*. Documentary 2020.
Gill, David W. "Jacques Ellul and Technology's Trade-Off." *Comment*, 11, no. 1 (Spring 2012). Online. https://www.cardus.ca/comment/article/jacques-ellul-and-technologys-trade-off/
Gillespie, Michael Allen. "The Future of Political Theory: Using the Canon to Prepare for Tomorrow," February 23, 2017. Online. https://humanitiesfutures.org/papers/future-political-theory-using-canon-prepare-tomorrow/
Goehr, Lydia. *The Imaginary Museum of Musical Works: An Essay in the Philosophy of Music*. Oxford: Oxford University Press, 1992.
Goethe, Johann Wolfgang von. *Faust. Der Tragödie zweiter Teil*. Stuttgart und Tübingen: J. G. Cotta'schen Buchhandlung, 1832. [1808]
Goethe, Johann Wolfgang von. *Faust – Eine Tragödie. Beide Theile in Einem Bande*. Stuttgart und Augsburg: J. G. Cotta'schen Buchhandlung, 1858.
Goethe, Johann Wolfgang von. "Der Zauberlehrling." *Ausgabe letzter Hand.Band 1–40 in 20 Bänden*. Stuttgart and Tübingen: Cotta, 1830 [1827], Vol. 1, 217–20.
Goodman, Steve. *Sonic Warfare: Sound, Affect, and the Ecology of Fear*. Cambridge: MIT Press, 2010.
Gordon, Don. *Electronic Warfare: Element of Strategy and Multiplier of Combat Power*. New York: Pergammon Press, 1981.
Gratzer, Wolfgang. "Listening to Music: An Art in It Self or Not." In Christian Thorau and Hansjakob Ziemer (eds.), *The Oxford Handbook of Music Listening in the 19th and 20th Centuries*, 462–76. Oxford: Oxford University Press, 2019.
Grandin, Greg. *The Empire of Necessity: Slavery, Freedom, and Deception in the New World*. New York: Metropolitan Books, 2014.
Grandin, Greg. "The Other 'Moby Dick': Melville's 'Benito Cereno' is an Analogy for American Empire." *Mother Jones*. Monday, January 27, 2014.
Gray, John. *Straw Dogs: Thoughts on Humans and Other Animals*. London: Granta Books, 2002.
Gray, John. "Why the Humanities Can't Be Saved." Online. https://unherd.com/2019/08/why-the-humanities-cant-be-saved/
Greffrath, Mathias. "Lob der Sturheit." *Die Zeit*, "Zeitläufte." 28/2002.
Groenewegen, Peter. "Thomas Carlyle, 'The Dismal Science,' and the Contemporary Political Economy of Slavery." *History of Economics Review*, 34 (Summer 2001): 74–94.
Griffin, David Ray. *The New Pearl Harbor Revisited: 9/11, the Cover-Up, and the Exposé*. Northampton: Olive Branch [Interlink Books], 2008.
Griffin, David Ray. *Evil Revisited: Responses and Reconsiderations*. Albany: State University of New York Press, 1991.
Grondin, Jean. "Heidegger und Augustine: Zur hermeneutischen Wahrheit." In Ewald Richter (ed.), *Die Frage nach der Wahrheit*, 161–73. Frankfurt am Main: Klostermann, 1997.
Günther, Rolf S. *Rainer Maria Rilke und Lou Andreas Salome: auf welches Instrument sind wir gespannt: Traumerzählung*. Würzburg: Königshausen & Neumann, 2005.
Hadot, Pierre. *Philosophy as a Way of Life*. Oxford: Blackwell, 1994.

Halbwachs, Maurice. "La mémoire collective chez les musiciens." *Revue philosophique* (March–April 1939): 136-65.
Haucke, Kai. "Anthropologie bei Heidegger. Über das Verhältnis seines Denkenszur philosophischen Tradition." *Philosophische Jahrbuch*, 105. Jahrgang / II (1998): 321-45.
Heelan, Patrick Aidan. *Quantum Mechanics and Objectivity*. The Hague: Nijhoff, 1965.
Hermann, Friedrich-Wilhelm von "Die 'Confessiones' des Heiligen Augustinus im Denken Heideggers." *Questio* 1 (2001): 113-46.
Han, Byung-Chul. The Burncut Society, trans. Erik Butler. Stanford: Stanford University Press, 2015.
Han, Byung-Chul. *In the Swarm: Digital Prospects*, trans. Erik Butler. Cambridge, MA: The MIT Press, 2017.
Harris, Bev and Kathleen Wynne. *Hacking Democracy*. HBO, 2006. Documentary. https://www.youtube.com/watch?v=iZLWPleeCHE
Hartmann, Frank (ed.), *Vom Buch zur Datenbank. Paul Otlets Utopie der Wissensvisualisierung* (= Forschung visuelle Kultur. 2). Berlin: Avinus, 2012.
Hauptman, Robert. "Surveillance: Ubiquitous and Oppressive." *Journal of Information Ethics*, 18, no. 2 (2009): 3-4.
Heelan, Patrick Aidan. *Quantum Mechanics and Objectivity*. The Hague: Nijhoff, 1965.
Heelan, Patrick Aidan. *Space-Perception and the Philosophy of Science*. Berkeley: University of California Press, 1983.
Heidegger, Martin. *Beiträge zur Philosophie (Vom Ereignis)*. Gesamtausgabe 65, ed. F. W. von Hermann. Frankfurt am Main: Klostermann, 1989.
Heidegger, Martin. *Bremer und Freiburger Vorträge GA 79*. Frankfurt: Vittorio Klostermann 1994.
Heidegger, Martin. *Die Frage nach der Technik* in: *Vorträge und Aufsätze*. Pfullingen: Neske, 1954.
Heidegger, Martin. "Die Frage nach der Technik." In Clemens Graf Podewils (ed.), *Die Künste im technischen Zeitalter*, 70-129. Munich: R. Oldenbourg, 1954.
Heidegger, Martin. "Einblick in das was ist, Bremer Vorträge 1949." In *Bremer und Freiburger Vorträge*, 5-80, GA 79. Frankfurt am Main: Klostermann, 1994.
Heidegger, *Introduction to Philosophy — Thinking and Poetizing*, trans. Phillip Jacques Braunstein. Bloomington: Indiana University Press, 2011.
Heidegger, Martin. *The Question Concerning Technology*, trans. William Lovitt. San Francisco: Harper Torchbooks, 1977. [1962]
Heidegger, Martin. *Sein und Zeit*. Tübingen: Max Niemeyer, 1984. [1927]
Heidegger, Martin. *On the Way to Language*." Trans. Peter Hertz. New York: Harper & Row, 1971.
Heidegger, Martin. *Überlegungen XII-XV (Schwarze Hefte 1939-1944)*, GA 96. Frankfurt am Main: Klosterman, 2014.
Heidegger, Martin. *What Is Philosophy*? Heidegger, *What Is Philosophy*, trans. William Kluback and Jean T. Wilde. New Haven: College and University Press, 1958.
Heidegger, Martin. "Wozu Dichter?" In *Holzwege*, 248-95. Frankfurt am Main: Vittorio Klostermann, 1950.
Helmholtz, Herman. *On the Sensations of Tone as a Physiological Basis for the Theory of Music*, trans. Alexander J. Ellis. New York: Dover Publications, Inc., 1954.
Hoeckner, Berthold. *Programming the Absolute: Nineteenth-century German Music and the Hermeneutics of the Moment*. Princeton: Princeton University Press, 2002.

Hölderlin, Friedrich. *Hyperion oder der Eremit in Griechenland*. 2nd Volume. Basel: Stroemfeld/Roter Stern: 1992. [1799]

Hölderlin, Friedrich. *Sämtliche Werke. Frankfurter Ausgabe*. Frankfurt am Main: Stroemfeld/Roter Stern, 1999.

Horkheimer, Max and Theodor Adorno. *Dialectic of Enlightenment*. London: Verso, 2016. [1976]

Horn, Eva. "Air Conditioning: Taming the Climate as a Dream of Civilization." In James Graham (ed.), *Climates: Architecture and the Planetary Imaginary, The Avery-Review*, 233–41. New York: Lars Müller Publishers, Columbia Books on Architecture and the City, 2016.

Huen, Wayne (ed.) *The Walking Dead and Philosophy: Zombie Apocalypse Now*. Bowling Green: Open Court, 2017.

Huie, William Bradford. *The Hiroshima Pilot: The Case of Major Claude Eatherly*. New York: Putnam, 1964.

Ihde, Don. *Husserl's Missing Technologies*. Oxford: Oxford University Press, 2016.

Ihde, Don. *Listening and Voice: Phenomenologies of Sound*. Albany: State University of New York Press, 1976.

Illich, Ivan. "An Address to 'Master Jacques.'" *The Ellul Forum*, 13 (July 1994): 16–17.

Illich, Ivan. "The Cultivation of Conspiracy." In Lee Hoinacki and Carl Mitcham (eds.), *The Challenges of Ivan Illich: A Collective Reflection*, 233–42. Albany: The State University of New York Press, 2002.

Illich, Ivan. *Deschooling Society*. London: Marion Boyars Publishers Ltd, 2000. [1971]

Illich, Ivan. *H2O and the Waters of Forgetfulness: Reflections on the Historicity of 'Stuff.'* Dallas: Dallas Inst Humanities & Culture, 1985.

Illich, Ivan. *The Powerless Church and Other Selected Writings, 1955–1985*. State College: Pennsylvania State Press, 2019.

Illich, Ivan. *In the Vineyard of the Text*. Chicago: University of Chicago Press, 1993.

Illich, Ivan. *Medical Nemesis: The Expropriation of Health*. New York: Pantheon Books, 1976.

Illich, Ivan. *Selbstbegrenzung: eine politische Kritik der Technik*. Munich: C.H.Beck, 1998. [1973]

Irigaray, Luce. *The Forgetting of Air in Martin Heidegger*, trans. Mary Beth Mader. Texas: University of Texas Press, 1999.

Iven, Mathias. "Spurensuche: Hannah Arendt und Günther Anders in Nowawes." *Mitteilungen der Studiengemeinschaft Sanssouci e.V.*, 18 Jg. (2013): 122–34.

James, Simon. "Phenomenology and the Problem of Animal Minds." *Environmental Values*, 18 (2009): 33–49.

Janicaud, Dominique. *Powers of Rationality: Science, Technology, and the Future of Thought*. Bloomington: Indiana University Press, 1994. [1985]

Jaspers, Karl. *The Atom Bomb and the Future of Man*, trans. E. B. Ashton. Chicago: University of Chicago Press, 1961.

Jenkins, Richard. "Disenchantment, Enchantment and Re-Enchantment: Max Weber at the Millennium." *Max Weber Studies*, 1, no. 1 (November 2000): 11–32.

Jerónimo, Helena M., José Luís Garcia, and Carl Mitcham (eds.). *Jacques Ellul and the Technological Society in the 21st Century*. Frankfurt am Main: Springer, 2013.

Jogschies, Rainer B. "Zur Chiffrierung von Atomkriegsängsten in Science-Fiction-Filmen und ihrer De-Chiffrierung in der Politik. Erzählweisen zwischen Überlieferung und Projektion in den USA, Großbritannien, und Japan – Wirklichkeitskonstruktion zwischen Fantasie und 'Nuclearismus.'" In Manfred Mai

and Rainer Winter (eds.), *Das Kino der Gesellschaft – die Gesellschaft des Kinos: Interdisziplinäre Positionen, Analysen und Zugänge*, 204–41. Cologne: Herbert von Halem Verlag, 2006.
Johannßen, Dennis. "Mensch und Dasein in Heideggers 'Sein und Zeit'." In Thomas Ebke and Caterina Zanfi (eds.), *Das Leben im Menschen oder der Mensch im Leben? Deutsch-Französische Genealogien zwischen Anthropologie und Anti-Humanismus*, 91–104. Potsdam: Universitätsverlag Potsdam, 2017.
Johannssenm Dennis. "Toward a Negative Anthropology." *Anthropology and Materialism*, 1 (2013): 1–14.
Johnson, Dorothy M. "The Man Who Shot Liberty Valance." *Cosmopolitan*, July 1949.
Johnson, Julian. "'The Elliptical Geometry of Utopia': New Music Since Adorno." In Berthold Hoeckner (ed.), *Apparitions: Essays on Adorno and Twentieth-Century Music*, 69–84. London: Routledge, 2013.
Jolly, Édouard. "Entre légitime défense et état d'urgence. La pensée andersienne de l'agir politique contre la puissance nucléaire." *Etica & Politica/Ethics & Politics: Rivista di filosofia*, 15, no. 2 (2013). Online.
Jolly, Édouard. *Étranger au monde: Essai sur la première philosophie de Günther Anders*. Paris: Classiques Garnier, 2019.
Jolly, Édouard. *Günther Anders. Une Politique de la Technique*. Paris: Decitre, 2017.
Jünger, Ernst. *The Worker: Dominion and Form*, trans. Bogdan Costea. Evanston: Northwestern University Press, 2017.
Jungk, Robert. *Strahlen aus der Asche. Geschichte einer Wiedergeburt*. Munich: Scherz, 1991. [1959]
Jungk, Robert. *Children of the Ashes: The Story of a Rebirth*. London: Heinemann, 1961.
Kaerlein, Timo. "Playing with Personal Media: On an Epistemology of Ignorance." *Culture Unbound*, 5 (2013): 651–70.
Kahn, Herman. *On Thermonuclear War*. Princeton: Princeton University Press, 1962.
Kane, Brian. "L'Objet Sonore Maintenant: Pierre Schaeffer, Sound Objects and the Phenomenological Reduction." *Organised Sound*, 12, no. 1 (April 2007): 15–24.
Kant, Immanuel. *Metaphysical Foundations of Natural Science*, trans. Michael Friedmann. Stanford: Stanford University, 2004.
Kant, Immanuel. *Foundations of the Metaphysicsl of Morals*, trans. Lewis White Beck. New York: Macmillan, 1990.
Kappeler, Susanne. "Speciesism, Racism, Nationalism or the Power of Scientific Subjectivity." In Carol Adams, et al. (eds.), *Animals and Women: Feminist Theoretical Explorations*, 320–66. Durham: Duke University Press, 1995.
Kateb, George. "Technology and Philosophy." *Social Research*, 64, no. 3 (1997): 1225–46.
Kimball, Roger. *Tenured Radicals: How Politics Has Corrupted Our Higher Education*. New York: Harper & Row, 1990.
Kittler, Friedrich. *Aufschreibesysteme 1800/1900*. Munich: Wilhelm Fink, 1985
Kittler, Friedrich. *Gramophone, Film, Typewriter*, trans. Geoffrey Winthrop-Young and Michael Wutz. Stanford: Stanford University Press, 1999.
Kittler, Friedrich. "*Weltatem* On Wagner's Media Technology." In Leroy R. Shaw, Nancy R. Cirillo and Marion Miller (eds.), *Wagner in Retrospect*, 203–12. Amsterdam: Rodopi, 1987.
Kleinberg, Ethan. *Generation Existential: Heidegger's Philosophy in France 1927–1961*. Ithaca: Cornell University Press, 2005.
Knelangen, Franz-Josef. "Günther Anders und die Musik oder 'Der Klavier-spieler mit dem Zeichenstift'." *Text & Kritik*, 115 (1992): 73–85.

Knott, Marie Luise Knott (ed.), *Hannah Arendt and Gershom Scholem, Der Briefwechsel.* Frankfurt am Main: Suhrkamp Verlag, 2010.

Koebler, Jason. "Martine Rothblatt Wants to Grow Human Organs in Pigs at This Farm." Motherboard. *Vice*, June 24, 2015.

Köhler, Joachim. *Zarathustras Geheimnis*. Nördlingen: Greno, 1989.

Kohn, Jerome. "Introduction." To: Hannah Arendt, *Understanding 1930–1954: Formation, Exile, and Totalitarianism*. New York: Schocken Books, 1994.

Kusch, Martin. *Language as Calculus vs. Language as Universal Medium*. Dordrecht: Springer, 1989.

Kusch, Martin. *Psychologism: A Case Study in the Sociology of Philosophical Knowledge*. New York: Routledge, 1995.

Kusch, Martin. "The Sociology of Philosophical Knowledge: A Case Study and a Defense." In Kusch (ed.), *The Sociology of Philosophical Knowledge*, 15–38. Dordrecht: Springer, 2000.

Latini, Michaela, Alessandra Sannella, and Alfredo Morelli (eds.), *La grammatica della violenza Un'indagine a più voci*. Milan: Mimesis Editioni, 2017.

Latour, Bruno. *Aramis or the Love of Technology*, trans. Catherine Porter. Cambridge: Harvard University Press, 1996.

Latour, Bruno. "Biography of an Investigation: On a Book about Modes of Existence," trans. Cathy Porter. *Social Studies of Science*, 43, no. 2 (2013): 287–301.

Latour, Bruno. *Facing Gaia: Eight Lectures on the New Climatic Regime*. London: Polity, 2017.

Latour, Bruno. *The Pasteurization of France*, trans. Alan Sheridan and John Law. Cambridge, MA: Harvard University Press 1988. [1984]

Latour, Bruno. *We Have Never Been Modern*, trans. Catherine Porter. New York: Harvester Books, 1993.

Latour, Bruno with Steve Woolgar. *Laboratory Life: The Construction of Scientific Facts*. Beverly Hills: Sage, 1979.

Leppert, Richard. "Commentary: Culture, Technology, and Listening." In *Theodor W. Adorno: Essays on Music*, 213–50. Berkeley: University of California Press, 2002.

Lessig, Lawrence. *Code and Other Laws of Cyberspace*. New York: Basic Books, 1999.

Lessing, Theodor. *Der Lärm. Eine Kampfschrift gegen die Geräusche unseres Lebens*. Wiesbaden: Bergmann, 1908.

Lester, Paul. "Gil Scott-Heron: The Revolution Lives On." *The Guardian*, August 26, 2015.

Levene, Mark. *The Meaning of Genocide: Genocide in the Age of the Nation State*. London: Tauris, 2005.

Levie, Françoise. *L'homme qui voulait classer le monde. Paul Otlet et le Mundaneum*. Brussels: Les Impressions Nouvelles, 2006.

Levin, Thomas Y. and Michael von der Linn. "Elements of a Radio Theory: Adorno and the Princeton Radio Research Project." *The Musical Quarterly*, 78, no. 2 (Summer 1994): 316–24.

Levine, Yasha. *Surveillance Valley: The Secret Military History of the Internet*. New York: Public Affairs, 2018.

Liessmann, Konrad Paul (ed.). *Günther Anders kontrovers*. Munich: Beck, 1992.

Liessmann, Konrad Paul (ed.). *Günther Anders: philosophieren im Zeitalter der technologischen Revolutionen*. Munich: C.H.Beck, 2002.

Liessmann, Konrad Paul (ed.). "Thought after Auschwitz and Hiroshima: Günther Anders and Hannah Arendt." *Enrahonar*, 46 (2011): 123–35.

Lin, Patrick. "Ethics of Hacking Back: Six arguments from Armed Conflict to Zombies. A Policy Paper on Cybersecurity." *U.S. National Science Foundation*, September 26, 2016.

Lin, Patrick. "Ethical Blowback from Emerging Technologies." *Journal of Military Ethics*, 9, no. 4 (2010): 313–31.

Löwith, Karl. "European Nihilism: Reflections on the Intellectual and Historical Background of the European War." In Löwith, *Martin Heidegger and European Nihilism*, trans. Gary Steiner, 173–284. New York: Columbia University Press, 1995.

Löwith, Karl. *Meaning in History: The Theological Implications of History*. Chicago: University of Chicago Press, 1949.

Lohmann, Margret. *Philosophieren in der Endzeit. Zur Gegenwartsanalyse von Günther Anders*. Munich: Fink, 1996.

Lohmar, Dieter. "On Some Motives for Husserl's Genetic Turn in His Research on a Foundation of the Geisteswissenschaften." *Studia Phaenomenologica*, 18 (2018): 31–48.

Long, Christopher P. "Art's Fateful Hour: Benjamin, Heidegger, Art and Politics." *New German Critique*, no. 83, Special Issue on *Walter Benjamin* (Spring–Summer 2001): 89–115.

Lorenz, Dagmar C. G. "Man and Animal: The Discourse of Exclusion and Discrimination in a Literary Context." In Sara Friedrichsmeyer and Patricia Herminghouse (eds.), *Women in German Yearbook 14*, 201–24. Lincoln: University of Nebraska Press, 1999.

Love, Jeff. *The Black Circle: A Life of Alexander Kojève*. New York: Columbia University Press, 2018.

Lucian, "Philosophies for Sale," trans. A. M. Harmon. In *Lucian: Volume II*. Loeb, 449–511. Cambridge, MA: Harvard University Press, 1915.

Luef, Wolfgang. "Leonard Cohen singt sein letztes Liebeslied." *Süddeutsche Zeitung*, 25. Oktober 2016.

Lütkehaus, Lüdger. "Der 'Atomphilosoph' scheitert an der Musik. Günther Anders will eine Theorie der Musik entwerfen, aber an Adorno kommt er nicht vorbei." *Neue Zürcher Zeitung*, November 29, 2017.

Lütkehaus, Lüdger. *Hannah Arendt–Martin Heidegger: Eine Liebe in Deutschland*. Marburg: Basilisken-Presse, 1999.

Lütkehaus, Lüdger. "In der Mitte sitzt das Dasein. Die Philosophen Günther Anders und Peter Sloterdijk lesen zweierlei Heidegger." *Die Zeit*, Januar 24, 2002.

Lütkehaus, Lüdger. *Philosophieren nach Hiroshima. Über Günther Anders*. Berlin: Fischer, 2018. [1992]

Lütkehaus, Lüdger. *Schwarze Ontologie. Über Günther Anders*. Lüneberg: Zu Klampen, 2002.

Lütkehaus, Lüdger. "Unversöhnte Dissonanz. Der Briefwechsel zwischen Hannah Arendt und Gershom Scholem." *Neue Zürcher Zeitung*, October 4, 2010.

Lütticken, Sven. "Shattered Matter, Transformed Forms: Notes on Nuclear Aesthetics, Part 1." *e-flux*, Journal #94 October 2018. Online. https://www.e-flux.com/journal/94/221035/shattered-matter-transformed-forms-notes-on-nuclear-aesthetics-part-1/

Luft, Sebastian. "Husserl's Concept of the 'Transcendental Person': Another Look at the Husserl–Heidegger Relationship." *International Journal of Philosophical Studies*, 13 (2006): 141–77.

Lynch, Michael P. *The Internet of Us: Knowing More and Understanding Less in the Age of Big Data*. New York: W.W. Norton and Company, 2016.

Lynch, Michael P. "The Philosophy of Privacy: Why Surveillance Reduces Us to Objects." *The Guardian*, Thursday, May 7, 2015.

MacIntyre, Alasdair. *After Virtue*. Notre Dame: University of Notre Dame Press, 1980.
MacLeish, Archibald. *Ars Poetica*. 1952.
Mahler, Horst. "'Ist Dein Mut zu töten wirklich so groß?' Offener Brief von Horst Mahler an den Philosophen Günther Anders." *die tageszeitung*, 3. July 1987: 16.
Maier-Katkin, Daniel. *Stranger from Abroad: Hannah Arendt, Martin Heidegger, Friendship and Forgiveness*. New York: Norton, 2010.
Manne, Kate. *Down Girl: The Logic of Misogyny*. Oxford: Oxford University Press, 2017.
Mannings, Robin. *Ubiquitous Positioning*. Norwood: Artech House, 2008.
Marcel, Gabriel. *Being and Having*, trans, Katharine Farrer. Glasgow: University Press, 1949.
Marcuse, Harold. *Günther Anders. (Guenther Anders, Gunther Anders). Journalist, Philosopher, Essayist, 1902–1992*. http://marcuse.faculty.history.ucsb.edu/anders.htm
Marcuse, Herbert. *One Dimensional Man*. Boston: Beacon Press, 1964.
Marian, Marco. "Günther Anders and the Modification of Reality." *Journal of Historical Archaeology & Anthropological Sciences*, 3/6 (2018): 789-92.
Maritain, Jacques. *Thomas and the Problem of Evil*. Milwaukee: University of Marquette Press, 1942.
Marx, Karl and Friedrich Engels. "Manifesto of the Communist Party." In *Collected Works, Volume 6 Marx and Engels, 1845–1848*, 477-96. New York: International Publishers, 1976.
Matlack, Samuel. "Confronting the Technological Society: On Jacques Ellul's Classical Analysis of Technique." *The New Atlantis*, 43 (Summer/Fall 2014): 45–64.
Matsakis, Louise. "How the Government Hides Secret Surveillance Programs." *Wired*, January 9, 2018.
Matassi, Elio. "Die Musik Philosophie bei Walter Benjamin und Günther Anders." In: Bernd Witte and Mauro Ponzi (eds.), *Theologie und Politik. Walter Benjamin und ein Paradigma der Moderne*, 212–2. 2 Berlin: Erich Schmidt, 2005.
Maus, Fred. "Listening and Possessing." In Christian Thorau and Hansjakob Ziemer (eds.), *The Oxford Handbook of Music Listening in the 19th and 20th Centuries*, 441–60. Oxford: Oxford University Press, 2019.
Mayor, Adrienne. "The Nessux Shirt in the New World: Smallpox Blankets in History and Legend." *The Journal of American Folklore*, 108, no. 427 (1995): 54–77.
McCormick, John. *Carl Schmitt's Critique of Liberalism: Against Politics as Technology*. Cambridge: Cambridge University Press, 1997.
McNulty, Michael Bennett. "What is Chemistry, for Kant?" *Kant Yearbook*, 9 (2017): 85–112.
Mccrae, Niall, Sheryl Gettings, and Edward Purssell, "Social Media and Depressive Symptoms in Childhood and Adolescence: A Systematic Review." *Adolescent Research Review*, 2, no. 4 (2017). Online. https://www.researchgate.net/publication/314172005_Social_Media_and_Depressive_Symptoms_in_Childhood_and_Adolescence_A_Systematic_Review
McMahon, Anne. "Tasmanian Aboriginal Women as Slaves." *Tasmanian Historical Research Association: Papers and Proceedings*, 23, no. 2 (June 1976): 44–9.
Mehlman, Jeffrey. *Émigré New York: French Intellectuals in Wartime Manhattan, 1940–1944*. New York: Plunkett Lake Press, 2019.
Meier, Bettina. *Goethe in Trümmern: Zur Rezeption eines Klassikers in der Nachkriegszeit*. Frankfurt am Main: Springer, 2019.
Merleau-Ponty, Maurice. "Indirect Language and the Voices of Silence." In Signs, trans. Richard McCleary. Chicago: Northwestern University Press, 1964.

Merry, Kay. "The Cross-Cultural Relationships between the Sealers and the Tasmanian Aboriginal Women at Bass Strait and Kangaroo Island in the Early Nineteenth Century," *Counterpoints*, 3, no. 1 (2003): 80–8.

Merton, Robert K. "The Unanticipated Consequences of Purposive Social Action." *American Sociological Review*, 1, no. 6 (1936): 895–904.

Meyer-Drawe, Käte. "Mit 'eiserner Inkonsequenz' fürs Überleben — Günther Anders." In Sven Kluge, Ingrid Lohmann, and Gerd Steffens (eds.), *Jahrbuch Für Pädagogik 2014. Menschenverbesserung Transhumanismus*, 105–19. Frankfurt a. M.: Lang, 2014.

Michael, M. G. and K. Michael (eds.), *Uberveillance and the Social Implications of Microchip Implants: Emerging Technologies*. Hershey: IGI Global, 2013.

Michaelis, Anthony R. "The Sorcerer's Apprentice." *Interdisciplinary Science Reviews*, 9, no. 1 (1984): 1–5.

Milchman, Alan and Alan Rosenberg. "Martin Heidegger and the Political: New Fronts in the Heidegger Wars." *Review of Politics*, 65, no. 3 (Summer 2003): 439–49.

Milner, Greg. *Pinpoint: How GPS Is Changing Technology, Culture, and Our Minds*. New York: Norton, 2017.

Moltmann, Jürgen. *The Coming of God: Christian Eschatology*. Minneapolis: Fortress Press, 2004.

Monod, Jean-Claude. "'L'interdit anthropologique' chez Husserl et Heidegger et sa transgression par Blumenberg." *Revue germanique internationale*, 10 (2009): 221–36.

Moneta, Giuseppina. "Profile." In Babich (ed.), *From Phenomenology to Thought, Errancy, and Desire: Essays in Honor of William J. Richardson, S.J*, 205–7. Dordrecht: Kluwer, 1995.

Montgomery, Marshall. "Hölderlin and 'Diotima.'" *The Modern Language Review*, 7, no. 2 (April 1912): 193–207.

Morozov, Evgeny. *The Net Delusion: The Dark Side of Internet Freedom*. New York: Public Affairs, 2011.

Morozov, Evgeny. *To Save Everything Click Here: Technology, Solutionism, and the Urge to Fix Problems that Don't Exist*. London: Penguin, 2014.

Müller, Christopher John. "Desert Ethics: Technology and the Question of Evil in Günther Anders and Jacques Derrida." *Parallax*, 21, no. 1 (2015): 87–102.

Müller, Christopher John. "Die Unangestellten. Ein Blick in die Zukunft der Arbeit." In Reinhard Ellensohn and Kerstin Putz (eds.), *Günther Anders-Journal, Jg. 1. Sonderausgabe zur Tagung „Schreiben für übermorgen". Forschungen zu Werk und Nachlass von Günther Anders*, Online. http://www.guenther-anders-gesellschaft.org/wp-content/uploads/2017/12/m%C3%BCller-2017.pdf

Müller, Christopher John. *Prometheanism: Technology, Digital Culture and Human Obsolescence*. Lanham: Rowman & Littlefield, 2016.

Müller, Christopher John and David Mellor. "Utopia Inverted: Günther Anders, Technology and the Social." *Thesis Eleven*, 153, no. 1 (2019): 3–8.

Müller, Marcel. *Von der "Weltfremdheit" zur "Antiquiertheit". Philosophische Antrhopologie bei Günther Anders*. Marburg: Tectum 2012.

Müller-Jung, Joachim. "Das Schwein, dein Spender. Vermenschlicht: gentechnisch veränderte Ferkel aus München." *Frankfurter Allgemeine Zeitung*, 19 (2009): 8.

Munster, Rens van and Casper Sylvest. "Appetite for Destruction: Günther Anders and the Metabolism of Nuclear Techno-Politics." *Journal of International Political Theory*, 15, Issue 3 (2019): 332–48.

Murphy, Rebekka. "Note, Routine Body Scanning in Airports: A Fourth Amendment Analysis Focused on Health Effects." *Hastings Const. L.Q.*, 39 (2012): 915.

Nehamas, Alexander. *Only a Promise of Happiness: The Place of Beauty in a World of Art.* Princeton: Princeton University Press, 2007.

Nehring, Holger. "Cold War, Apocalypse and Peaceful Atoms: Interpretations of Nuclear Energy in the British and West German Anti-Nuclear Weapons Movements, 1955-1964)." *Historical Social Research*, 29, no. 3 (2004): 150–70.

Neiman, Susan. *Evil in Modern Thought: An Alternative History of Philosophy.* Princeton: Princeton University Press, 2002.

Nettling, Astrid. "Halsstarriger Streiter in Sachen Vernunft." *Deutschlandfunk*, November 07, 2012. Online. https://www.deutschlandfunk.de/halsstarriger-streiter-in-sachen-vernunft.700.de.html?dram:article_id=214615

Neumann, Bernd. *Hannah Arendt und Heinrich Blücher.* Berlin: Rowohlt, 1998.

Neumann, Bernd. "Noch Einmal: Hannah Arendt, Günther Stern/Anders mit bezug auf den jüngst komplettierten Briefwechsel zwischen Arendt und Stern und unter Rekurs auf Hannah Arendts unveröffentlichte Fabelerzählung Die weisen Tiere." In Bernd Neumann, Helgard Mahrdt, and Martin Frank (eds.), "The angel of history is looking back": *Hannah Arendts Werk*, 107–26. Würzbach: Königshausen u. Neumann, 2001.

Nietzsche, Friedrich. *Also Sprach Zarathustra. Ein Buch für Alle und Keinen.* Chemnitz: Verlag von Ernst Schmeitzner, 1883. Online. http://www.zeno.org/Philosophie/M/Nietzsche,+Friedrich/Also+sprach+Zarathustra/Zarathustras+Vorrede

Nietzsche, Friedrich. *Beyond Good and Evil/Genealogy of Morality*, trans. Adrian del Caro. Stanford: Stanford University Press, 2014.

Nietzsche, Friedrich. *Die fröhliche Wissenschaft.* Leibniz: Verlag von Ernst Schmeitzner, 1882.

Nietzsche, Friedrich. *Ecce Homo. Wie man wird was man ist.* Leipzig: Insel, 1908.

Nietzsche, Friedrich. *Götzen-Dämmerung oder Wie man mit dem Hammer philosophirt.* Leipzig: Verlag von C. G. Naumann, 1889.

Nietzsche, Friedrich. "On the Theory of Quantitifying Rhythm," trans James W. Halporn. *New Nietzsche Studies*, 10, nos. 1 and 2 (2016): 69–78.

Nietzsche, Friedrich. *Selected Letters of Friedrich Nietzsche*, ed. Christopher Middleton. Notre Dame: University of Notre Dame, 1996 [1969].

Nietzsche, Friedrich. "Über Wahrheit und Lüge im Aussermoralischen Sinne." [1873] Online. http://www.zeno.org/Philosophie/M/Nietzsche,+Friedrich/%C3%9Cber+Wahrheit+und+L%C3%BCge+im+au%C3%9Fermoralischen+Sinn

Norberg-Schulz, Christian. *Genius Loci: Towards a Phenomenology of Architecture.* New York: Rizzoli, 1979.

Nunes, Mark. "Jean Baudrillard in Cyberspace: Internet, Virtuality, and Postmodernity." *Style*, 29, no. 2, *From Possible Worlds to Virtual Realities: Approaches to Postmodernism* (Summer 1995): 314–27.

Nunez, Tyke. "Logical Mistakes, Logical Aliens, and the Laws of Kant's Pure General Logic." *Mind*, 128, Issue 512 (October 2019): 1149–80.

Oberschlick, Gerhard. "Editorische Notiz." In: Anders, *Die Kirschenschlacht. Dialoge mit Hannah Arendt und ein akademisches Nachwort*, 61–72. Munich: Beck, 2012.

Ogden, Daniel. *In Search of the Sorcerer's Apprentice: The Traditional Tales of Lucian's Lover of Lies.* Swansea: Classical Press of Wales, 2007.

Ogden, Daniel. "The Apprentice's Sorcerer: Pancrates and his Powers in Context (Lucian, 'Philopseudes'), 33–36." *Acta Classica*, 47 (2004): 101–26.

O'Connell, Mark. *To Be a Machine: Adventures Among Cyborgs, Utopians, Hackers, and the Futurists Solving the Modest Problem of Death.* London: Granta, 2018.

O'Neil, Timothy P. "Two Concepts of Liberty Valance: John Ford, Isaiah Berlin, and Tragic Choice on the Frontier." *Creighton Law Review*, 37 (2004): 471–92.
Oppenheimer, J. Robert. *The Decision to Drop the Bomb*. NBC Documentary, 1965.
Orwell, George. *Nineteen Eighty-four*. London: Secker & Warburg, 1949.
Østergaard, Edvin. "Echoes and Shadows: A Phenomenological Reconsideration of Plato's Cave Allegory." *Phenomenology & Practice*, 13, no 1 (2019): 20–33.
Østern, Anna-Lena. "Norwegian Perspectives on Aesthetic Education and the Contemporary Conception of Cultural Literacy as Bildung ('Danning')." *Zeitschrift für Erziehungswissenschaft*, 16 (2013): 43–63.
Oxley, Alan. *Uncertainties in GPS Positioning: A Mathematical Discourse*. London: Elsevier, 2017.
Paddison, Max. *Adorno's Aesthetics of Music*. Cambridge: Cambridge University Press, 1997.
Paddison, Max. "Review Article: Adorno's Aesthetic Theory." *Music Analysis*, 6, no. 3 (1987): 355.
Palaver, Wolfgang. "The Respite: Günther Anders' Apocalyptic Vision in Light of the Christian Virtue of Hope." In Günter Bischof, Jason Dawsey and Bernhard Fetz (eds.) *The Life and Work of Günther Anders: Émigré, Iconoclast, Philosopher, Man of Letters*, 83–92. London: Taylor & Francis, 2014.
Pan, David (ed.), *Carl Schmitt and the Critique of Technical Rationality*. Telos, No. 187 (2019).
Patrikarakos, David. *War in 140 Characters: How Social Media Is Reshaping Conflict in the Twenty-First Century*. New York: Basic Books, 2017.
Patterson, Kristine B. and Thomas Runge. "Smallpox and the Native American." *American Journal of Medical Science*, 342, no. 4 (April 2002): 216–22.
Patterson, Charles. *Eternal Treblinka: Our Treatment of Animals and the Holocaust*. New York: Lantern Books, 2002.
Perlman, Lawrence. *The Eclipse of Humanity: Heschel's Critique of Heidegger*. Berlin: Walter de Gruyter, 2016.
Perreau, Laurent. *Günther Anders à l'école de la phénoménologie*. Paris: Kimé, 2007.
Pietsch, Wolfgang, Jörg Wernecke, and Max Ott (eds.), *Berechenbarkeit der Welt?: Philosophie und Wissenschaft im Zeitalter von Big Data*. Frankfurt: Springer, 2017.
Pilkey, Orrin with Linda Pilkey-Jarvis. *Useless Arithmetic: Why Environmental Scientists Can't Predict the Future*. New York: Columbia University Press, 2007.
Pinker, Steven. *Enlightenment Now: The Case for Reason, Science, Humanism, and Progress*. London: Penguin, 2019.
Postman, Neil. *Amusing Ourselves to Death in the Age of Show Business*. New York: Penguin, 1985.
Postman, Neil. *Technopoly: The Surrender of Culture to Technology*. New York: Vintage Books, 1993.
Putz, Kerstin. "Nachwort: Korrespondenzen Hannah Arendt and Günther Anders." In Anders/Arendt, *Schreib doch mal* hard facts *über Dich*, ed. Kertin Putz, 229. Munich: Piper, 2016.
Putz, Kerstin. Trägt ihn mit stolz, den gelben Fleck!" In F. A. Krummacher (ed.), *Die Kontroverse, Hannah Arendt: Eichmann – und die Juden*. 152–60. Munich: Nymphenburger Verlagshandlung, 1964.
Rasini, Vallori. "Il potere della violenza. Su alcune riflessioni di Günther Anders." *Etica & Politica/Ethics & Politics: Rivista di filosofia*, 15, no. 2 (2013): 258–70.
Rath, Matthias. Der *Psychologismusstreit in der deutschen Philosophie*. Freiburg i. B.: Karl Alber Verlag, 1994.
Reemtsma, Jan Philipp. *Vertrauen und Gewalt. Versuch über eine besondere Konstellation der Moderne*. Hamburg: HIS Verlages, 2007.

Rehberg, Karl-Siegbert. "Philosophical Anthropology from the End of World War I to the 1940s and in a Current Perspective." *Iris*, 1, no. 1 (2009): 131–52.

Reid-Cunningham, Allison Ruby. "Rape as a Weapon of Genocide." *Genocide Studies and Prevention*, 3, no. 3 (December 2008): 279–96.

Reinalter, Helmut and Andreas Oberprantacher (eds.), *Außenseiter der Philosophie*. Würzburg: Köngishausen und Neumann, 2012.

Reinthal, Angela. "Mich hält ein reines Intervall". *Carl Schmitt und die Musik*. Berlin: Duncker & Humblott, 2019.

Rilke, Rainer Maria. *Das Stundenbuch*. Leipzig: Insel Verlag, 1905.

Rilke, Rainer Maria. *The Book of Hours*, trans. Babette Deutsch. New York: New Directions Books, 1941.

Riskin, Jessica. "Pinker's Pollyannish Philosophy and Its Perfidious Politics." *Los Angeles Review of Books*, December 15, 2019.

Roberts, Michael James. *Tell Tchaikovsky the News: Rock 'n' Roll, the Labor Question, and the Musicians' Union, 1942–1968*. Durham: Duke University Press, 2014.

Robison, John Elder. "Is the Internet Making People a Little Bit Autistic?" *Psychology Today*, November 30, 2008. https://www.psychologytoday.com/us/blog/my-life-aspergers/200811/is-the-internet-making-people-little-bit-autistic

Roholt, Tiger. "On the Divide: Analytic and Continental Philosophy of Music." The American Society for Aesthetics, 71, Issue 1 (Winter 2017): 49–58.

Rohrlich, Elisabeth. "'To Make the End Time Endless:' The Early Years of Gunther Anders' Fight against Nuclear Weapons." In Günter Bischof et al. (eds.), *The Life and Work of Günther Anders: émigré, iconoclast, philosopher, man of letters*, 45–57. Innsbruck: Studienverlag, 2014.

Rolls, Mitchell. "The Changing Politics of Miscegenation." *Aboriginal History*, 29 (2005): 64–76.

Rose, Ellen. "Errors of Thamus: An Analysis of Technology Critique." Bulletin of Science, Technology, and Society, 23 (2003): 147–56.

Rose, Gillian. *Love's Work: A Reckoning with Life*. New York: Schocken Books, 1998.

Ross, Alex. "Searching for Silence: John Cage's Art of Noise." *The New Yorker*, October 4, 2010.

Rothblatt, Martine. "Biology Is Technology." Youtube Lecture. Online. https://www.youtube.com/watch?v=wSZgrEtakz8.

Rougement, Denis de. *The Devil's Share: An Essay on the Diabolic in Modern Society*, trans. Haakon Chevalier. New York: Pantheon, 1945. [1942]

Rougement, Denis de. "On the Devil and Politics." *Christianity and Crisis*, Vol 1 (June 2, 1941): 2–5.

Ruin, Hans. *Being with the Dead: Burial, Ancestral Politics, and the Roots of Historical Consciousness*. Stanford: Stanford University Press, 2019.

Rougement, Denis de. "The Inversion of Mysticism: *Gelassenheit* and the Secret of the Open in Heidegger." *Religions* 10, no. 1 (2019). Online. https://www.mdpi.com/2077-1444/10/1/15/htm

Rougement, Denis de. "Saying *Amen* to the Light of Dawn: Nietzsche on Praise, Prayer, and Affirmation." *Nietzsche-Studien*, 48 (2019): 99–116.

Russell-Brown, Sherrie L. "Rape as an Act of Genocide." *Berkeley Journal of Law*, 20 (2003): 350–74.

Ryan, Lyndall. *The Aboriginal Tasmanians*, 2nd ed. St. Leonards: Allen & Unwin, 1996.

Saap, Jan. *Where the Truth Lies: Franz Moewus and the Origins of Molecular Biology*. Cambridge: Cambridge University Press, 1999.

Sarat, Austin D. et al. (eds.). *The Time of Catastrophe: Multidisciplinary Approaches to the Age of Catastrophe*. Surrey: Ashgate 2015.

Saito, Yuichi. "The Way of the Reduction via Anthropology: Husserl and Lévy-Bruhl, Merleau-Ponty and Lévi-Strauss." *Bulletin d'analyse phénoménologique*, X, no. 1 (2014): 1–18.

Sare, Laura. "A Comparison of HathiTrust and Google Books Using Federal Publications." *Practical Academic Librarianship: The International Journal of the SLA Academic Division*, 2, no. 1 (2012): 1–25.

Schandl, Franz. "Work Will Not Set You Free: Notes on Günther Anders." https://libcom.org/library/work-will-not-set-you-free-notes-g%C3%BCnther-anders-%E2%80%93-franz-schandl

Schmid, Holger. (ed.). *Kant Sur les extraterrestres: Théorie du ciel*. Paris: Editions Manucius, 2019.

Schmid, Holger. (ed.). *Kunst des Hörens. Orte und Grenzen philosophischer Spracherfahrung*. Cologne: Böhlau, 1999.

Schmidt, Christoph. "The Leviathan Crucified: A Critical Introduction to Jacob Taubes' 'The Leviathan as Mortal God.'" *Political Theology*, 19 (2018): 172–92.

Schmitt, Carl. *Über Schuld und Schuldarten. Eine terminologische Untersuchung*. Berlin: Duncker & Humblott, 1977 [1910].

Schneier, Bruce. "Why are we spending $7 billion on TSA?" CNN. https://www.cnn.com/2015/06/05/opinions/schneier-tsa-security/index.html. Last modified June 05, 2015.

Scholtes, Jennifer. "Price for TSA's Failed Body Scanners: $160 Million." *Politico*, August 17, 2015.

Schonfeld, Martin. "Kant on Animals." In J. Vonk and T. K. Shackelford (eds.), *Encyclopedia of Animal Cognition and Behavior*. Frankfurt am Main: Springer, 2018.

Schrader Jr., George A. "Heidegger's Ontology of Human Existence." *The Review of Metaphysics*, 10, no. 1 (1956): 35–56.

Schraube, Ernst. "'Torturing Things Until They Confess': Günther Anders' Critique of Technology." *Science as Culture*, 14, no. 1 (March 2005): 77–85.

Schürmann, Reiner. "Concerning Philosophy in the United States." *Social Research*, 61, no. 1 (Spring 1994): 89–113.

Schürmann, Reiner. "De la philosophie aux Etats-Unis." *Le Temps de la Réflexion*, 6 (1985): 303–21.

Schürmann, Reiner. *Heidegger on Being and Acting: From Principles to Anarchy*. Bloomington: Indiana University Press, 1987.

Schürmann, Reiner. *On Heidegger's Being and Time*. London. Routledge, 2008.

Schütz, Alfred. "Fragments Toward a Phenomenology of Music." In Helmut Wagner and George Psathas (eds.), *Collected Papers IV*, 243–75. Dordrecht: Kluwer Academic Publishers, 1996.

Schütz, Alfred. "Making Music Together." In: *Collected Papers II: Studies in Social Theory*. Edited by Arvid Brodersen. The Hague: Martinus Nijhoff, 1964. 159–178.

Schütz, Alfred. "Phenomenology and the Social Sciences." In Maurice Natanson (ed.), *Collected Papers I: The Problem of Social Reality*, 118–39. Dordrecht: Kluwer, 1972.

Schutz, Alfred. "Phenomenology and the Social Sciences." In Maurice Natanson (ed.),*Collected Papers I: The Problem of Social Reality*, 118–39. . Dordrecht: Kluwer, 1972.

Schwartz, Regina M. *The Curse of Cain: The Violent Legacy of Monotheism*. Chicago: University of Chicago Press, 1997.

Scott-Heron, Gil. *The Revolution Will Not Be Televised*. 1971. https://www.youtube.com/watch?time_continue¼3&v¼vwSRqaZGsPw

Scruton, Roger. "Is Adorno a Dead Duck?" Lecture: *Royal Musical Association Music and Philosophy Study Group* 2nd Annual Conference, King's College London. 20 J.

Sebald, W. G. *On the Natural History of Destruction*. New York: Modern Library, 2004.
Selinger, Evan (ed.), *Postphenomenology: A Critical Companion to Ihde*. Albany: State University of New York Press, 2006.
Sell, Annette. "Leben führen – Dasein entwurfen: Zur systematischen und gesellschaftspolitischen Bedeutung von Plessners anthropologischem und Heideggers fundamental-ontologischem Konzept des Menschen." In Kevin Liggieri and Julia Gruevska (eds.), *Vom Wissen um den Menschen: Philosophie, Geschichte, Materialität*, 46–61. Freiburg im Briesgau: Alber, 2018.
Seymour, Richard. *The Twittering Machine*. London: Indigo, 2019.
Sharma, Ved Parkash and Neelima R. Kumar. "Changes in honeybee behaviour and biology under the influence of cellphone radiations." *Current Science*, 98, no. 10 (May 25, 2010): 1376–8.
Shelley, Mary. *Frankenstein or a Modern Prometheus*. London: Lackington, Hughes, Harding, Mavor & Jones, 1818.
Shiva, Vandana. *Biopiracy: The Plunder of Nature and Knowledge*. New Delhi: Natraj Publishers, 1997.
Simonelli, Thierry. Le monde, "Vu à la télé". *Dogma. Revue électronique de Sciences Humaines*. 2004. http://1libertaire.free.fr/GAnders25.html [Archive version.]
Singer, Isaac Bashevis. "The Slaughterer," trans. Mirra Ginsburg. *The New Yorker*, November 25, 1967: 60–5.
Slade, Giles. *Big Disconnect: The Story of Technology and Loneliness*. Amherst: Prometheus, 2012.
Slade, Giles. *Made to Break: Technology and Obsolescence in America*. Cambridge, MA: Harvard University Press, 2006.
Sloterdijk, Peter. *Critique of Cynical Reason*. Minneapolis: University of Minnesota Press, 1987.
Sloterdijk, Peter. *Luftbeben. An den Wurzeln des Terrors*. Frankfurt am Main: Suhrkamp, 2002.
Sloterdijk, Peter. *Terror from the Air*. Amy Patton and Steve Corcoran, trans. Los Angeles: Semiotext(e), 2009. [2002]
Smith, Dominic. *Exceptional Technologies: A Continental Philosophy of Technology*. London: Bloomsbury, 2018.
Smith, F. Joseph. *Jacobi Leodiensis Speculum musicae / A Commentary in Three Volumes*. Brooklyn: Institute of Mediaeval Music, [1966–1983].
Smith, F. Joseph. *The Experiencing of Musical Sound: Prelude to a Phenomenology of Music*. New York/Montreux: Gordon and Breach, 1979.
Smith, F. Joseph (ed.). *Understanding the Musical Experience*. New York/Montreux: Gordon and Breach, 1989.
Smythe, Dallas. "Communications: Blindspot of Western Marxism." *Canadian Journal of Political and Social Theory/Revue canadienne de theorie politique et sociale*, 1, no. 3 (Fall/Automne 1977): 1–27.
Smythe, Dallas. *Dependency Road: Communications, Capitalism, Consciousness, and Canada*. Norwood: Ablex, 1981.
Snelly, John L. *The Nazi Revolution*. Boston: B.C. Heath, 1959.
Soffer, Gail. "Heidegger, Humanism, and the Destruction of History." *The Review of Metaphysics*, 49, no. 3 (1996): 547–76.
Sommer, Christian. "'Ni homme, ni capucin, c'est un Dasein' Remarques sur Über Heidegger." *Tumultes* 1, nos. 28–9 (2007): 51–68.

Sonolet, Daglind. *Günther Anders: phénoménologue de la technique*. Bourdeaux: Presses Universitaires de Bordeaux, 2006.
Sonolet, Daglind. "Literature and Modernity: Günther Anders, Hannah Arendt, and Theodor W. Adorno – Interpreters of Kafka." In Jeffrey A. Halley and Sonolet (eds.), *Bourdieu in Question*, 426–41. Amsterdam: Brill, 2018.
Sorgner, Stefan Lorenz. "From Nietzsche's Overhuman to the Posthuman of Transhumanism." *English Language and Literature*, 62, no. 2 (2016): 163–76.
Spigel, Lynn. *Make Room for TV: Television and the Family Ideal in Postwar America*. Chicago: The University of Chicago Press, 1992.
Stern, William and Clara Stern. *Die Kindersprache. Eine psychologische und sprachtheoretische Untersuchung*. Leipzig: Barth, 1907.
Stern, William and Clara Stern. *Erinnerung, Aussage und Lüge in der ersten Kindheit*. Leipzig: Barth, 1909.
Stern, Clara. *Aus einer Kinderstube: Tagebuchblätter einer Mutter*. Leipzig: B.G. Teubner, 1914.
Stewart, Garrett. *Closed Circuits: Screening Narrative Surveillance*. Chicago: University of Chicago Press, 2015.
Stingl, Alexander I. *The Digital Coloniality of Power: The Epistemic Disobedience in the Social Sciences and the Legitimacy of the Digital Age*. Lanham, MD: Lexington Book, 2015.
Stoelker, Thomas. "Babette Babich on Love, Social Media, and Megxit." Video Interview, January 16, 2020. Online. https://www.youtube.com/watch?v=9Emj6JAEVKs
Størmer, Carl. "Shortwave Echoes and the Aurora Borealis." *Nature*, 122, no. 3079 (November 3, 1928): 681.
Størmer, Carl Christian Lein. *The Ghost Radio Hunter*. Documentary, 2013.
Straus, Scott. "Contested Meanings and Conflicting Imperatives: A Conceptual Analysis of Genocide." *Journal of Genocide Research*, 3, no. 3 (2011): 349–75.
Street, John. *Politics and Technology*. New York: Guilford Press, 1992.
Strong, Tracy B. "At Home Alone: The Problems of Citizenship in Our Age" of Strong, *Learning One's Native Tongue: Citizenship, Contestation, and Conflict in America*. Chicago: University of Chicago Press, 2019.
Strong, Tracy B. "Follow Your Leader: Benito Cereno and the Case of Two Ships." In Jason Frank (ed.), *A Political Companion to Herman Melville*, 281–309. Lexington: University Press of Kentucky, 2013.
Strong, Tracy B. *Politics Without Vision: Thinking Without a Bannister in the Twentieth Century*. Chicago: University of Chicago Press, 2012.
Styfhals, Willem. "Evil in History: Karl Löwith and Jacob Taubes on Modern Eschatology." *Journal of the History of Ideas*, 76, no. 2 (April 2015): 191–213.
Szendy, Peter. *Kant in the Land of Extraterrestrials: Cosmopolitical Philosofictions*, trans. Will Bishop. New York: Fordham University Press, 2013.
Taminiaux, Jacques. "Arendt, disciple de Heidegger?" *Études Phénoménologiques*, 1, no. 2 (1985): 111–36.
Taminiaux, Jacques. 'On Heidegger's Interpretation of the Will to Power as Art'. *New Nietzsche Studies*, vol. 2, nos. 1 and 2 (Winter 1999): 1–22.
Tanzer, Mark Basil. *Heidegger: Decisionism and Quietism*. Amherst: Humanity Books, 2002.
Taubes, Jacob (ed.), *Der Fürst dieser Welt. Carl Schmitt und die Folgen*. Munich: Fink, 1983.
Taubes, Jacob. *Die politische Theologie des Paulus: Vorträge gehalten an der Forschungsstätte der evangelischen Studiengemeinschaft in Heidelberg, 23–27 Februar 1987*. Munich: Wilhelm Fink 1995. [1993]
Taubes, Jacob. *La théologie politique de Paul*. Paris: Seuil, 1999.

Taubes, Jacob. *Occidental Eschatology*. David Ratmoko, trans. Stanford: Stanford University Press, 2009. [1947]
Taubes, Jacob. *To Carl Schmitt*. New York: Columbia University, 2013. [1987]
Taupitz, Jochen and Marion Weschka (eds.), *Chimbrids – Chimeras and Hybrids in Comparative European and International Research: Scientific, Ethical, Philosophical and Legal Aspects*. Frankfurt: Springer Science & Business Media, 2009.
Terpstra, Marin. "The Management of Distinctions: Jacob Taubes on Paul's Political Theology." In Gert Jan van der Heiden, George Henry van Kooten, and Antonio Cimino (eds.), *Saint Paul and Philosophy*, 251–68. Berlin: De Gruyter, 2017.
Tewinkel, Christiane. "'Everybody in the concert hall should be devoted entirely to the music': On the Actuality of Not Listening to Music in Symphonic Contexts." In Christian Thorau and Hansjakob Ziemer (eds.), *The Oxford Handbook of Music Listening in the 19th and 20th Centuries*, 477–99. Oxford: Oxford University Press, 2019.
Theunissen, Michael. *The Other: Studies in the Social Ontology of Husserl, Heidegger, Sartre, and Buber*, trans. Christopher Macann. Cambridge: MIT Press, 1984. German: *Der Andere: Studien zur Sozialontologie der Gegenwart*. Berlin: de Gruyter, 1977.
Thadeusz, Frank. "Nazi-Labor in Oberfranken. Geheimwaffen aus dem Burgverlies." *Spiegel*, April 21, 2011.
The New Shorter Oxford English Dictionary. Vol. 2, N–Z. ed. Lesley Brown. Oxford: Clarendon Press, 1993. [1973]
Thieme, Nick. "The Gruesome Truth About Lab-Grown Meat: It's Made by Using Fetal Cow Blood." *Slate*, July 11, 2017.
Thielens, Arno, Mark K. Greco, Leen Verloock, Luc Martens, amd Joseph Wout, "Radio-Frequency Electromagnetic Field Exposure of Western Honey Bees." *Scientific Reports* 10, 461 (2020). Online.
Thwaites, Trevor. "Heidegger and Jazz: Musical Propositions of Truth and the Essence of Creativity." *Philosophy of Music Education Review*, 21, no. 2 (Fall 2013): 120–35.
Tiedemann, Rolf. "'Do you know what it will look like?' On the Relevancy of Adorno's Theory of Society," trans. Sean Nye. *Cultural Critique*, 70 (Fall 2008): 123–36.
Tirosh, Yofi and Michael Birnhack. "Naked in Front of the Machine: Does Airport Scanning Violate Privacy?" *Ohio State Law Journal*, 74, no. 6 (2013): 1263–306.
Trafton, Anne. "Storing Medical Information Below the Skin's Surface: Specialized Dye, Delivered Along With a Vaccine, Could Enable 'on-patient' Storage of Vaccination History." *MIT News Office*, December 18, 2019. http://news.mit.edu/2019/storing-vaccine-history-skin-1218. Accessed May 19, 2020.
Trawny, Peter. *Irrnisfuge. Heideggers An-Archie*. Berlin: Matthes & Seitz Verlag, 2014.
Trepca, Amalia. "The Utopia of Eidetic Intuition: A Phenomenological Motif in Adorno." *Meta: Research in Hermeneutics, Phenomenology, and Practical Philosophy*, XII, 1 (June 2020): 102–126.
Tschasslaw D. Kopriwitza. "Heidegger und der Anthropozentrismus." In Holger Zaborowsky and Alfred Denker (eds.), *Heidegger Jahrbuch 10*. 178–90. Freiburg im Briesgau: Alber, 2017.
Tuncel, Yunus (ed.). *Nietzsche and Transhumanism: Precursor or Enemy?* Cambridge: Cambridge Scholars, 2017.
Turkle, Sherry. *Alone Together: Why We Expect More from Technology and Less from Each Other*. New York: Basic Books, 2011.
Tyler, Don. *Hit Songs, 1900–1955: American Popular Music of the Pre-Rock Era*. Jefferson: Macfarland, 2007.

Van Brakel, Jaap. "Kant's Legacy for the Philosophy of Chemistry." In Davis Baird, Eric Scerri, and L. McIntyre (eds.), *Philosophy of Chemistry: Synthesis of a New Discipline*. 69–91. Dordrecht: Springer, 2006.
Vandenberghe, Frédéric. "Empathy as the Foundation of the Social Sciences and of Social Life: A Reading of Husserl's Phenomenology of Transcendental Intersubjectivity." *Sociedade e Estado, Brasília*, 17, no. 2 (2002): 563–85.
van Dijk, Paul. *Anthropology in the Age of Technology*. Amsterdam: Rodopi, 2000. [1998]
van Schalkwyk, G. I., C. E. Marin, M. Ortiz et al. "Social Media Use, Friendship Quality, and the Moderating Role of Anxiety in Adolescents with Autism Spectrum Disorder." *Journal of Autism and Developmental Disorders* 47 (2017): 2805–13.
Verbeek, Peter-Paul. *What Things do: Philosophical Reflections on Technology, Agency, and Design*. College Station: Pennsylvania State University Press, 2005.
Vetlesen, Arne Johan. *Cosmologies of the Anthropocene: Panpsychism, Animism, and the Limits of Posthumanism*. London: Routledge, 2019.
Vetlesen, Arne Johan. *The Denial of Nature: Environmental Philosophy in the Era of Global Capitalism*. London: Routledge, 2015.
Vézin, François. "L'étendue du désastre ". 8 août 2014. Online.
Virilio, Paul. *Art & Fear*, trans. Julie Rose. London: Continuum, 2003. [2000]
Virilio, Paul. *Desert Screen: War at the Speed of Light*, trans. Michael Degener. London: Continuum, 2002.
Virilio, Paul. *La procédure silence*. Paris: Galilée, 2000.
Visvanathan, Shiv. "On the Annals of the Laboratory State." In Ashis Nandy (ed.), *Science, Hegemony, and Violence: A Requiem for Modernity*, 257–88. Oxford: Oxford University Press, 1988.
von Bonsdorff, Pauline. "Aesthetics and Bildung." *Diogenes*, 59, nos. 1–2 (2013): 127–37.
von Bülow, Hans. Letter to Nietzsche, July 20, 1872. In Nietzsche, *Kritische Briefausgabe*, Vol. 4, 26–7. Berlin: de Gruyter, 1986.
von Hermann, Friedrich-Wilhelm. ",,'»Confessiones'« des Heiligen Augustinus im Denken Heideggers." *Questio*, 1 (2001): 113–46.
Vonnegut, Kurt. "Harrison Bergeron." *Magazine of Science and Science Fiction* (October 1961): 5–10.
Vovolis, Thanos. *Prosopon: The Acoustical Mask in Greek Tragedy and in Contemporary Theatre: Form, Function and Appearance of the Tragic Mask and Its Relation to the Actor, Text, Audience and Theatre Space*. Stockholm: PUBLISGHER, 2009.
Vowinckel, Annette. *Geschichtsbegriff und Historisches Denken bei Hannah Arendt*. Cologne: Böhlau, 2001.
Waelbers, Katinka. "Technological Delegation: Responsibility for the Unintended." *Science and Engineering Ethics*, 15, Issue 1(March 2009): 51–68.
Wagner, Kate. "How Restaurants Got So Loud. Fashionable Minimalism Replaced Plush Opulence. That's a Recipe for Commotion." *The Atlantic* (November 27, 2018) Online. https://www.theatlantic.com/technology/archive/2018/11/how-restaurants-got-so-loud /576715/.
Waite, Geoff. "Radio Nietzsche, or, How to Fall Short of Philosophy." In Bruce Krajewski (ed.), *Gadamer's Repercussions: Reconsidering Philosophical Hermeneutics*, 169–211. Berkeley: University of California Press, 2004.
Wallenstein, Sven-Olov. "The Historicity of the Work of Art in Heidegger." In Marcia Sá Cavalcante Schuback, and ans Ruin (eds.), *The Past's Presence: Essays on the Historicity of Philosophical Thinking*. Huddinge: Södertörns högskola, 2005.

Wallrup, Erik. *Being Musically Attuned: The Act of Listening to Music*. London: Routledge, 2018.
Wallrup, Erik. "Music, Truth and Belonging: Listening with Heidegger." In Frederik Pio and Øivind Varkøy (eds.), *Philosophy of Music Education Challenged: Heideggerian Inspirations*, 131–46. Frankfurt am Main, 2014.
Walter-Busch, Emil. *Geschichte der Frankfurter Schule. Kritische Theorie und Politik*. Munich: Fink, 2010.
Wang, Sheng-Chih. *Transatlantic Space Politics: Competition and Cooperation Above the Clouds*. London: Routledge, 2013.
Ward, Deborah M., Karen E. Dill-Shackleford, and Micah O. Mazurek, "Social Media Use and Happiness in Adults with Autism Spectrum Disorder." *Cyberpsychology, Behavior, and Social Networking* 21(3) (2018 March): 205–9.
Weinberg, Alvin. "A Wake-Up Call for Technological Somnambulists." The Scientist, January 12, 1987.
Weiner, James F. "Anthropology contra Heidegger Part I: Anthropology's Nihilism." *Critique of Anthropology*, 12, no. 1 (1992): 75–90
Weiner, James F. "Anthropology contra Heidegger Part II: The Limit of Relationship." *Critique of Anthropology*, 13, no. 3 (1993): 285–301.
WHO. *Electromagnetic Fields and Public Health: Mobile Phones, Fact sheet N°193*. May 2010.
WHO. *Establishing a Dialogue on Risks from Electromagnetic Fields: A Handbook*. Geneva: World Health Organization, 2002.
Wiggershaus, Rolf. "The Frankfurt School's 'Nietzschean Moment,'" trans. Gerd Appelhans. *Constellations*, 8, no. 1 (2001): 144–7.
Wilamowitz-Möllendorf, Ullrich von. "Future Philology.", trans. Gertrude Postl, Babette Babich, and Holger Schmid. *New Nietzsche Studies*, 4, nos. 1 and 2 (2000): 1–32.
Williams, Charles Francis Abdy. *The Aristoxenian Theory of Musical Rhythm*. Cambridge: Cambridge University Press, 1911.
Windham, Elizabeth. "The TSA Opting-Out of Opt-Outs: The New TSA Full-Body Scanner Guidelines and Travelers' Right to Privacy." *North Carolina Journal of Law & Technology*, 17 (2016): 329–67.
Windshuttle, Keith. *The Fabrication of Aboriginal History: Volume One: Van Diemen's Land 1803–1847*. Paddington: Macleay Press, 2002.
Winner, Langdon. *Autonomous Technology. Technics-out-of-Control as a Theme in Political Thought*. Cambridge, MA: MIT Press, 1977.
Winner, Langdon. *The Whale and the Reactor. Search for Limits in an Age of High Technology* Chicago: University of Chicago Press, 1988.
Wittulski, Eckhard. "Der tanzende Phänomenologe." In: Konrad Paul Liessmann (ed.) *Günther Anders kontrovers*. Munich: Beck, 1992. 17–33.
Woessner, Martin. "Hermeneutic Communism: Left Heideggerian's Last Hope." In Silvia Mazzini and Owen Glyn-Williams (eds.), *Making Communism Hermeneutical*, 35–49. Frankfurt am Main: Springer, 2017.
Wolff, Ernst. "From Phenomenology to Critical Theory: The Genesis of Adorno's Critical Theory from His Reading of Husserl." *Philosophy & Social Criticism*, 32, Issue 5 (2008): 555–72.
Worster, Donald. *Nature's Economy: A History of Ecological Ideas*. Cambridge: Cambridge University Press, 1977.
Wu, Tim. *The Attention Merchants*. London: Penguin, 2016.

Yeats, William Butler. *The Collected Works of W. B. Yeats: Volume I: The Poems*, ed. Robert Finneran. New York: Simon and Schuster, 1997.
Zaborowski, Holger. "Bedingungen und Möglichkeiten des Humanismus – heute. Jaspers, Heidegger und Levinas zur Frage nach dem Menschen." In Holger Zaborowsky and Alfred Denker (eds.), *Heidegger Jahrbuch 10*, 251–64. Freiburg im Briesgau: Alber, 2017.
Zaborowski, Holger and Alfred Denker (eds.). *Heidegger Jahrbuch 10*. Freiburg im Briesgau: Alber, 2017.
Zadka, Saul. *Blood in Zion: How the Jewish Guerrillas Drove the British out of Palestine*. London: Brassey's, 1995.
Zanotti-Fregonara, P. and E. Hindie. "Radiation Risk from Airport X-ray Backscatter Scanners: Should We Fear the Microsievert?" *Radiology*, 261, no. 1 (2011): 330–1.
Zuboff, Shoshana. *The Age of Surveillance Capitalism: The Fight for a Human Future at the New Frontier of Power*. London: Profile, 2019.
Zuboff, Shoshana. "Big Other Surveillance Capitalism and the Prospects of an Information Civilization." *Journal of Information Technology*, 30, no. 1 (2015): 75–89.
Zwonitzer, Mark and Charles Hirshberg. *Will You Miss me When I'm Gone? The Carter Family & their Legacy in American Music*. New York: Simon & Schuster, 2004.

Name Index

Abelard 58
Adorno, Theodor Wiesengrund 1, 4,
  6, 8, 11, 15, 21f, 28, 36f, 37, 39f, 42,
  46f, 51, 56, 65, 68, 69f, 72, 83, 92, 102,
  110f, 114, 137f, 139f, 143, 153f, 158,
  161, 163, 165, 176, 180, 182f, 187, 190,
  194, 209
Aeschylus 10, 21, 24, 71, 207, 214
Agamben, Giorgio 2, 57f, 100, 123f, 192,
  211, 214, 217, 220, 228n, 241n, 249n,
  257n, 271n
Albrechtsberger, Johann Georg 192
Alexander 59
Alford, C. Fred 26-7
Allen, Valerie 84
Allen, Woody 41
Alvis, Jason 231n
Anacreon 59
Anaximander 24
Anderson, Laurie 146
Archilochus 59
Arendt, Hannah 1, 4, 7f, 14ff, 21f, 26f,
  29, 32, 36, 51, 55f, 58f, 66ff, 69ff, 78ff,
  89, 102, 118, 124, 152, 159, 162, 174,
  248n, 230n, 251n
Aristotle 43, 44, 49, 58ff, 65f, 155, 162
Armon, Adi 227n
Arnheim, Rudolf 17, 139, 272n
Aronowitz, Stanley 27, 236n, 237n
Augustine 28, 44, 50, 52, 61, 63, 67, 78,
  82, 92, 100, 128, 154f, 157, 164, 182f,
  193, 243n, 244n
Ausländer, Rose 83
Axelos, Kostas 53
Axiotis, Are D. 83

Bachelard, Gaston 25
Bachmann, Ingeborg 69
Baez, Joan 61, 137
Bardot, Brigitte 182
Barth, Ferdinand 173, 204

Bataille, Georges 124
Bateman, Chris 228n, 263n, 268n
Baudelaire, Charles 2, 56, 248-9n
Baudrillard, Jean 2, 4, 10ff, 24f, 39f, 89,
  105, 108f, 123, 127, 131, 143f, 148ff,
  151, 162, 169, 198, 201, 203, 207,
  215f, 220, 232n, 233n, 236n, 237n,
  238n, 241n, 260n, 263n, 265n, 269n,
  273n, 276n
Becker, Howard 234n
Becker, Paul 148
Beckett, Samuel 2, 9, 17, 89, 175ff,
  214, 232n
Beethoven, Ludwig van 37, 102, 160,
  166f, 186f, 189, 192, 255n, 265n
Beinsteiner, Andreas 242-3n
Benjamin, Walter 1, 3, 8, 14f, 22, 26f, 36,
  65, 69f, 76, 88f, 91f, 97, 102, 106, 120,
  164, 183, 188, 229n, 233n, 235n
Bennett, H. Stith 158, 190ff, 234n,
  244n, 270n
Benny, Jack 185
Berg, Alban 161
Bergson, Henri 185
Berlioz, Hector 166, 181
Bernasconi, Robert 120, 239n, 260n
Bernays, Edward 171, 184, 206
Berry, David 26, 233n
Bertolotti, David S. 106f, 206, 268n,
  275n
Beuys, Joseph 130
Birkin, Jane 182
Bizet, Georges 37
Blitz, Mark 236n
Bloch, Ernst 9, 56, 249n
Blok, Vincent 245n, 267n
Blondel, Maurice 217f
Blücher, Heinrich 59, 63f, 65ff, 251n, 252n
Bogart, Humphrey 7
Böhme, Hartmut 239n
Bostrom, Nick 133, 243n, 275n

Bourdieu, Pierre   2, 7, 24, 153, 238n, 239n, 244n, 264n, 266n, 269n
Brecht, Bertolt   139
Brendel, Alfred   191
Brock, Bazon [Jürgen Johannes Hermann]   66, 252n
Brown, Norman O.   41
Buber, Martin   16, 56, 249n
Busch, Briton Cooper   118f
Busoni, Ferrucio   159, 192, 254n
Bussolini, Jeffrey   247n, 258n

Cage, John   192f, 271n
Cain   99, 259n
Cale, John   189
Callas, Maria   191
Camus, Albert   16
Čapek, Karel   29
Carnap, Rudolf   161
Carroll, Lewis   102, 121f
Carson, Anne   59
Carson, Gerald   269n
Carter, Jimmy   116
Caruso, Enrico   191
Casals, Pablo   191
Cassirer, Ernst   85
Castor   65
Catullus   59
Cavell, Stanley   11
Caygill, Howard   253n
Celan, Paul   83
Chargaff, Erwin   205, 240–1n
Clausewitz, Carl von   170
Cohen, Leonard   5, 181f, 185f, 189, 193, 234n, 271n
Collins, Judy   185
Copernicus, Nicolaus   51, 85f, 156, 260n
Corbusier   170
Cornelius, Hans   230n

Dahlhaus, Carl   187
Dahlstrom, Dan   48, 53
Dali, Salvador   101
Dante   103
Danto, Arthur   192
Darling, David   190f
Dastur, Françoise   48, 245n
David   116
David, Christophe   240n, 243n, 256n

Dawsey, James.   229n, 235n
de Beauvoir, Simone   16, 58, 110, 118, 185
Debord, Guy   12, 148, 179
Debussy, Claude   181
de Certeau, Michel   2, 7, 16, 237n, 239n, 242n
Degerman, Stig   107
del Caro, Adrian   85
Deleuze, Gilles   25, 123, 263n
de Rougement, Denis   55f, 224, 248–9n, 277n
Derrida, Jacques   25, 51, 106, 114
Descartes, René   2, 32f, 43f, 49, 67, 243n
De Toqueville, Alexis   151
Dick Tracy   5
Dilthey, Wilhelm   156f, 264n
Dionysus   32, 128, 166, 226n
Diotima   58
Doré, Gustav   103f
Downes, Hugh   190
Dreyfus, Bert   11
Dries, Christian   83, 214, 227n, 235n, 240n, 251n
Drouillard, Jill   260n
Dücker, Bürckhard   82f
Duhem, Pierre   130
Dupuy, Pierre   56
Durbin, Paul   237–8n
Dylan, Bob   184

Eatherly, Claude   89, 94ff, 97f, 99f, 106, 108, 197, 232n, 256n, 257n, 258n, 260n
Eichmann, Adolf   56, 68, 89, 99
Eichmann, Klaus   68, 89, 197, 232n
Einstein, Albert   67, 102
Eisler, Hanns   155, 270n
Ellensohn, Reinhard   167, 186, 230n, 241n, 262n
Ellul, Jacques   2, 7, 10ff, 23f, 95f, 127, 149, 169, 172, 234n, 236n, 237n, 256n, 268n
Engels, Friedrich   4
Enkidu   99
Epictetus   200, 204, 207, 273n
Epicurus   254n

Fanon, Franz   16, 25
Fetz, Bernard   238n
Figal, Günther   103, 257n

Filk, Christian   230n
Fink, Eugen   157, 265n
Fleck, Ludwik   237n
Forman, Paul   171
Foucault, Michel   100f, 174f, 263n
Frege, Gottlob   45
Freire, Paolo   16
Freud, Sigmund   64, 83, 111, 137, 184
Freundlich, Elisabeth   28, 64, 66
Frith, Jordan   267n, 272n
Fuchs, Christian   7, 229n, 235n, 238n, 242n
Fuller, Steve   12, 25, 50, 133, 215, 217, 220, 232n, 239n, 241n, 242n, 245n, 268n, 275n

Gadamer, Hans-Georg   23, 26, 38, 72, 84f, 103, 105, 154, 156f, 158, 172, 181, 254n
Gainsborough, Serge   182
Gehlen, Arnold   26
Gellen, Kata   69f
Germaine, Gilbert   236n, 237n
Geuss, Raymond   22
Gide, André   55
Gillespie, Michael Allen   236n
Goehr, Lydia   153
Goethe, Johann Wolfgang von   5, 15, 24, 61, 65, 71, 78f, 90, 111f, 133, 137f, 164, 172f, 178, 204, 215, 222, 228n, 238n, 249n, 258n, 261n, 268n
Goodman, Steve   171, 199
Gore, Lesley   184
Gould, Glenn   191
Grandin, Greg   119
Gray, John   27, 239–40n
Grien, Hans Baldung   60
Griffin, David Ray   219, 276n
Grondin, Jean   244n
Grönemeyer, Herbert   182
Guthrie, Arlo   188

Habermas, Jürgen   22, 26, 38, 88, 91, 98, 103, 111, 239n
Hadot, Pierre   72, 244n
Halbwachs, Maurice   155
Han, Byung-Chul   237n
Hanslick,   152
Hardy, Thomas   60

Harré, Rom   45
Harris, Sir Arthur   107
Heelan, Patrick Aidan   171f, 278n
Hegel, Georg Wilhelm Friedrich   2f, 4, 28, 50, 55, 63, 69, 87, 91, 231n
Heidegger, Elfriede   66, 251n, 252n
Heidegger, Martin   1f, 8f, 11f, 15f, 21f, 24, 26, 28f, 32f, 42f, 45ff, 47f, 53, 56, 59f, 61, 64f, 70f, 76, 83f, 88, 90, 92, 94, 97, 100, 107f, 121, 123f, 129, 132, 138, 143, 153, 155f, 158f, 161, 166, 172, 174, 180, 184f, 193f, 198, 292, 204f, 208, 210ff, 213f, 216, 220, 222f, 227n, 230n, 231n, 235n, 238n, 241n, 242n, 245n, 231n, 252n, 255n, 256n, 260n, 266n, 269n, 273n, 277n
Heinlein, Robert   4
Heisenberg, Werner   53f
Helm, Birgitte   30ff
Helöise   58
Hempel, Hans-Peter   237n
Hepburn, Katherine   7
Herzl, Theodor   67
Hirst, Damien   130
Hitler, Adolf   58, 106, 139f, 150, 152, 220, 246n, 249n, 261n
Hoeckner, Berthold   102
Hoffman, Anne Golomb   89
Hoffmann, E.T.   166
Hölderlin, Friedrich   36, 53f, 58, 71ff, 76, 81, 90, 108, 115, 158, 163, 176, 202f, 214, 249n, 273n, 274n
Homer   42, 116
Honneth, Axel   97, 111
Hook, Sidney   67
Horkheimer, Max   4, 8, 11, 21, 37, 39f, 83, 143
Huck Finn   65, 188
Hullot-Kentor, Robert   83, 230n
Hume, David   172, 183, 244n
Husserl, Edmund   2, 8, 22f, 28, 45, 48f, 59, 111, 124, 137, 154f, 156, 159, 166, 172, 174, 227n, 231n
Huxley, Aldous   141

Ihde, Don   10f, 23, 25ff, 133, 172, 219, 232n, 241n, 239n, 268n
Illich, Ivan   2, 7, 11f, 16, 24f, 27f, 47, 53, 63, 148, 179, 193, 198, 201, 205, 217f,

230n, 234n, 237n, 239n, 260n, 261n, 263n, 272n, 273n, 276n, 277n
Irigaray, Luce   124, 185, 220, 277n

Janicaud, Dominique   53f, 237n
Jaspers, Karl   15, 63, 66f, 107, 169, 185, 227n, 240n, 246n, 250n, 258n
Jesus   53, 207
Johnson, Julian   161
Jolly, Édouard   227n, 235n
Jonas, Hans   17, 26, 32, 63, 91, 93f, 227n, 251n
Jünger, Ernst   96

Kafka, Franz   2, 10, 14f, 24, 69ff, 176, 221, 224, 234n, 238n, 253n
Kahn, Herman   107, 258n
Kant, Immanuel   2, 8, 22, 26, 28, 34, 42ff, 45, 48ff, 52, 70, 93, 120, 123, 132, 155, 186, 189f, 244n, 253n, 259n, 271n
Kateb, George   23
Kaufmann, Walter   186
Kayser, Hans   160
Kendrick, Anna   187
Kepler, Johannes   86, 156
Kierkegaard, Søren   230n
Kimball, Roger   27, 239n
Kircher, Athanasius   86
Kissinger, Henry   116
Kittler, Friedrich   26f, 40, 70, 138f, 140f, 145f, 149f, 162, 170f, 176, 178, 199, 222, 233n, 239n, 262n, 263n, 267n, 269n, 272n
Klee, Paul   5, 74, 89f, 106
Köhler, Wolfgang   8
Kohn, Jerry   61, 75
Kojève, Alexander   69f, 231n
Kracauer, Siegfried   22
Krauss, Karl   76, 88
Kreisler, Georg   152
Kroker, Arthur   237n
Kuhn, Thomas   7, 237n
Kusch, Martin   47, 246n

Lacan, Jacques   41, 64, 88, 91, 99, 106, 125, 173, 205
Laine, Frankie   184
Lang, Fritz   29ff
Lanier, Jaron   23

Latour, Bruno   24f, 27, 91f, 121, 215, 218, 236n, 238n, 256n, 260n, 275n, 276n
Lazare, Bernard   67
Le Bon, Gustave   140
Leibniz   2, 84ff
Levarie, Sigmund   191
Levene, Mark   112, 119
Levinas, Emmanuel   16, 25, 29, 51f, 114, 246n
Levinson, Jerrold   153
Levy, Ernst   191
Lewis, C. S.   56, 102
Liessmann, Konrad Paul   239n
Lincoln, Abraham   140
Lingis, Alphonso   277n
Locke, John   120, 175
Lonergan, Bernard   38
Lovitt, William   40
Löwith, Karl   68, 92, 107, 248n, 258n
Lubitsch, Ernst   185
Lucian   5, 55, 160, 173, 268n
Lühman, Niklas   26
Lukács, György   96
Luther, Martin   56
Lütkehaus, Lüdger   38, 59, 154f, 253n, 231n, 250n, 252n, 256n, 264n

McClain, Ernest   187, 191, 262n, 265n
McCormick, John   23
McLuhan, Marshall   23, 95, 237n
Maier-Katkin, Daniel   64, 66f, 251n, 252n
Manne, Kate   6, 41, 228n
Mannheim, Karl   22
Marcel, Gabriel   8, 15, 110f, 230n, 258n
Marcus Aurelius   97, 107
Marcuse, Harold   232n, 268n
Marcuse, Herbert   11, 17, 39ff, 55, 125, 143, 169, 172, 240n, 242n, 268n, 270n, 272n
Marcuse, Ludwig   8, 17, 21
Marinetti, Filippo Tommaso   129
Maritain, Jacques   55
Martin, George R. R.   37
Martin, Rickie   188f
Marx, Karl   4, 28, 53, 71, 94, 116, 228n, 231n
Matassi, Elio   238n
Meier, Bettina   238n
Mellor, David   7

Melville, Herman 120
Mephisto 56, 110, 268n
Merleau-Ponty, Maurice 16, 61f
Merton, Robert K. 173, 237n, 268n
Michaelis, Anthony R. 228n
Michaelis, David 68, 228n, 229n, 241n, 253n
Michaelis-Stern, Eva 8, 68, 229n, 253n
Milner, Greg 241n
Milton, John 56, 133
Minelli, Liza 184
Moltmann, Jürgen 91
Moneta, Giuseppina 61f
Morozov, Evgeny 13, 102, 233n
Moses 4
Müller, Christopher John 7, 214, 228n, 229n, 231n, 233n, 235n, 242n, 275n
Müller, Marcel 230n

Napoleon 116
Nehamas, Alexander 55
Nehring, Holger 106
Neiman, Susan 55
Nettling, Astrid 24
Nietzsche, Friedrich 2, 4ff, 8, 15ff, 27, 29, 37f, 43, 48, 52, 58, 62, 70f, 73, 78f, 83f, 85f, 87, 90f, 96, 99, 102, 108, 116, 128, 155, 158, 163f, 179, 182, 185f, 191f, 193f, 227n
Nono, Luigi 177, 192

Ogden, Daniel 228n
Orwell, George 6, 128, 168
Østergaard, Edvin 142m, 262n
Otlet, Paul 170

Paddison, Max 153, 234n, 254n
Palaver, Wolfgang 56
Pascal, Blaise 156
Patterson, George 118, 140
Pericles 140
Pinker, Steve 25, 238n, 239n
Plato 43, 58, 67, 83, 161ff, 212f, 256n, 262n
Plessner, Helmut 8, 15
Pöggeler, Otto 83
Pollux 65
Postman, Neil 11f, 23, 232n, 237n
Presley, Elvis 182, 191

Prometheus 99
Ptolemy 86
Putz, Kerstin 60
Pythagoras 164

Rançiere, Jacques 16
Raulff, Helga 83
Rée, Paul 60
Reemtsma, Jan Philip 83
Richardson, William J., S.J. 16, 49, 61
Rickman, Alan 56, 102, 107, 182
Ricoeur, Paul 138
Rilke, Rainer Maria 15, 24, 58, 71f, 78f, 83, 89, 97, 154, 157, 159, 162, 176, 210ff, 215, 224, 254n
Ritschl, Friedrich 65
Roberts, Mike 192
Rohde, Erwin 65
Ropohl, Günther 26
Rose, Gillian 63
Rossini, Gioachino 37
Rothblatt, Martine 130
Rousseau, Jean-Jacques 34, 43, 53, 164
Rowling, J. K. 102
Ruin, Hans 237n, 242n, 243n

Salomé, Lou von 58, 249n
Sappho 59f
Sartre, Jean-Paul 16, 25, 29, 32, 47f, 56, 58, 132, 185, 231n
Satan 56
Saul 115f
Schandl, Franz 233n
Scheler, Max 1, 8, 15, 43, 48f, 55, 87, 107, 123, 131f, 156, 245n
Schelling, Friedrich Wilhelm Joseph 48, 76, 155
Schiller, Friedrich 34, 46, 53, 64f, 186, 189, 249n, 251n
Schlegel, Friedrich 161
Schmid, Holger 245n, 264n
Schmitt, Carl 23, 57, 67f, 155
Schoenberg, Arnold 192
Scholem, Gershom 67f, 89, 252-3n
Schopenhauer, Arthur 158, 163f, 166
Schraube, Ernst 70, 227n, 235n, 253n, 258n
Schürmann, Reiner 47ff, 53, 240n
Schwartz, Regina M. 259n

Name Index

Scott-Heron, Gil 150, 263n
Scruton, Roger 153
Sebald, Winfried 105, 107, 258n
Seymour, Richard 5f, 233n
Shakespeare, William 5
Sibelius 231n
Simmel, Georg 55, 87
Simondon, Gilbert 2, 237n, 263n
Singer, Peter 173
Sloterdijk, Peter 24f, 57, 91, 94, 103, 108f, 198, 212f, 214f, 216f, 219, 221f, 223, 256n, 258n, 274n, 275n, 276n
Smith, F. Joseph 72, 158, 231n, 242n, 268n
Smith, Jonathan 29
Smythe, Dallas 23, 39, 232n, 233n, 272n
Socrates 14, 58
Sommer, Christian 260n
Sonolet, Daglind 24, 230n, 237n
Sophocles 123, 257n
Sorgner, Stefan 12
Speer, Albert 139
Spinoza, Baruch 2
Steiner, Rudolf 164
Stern, Carla 6, 8, 229n
Stern, Günther 14f, 59, 61, 65, 76, 110, 154f, 197, 229n, 231n
Stern, William 8f, 14, 44, 142, 229n, 231n
Stewart, Garrett 220
Stiegler, Bernard 25, 237n
Stoneman, Rod 264n
Strauss, Leo 17, 67
Street, John 169ff, 172, 236n, 237n, 268n
Strong, Tracy 137, 236n, 252n, 259n, 261n
Stumpf, Carl 9, 72

Tanzer, Mark 240n
Taubes, Jacob x, 27, 57, 64, 67f, 89, 216, 248n, 249n, 252n, 256n
Taubes, Susan 64
Thales 46
Theunissen, Michael 16, 34
Theweleit, Klaus 29
Thoreau, Henry David 11
Thrasymachus 58, 93
Tiedemann, Rolf 114

Tillich, Paul 8, 16
Tolkein, J. R. R. 56
Tom Sawyer 65
Trawny, Peter 240n
Treitler, Leo 161
Truman, Harry 98, 116
Tudor, David 192
Turkle, Sherry 12, 23, 232n

van Dijk, Paul 11, 167, 169, 227n, 235n, 238n, 240n
Verbeeck, Peter-Paul 11, 25, 236n, 237n, 238n
Verne, Jules 102, 257n
Vetlesen, Arne Johan 26, 238n, 239n
Vézin, François 241n
Vietta, Silvio 237n
Virilio, Paul 2, 4, 6, 11, 24f, 70, 108, 117, 120f, 129f, 130f, 133, 171, 174, 205, 223, 236n, 238n, 241n, 260n, 261n, 265n
Vogel, Steve 239n
von Brentano, Margherita 69
von Bülow, Hans 191
von Hildebrandt, Dietrich 17
Vonnegut, Kurt 240n

Waelbers, Katinka 236n
Wagner, Cosima 192
Wagner, Richard 16, 37, 70, 152, 166, 186f, 199
Waite, Geoff 254n
Wallenstein, Sven-Olov 234n
Weber, Max 194
Weil, Simone 227n
Weill, Kurt 184
Wieland, Christopher Martin 5, 268
Winner, Langdon 23, 233n, 236n, 237n
Wittgenstein, Ludwig 54, 162
Wohlfart, Günter 250–1n
Wölfflin, Heinrich 8
Worster, Donald 259n

Yeats, William Butler 98, 257n

Zelka, Charlotte 28, 61, 66
Žižek, Slavoj 41, 91, 109, 127
Zuboff, Shoshana 233n
Zuckerman, Lord Solly 107

Subject Index

4'33" 192f, 271n
9/11 127, 169, 170, 206, 210, 217, 219, 276n

acoustic 72ff, 76ff, 139f, 141ff, *see also* stereoscope
    double 138f
    lifeworld 141
    notification/signal 148, 168, 184
    phenomenology 137f, 142f, 182, 231n, 241n, 265n
    reproduction 17
acroamatic 160, 265n
Adorno Prize 83, 230n
advertising 13, 35, 39, 93f, 141, 150, 272n
aesthetics, artist's 125, 158, 192
    spectator's 4, 130, 150, 158, 185, 190
Afghanistan 108, 112, 127
air 3, 62, 94ff, 109, 137, 144f, 146, 159f, 170, 175, 205, 215ff, 220, 222ff
    dead 192f
Alexandrian 192
American Gods 56
American Pie 188
analytic philosophy 92, 111f, 153f, 172
    dominance of 92, 112, 264n
Animal Farm 6
animals 46, 49ff, 73f, 100, 109, 114ff, 130ff, 246n, 247n, 248n, 261n
    abused in art 130
angel 14, 73f, 77f, 82, 88f, 90, 97, 102, 107, 124, 224, 252n, 254n
Angelus Novus 14, 89f, 102
annihilation 93, 96, 115ff, 132ff, 144f, 245n
Anthropocene 4, 50, 53, 219, 239n, 242n
anthropocentrism 51f, 93
anthropologism 42, 45ff, 48, 244n
anthropology, anthropological 7, 22, 45ff, 48, 52f, 138, 244n, 245n, 246n, *see also* ethnography
    media 230n
    musical 164

phenomenology 2f, 107, 121, 222, 227n
    philosophical 8, 10, 22, 26f, 42f, 37, 82f, 230n, 254n
    physical 242n
    sociological 107
Antigone 123
anti-Semitism 51, 92, 108
apocalypse 9, 24, 56, 88ff, 93, 97, 99, 101, 105f, 109, 128, 203ff, 209ff, 214, 223f
Apocalypse Now 199
Archaic Torso of Apollo 72
art, art culture 15f, 47, 88, 130, 133
arthritis 255n
ascetic, ascesis 58, 85, 144, 194, 207
ASMR 143, 166, 184, 266n
assignment 71, 75, 78ff
atheism 52f, 85, 238n
Athens 32, 140
atomic bomb, atomic power, Atomic Age 3, 9, 17, 28f, 83f, 88, 94ff, 98, 103
    manufacture of 106
    neutron 117f
    war 126
Augustinians 55, 124
Auschwitz 47, 57, 69, 83, 106, 111, 113f, 117, 121, 126f, 129, 133, 197, 204, 245n
    -Birkenau 121, 129f
authenticity 42, 44, 97, 124, 151, 163, 194, 209
autonomy 145, 150, 202

banality 89, 157, 220, 248n
being-in-music 74, 142, 147, 157f
big data 12, 203, 229n
Bilderverbot 106, 185
biotech 125ff, 129, 130ff, 133
Blue Hawaii 182
body 32f, 53, 58, 101, 110f, 123f
    naked 91
    parts, spare 127ff

Subject Index

bomb, *see* atomic bomb
Brave, New World 141
broadcasting 160, 176f, 184f, 188, 199
Brot und Wein 72

can do, can do-ability 90, 125
  obligatory, compulsory 131
capitalism, capital 2, 33, 56, 125, 150, 179, 229n, 272n
  disaster 56
  surveillance 12, 233n, 269n
cargo-cult 169, 214
Carnivàle 56
causality 112, 183f
cell phones, mobile phones 5f, 34f, 36f, 116, 148ff, 168, 180, 199f
censorship 24, 206, 216f
Chernobyl 57, 129, 170, 175, 197f, 199, 205, 232n, 274n, 275n
chimeras, chimbrids, animal-human 116f, 120f, 130, 224, 261n
City of New Orleans 188
climate control/crisis 3, 87, 93, 109, 116, 128, 133, 209, 214, 222ff, *see also* geoengineering
cold war 88, 91, 126f, 258n
commodity 144, 149, 159, 180, 233n, 256n
Communist Manifesto 4
Confessions 44, 82, 154, 182, 243n
conformism 10, 151, 203
consumers 4, 12ff, 37, 43, 93, 116f, 125ff, 131, 147, 149, 160, 192, 200f, 217, 221
  society 35
continental philosophy 92, 112, 264n, 238n
corporeality, *see* body
cosmos, cosmology 84ff, 93, 116, 160
covers, cover culture 13, 15, 30, 153, 177, 181ff, 185f, 187ff, 190, 234n
critical theory ix, 1f, 21f, 25, 38f, 40, 55, 88, 106, 111f, 138, 152f, 154, 167, 220, 233n
  thinking 28, 111
criticism, negative, revolutionary 22, 34, 38, 88f, 94, 113, 153
Critique of Cynical Reason 108, 109
crystallization, political 141, 171, 184, *see also* propaganda

culinary 93, 124, 144, 159, 183f, 187, 192, 230n
culture industry 12, 15, 37, 40, 93, 107, 144f, 149f, 160, 165ff, 171, 183, 192f, 194, 200, 221f, 222, 233n, 270n, 273n
Cup Song 187f, 190
Current of Music 138, 188, 200
cyborg 5, 34f, 47, 221, 261n

Darling, David 190
Darwinism 51, 131
Daseinsanalyse 163
data mining, datification 150, 201f, 203, 233n, 273n
death 47, 97, 109, 183f, 193f
  of God 29, 97, 198
democracy impossible 200
desire 41, 63, 154
Deutsche Menschen 14, 233n
devil, 'deviltry' 24, 50, 55f, 57, 248–9n, 277n
Dialectic of Enlightenment 21, 40, 209, 242n
Diamonds and Rust 61, 137
Die Kirschenschlacht 51, 61, 64, 78, 81ff, 93, 247n
Die molussische Katakombe 4, 24, 30, 110, 124, 238n
digital hacking 272n
dignity 46
diner culture, *see* jukebox culture
Dionysus, dionysian, dionysiac 32, 166, 187, 266n
Dionysus Dithyrambs 128
Dresden 105, 107f, 109, 212, 215f
Duino Elegies 71ff, 76, 89f, 142, 162, 254n

earphones, -buds 39, 140, 146, 149
eavesdropping 148, *see also* surveillance
echo 2, 72, 78, 176, 178
  radio 137, 139, 141f, 156f, 261n
economics 72, 109, 123f, 127f, 132f, 192, 212, 233n, 259n
Eichmann in Jerusalem 68, 71, 248n
'empirical turn' 246n
end-time, *Endzeit* 89f, 105
energeia 155, 157
erotic 41, 58, 65, 70, 79, 124f, 199, 263n

eschatology   57, 92f, 248n, 256n
ethnography   ix, 2, 45, 48, 121, 227n, 245n, *see also* anthropology
evil   10, 28, 55ff, 248n, 88, 108f, 111, 116, 173, 175, 201, 204, 214, 242n, 248n, 275n
  banality   96
existentialism   45, 49, 53, 132, 163, 230n
extraterrestrials   46, 243n

Facebook   5, 13, 39f, 65, 144, 148f, 174ff, 179, 198ff, 201f
facts   48, 99, 236n
'fake news'   12, 39, 141, 205, 215, 217, 249n, 277n
Fantasia   90
Faust   78f, 137, 258n, 261n, 268n
Fear of Missing Out [FOMO]   146f, 168f
First World War   81f, 105
Frankfurt School   1f, 8, 22, 26, 38, 47, 111, 125, 133, 183
free opinion   200, 202f
free variation   141f
friendship   62ff, 67, 123
Fukushima   57, 199, 205, 274n
'fun'   92, 150, 220, 263n

gadget/*Gerät*   2, 5, 33f, 40f, 43, 93, 96, 100, 117, 167f, 266n
gain of function   3, 100, 133
Garden Party   188
Gaza   112, 127, 131
Gelassenheit   101, 127
Genealogy of Morals   52, 85, 144, 200f, 227n
Genesis   10, 84, 99
genocide   110ff, 114ff, 197ff, 223, 224
  animal   118ff
  genocidal rape   259n
  slave trade   119f
geoengineering   3, 57, 87, 109, 128, 172, 205, 209ff, 215, 219, 224, 247n, 258n
Ge-Stell   2, 35, 43, 54
ghosts   2, 42, 82, 137ff, 177f, *see also* radio, ghosts
  effect, acoustic   139f
  television   141
global destruction, globicide   13, 50, 86f

GMO   3, 101, 129, 213
gnostic, gnosticism   x, 2, 56, 93
Good Omens   56
GPS   39, 147, 169, 176, 179, 200, 241n, 263n, 272n
guilt   99, 155, 169, 209, 264n
guiltlessly guilty   95, 97, 105, 108f, 121
Gulf War   3, 108, 131, 220

HAARP   101, 217, 219, 222
Hallelujah   181ff, 185f, 188ff, 193, 234n
'Hallelujah effect'   39f, 63, 137, 167, 183f, 185, 192
happiness   194
Happy Days   39
"Harrison Vergeron"   240n
having   8f, 50, 96f, 101f, 105, 230n, *see also* property
  been   197
  been born   51, 66
  done   110
headphones   146, 179, 199
hearing   74f, 159, 161, 163, 266n, *see also* listening
hearkening   159, 162
hear stripe   190
Hearts of Space   166
hermeneutical phenomenology   49, 98, 153, 224
hermeneutics   x, 1, 21, 16, 38, 49, 71f, 96, 103, 138, 111f, 154, 163, 203
  material   239
Hey, Jude   189f
Hiroshima, Mon Amour   106
Hiroshima/Nagasaki   9, 26, 28, 47, 56, 68, 82f, 94f, 96, 100, 105f, 108f, 111, 116, 126, 170, 197ff, 204ff, 215f, 235n, 245n, 256n, 258n, 271n, 274n, 275n
history   165, 194, 248n
Holocaust   28, 55f, 68, 115ff
  animal   118
homework, homeworker   ix, 12f, 117, 145, 147, 150, 160, 198
Honolulu (love is only in)   182ff, 189f
human, humanism   42ff, 47ff, 132f, 197, 246n
  condition   50, 65f, 99, 121

Subject Index

destruction of 98f, 100f, 114f, 118, 120, 121, 205, 245n
engineering 33, 66f, 102, 121, 129, 132f, 205, 245n, 263n, 268n
error 33
exceptionalism 54
 as indeterminate/unfinished animal 43, 51, 254n
 as irrelevant 84, 93
 musical situation of 164, 166
 total mobilization of 47f, 114, 134
humanity as a problem 29, 34, 43, 46, 48, 52ff, 57, 98, 109
Hyperion 58

ideology 27, 49, 106, 116, 222, 235n
inauthenticity 44
incarnationist 57
industrial Platonism 138
Introduction to the Reading of Hegel 69
Iraq 105, 112, 114, 127, 131, 218
isolation ix, 13, 39, 149, 180, 199
Israel 67f, 76, 112, 114f, 124, 274n

jazz 32, 37, 142, 153, 154, 161, 181ff, 190, 192, 265n, 271n
Je t'aime... moi non plus 182
Jewish science 130
Judaism, Judaic tradition 1, 28, 50f, 67f, 74, 76, 99
jukebox culture 39, 148, 179f, 184, 241n

Korea 3, 82, 127, 212

Lake Woebegone 6
Lebensphilosophie, life philosophy 123f, 185
listening-to 1, 9, 61f, 77, 138f, 141ff, 145, 148, 153, 160f, 162, 178, 181f, 184, 265n
 active, attuned 158, 161, 189f, 265n
lockdown 5f, 7, 39, 149, 168
Lost in Space 29
loudspeakers 139f, 140f, 145, 147, 165, 191, 200, 262n, 272n
love 58, 62ff, 65, 71, 78ff
 affair(s) 63ff, 85f
 erotic 70, 124f
 triangles 58f

Luddite, luddism 11, 22f, 35, 88, 95, *see also* criticism
lyric 16, 189
 subjectivity 189f

McCarthyism 169
Manhattan Project 106
'mania for interpretation' 71
'manufacture of corpses' 52, 121, 131
marketing 93f, 184f
mass, masses 10, 16, 141f, 148, 150f, 179, 202f
media ixf, 107, 126, 135f, 143, 161, 168f, 197f, 203f, *see also* social media
 digital 141, 143, 184, 192, 203, 221
Metropolis 29ff
Minima Moralia 51, 144, 159, 176, 214, 221
mobbing 92, 219, 276n
monads 6, 84, 86f
mortality 81, 89, 97, 188
Mündigkeit 44, 145, 202, *see also* autonomy
music 5, 37, 71f, 84, 152, 158, 165, 190, 192, 257n
 aesthetics 155, 231n
 appreciation 144, 153, 265n
 background 159, 184, 241n
 criticism 152f
 electronic 146, 199, 261n
 Greek 16, 37, 75, 140, 160f, 186, 262n
 in- 9, 74, 158f
 live 145, 178
 machine- 32
 making, musicking 15, 153f, 192
 modes 161
 new 190, 192, 265n, 266n
 phenomenology 139, 190f, 265n
 pop 182ff, 186ff, 190f, 192f, 270n, 271n
 postmodern 166
 sentences 156
 situation 138f, 153ff, 160f
 sociology 8, 37f, 138f, 152f, 154, 157f, 158f, 165f, 178f, 180f, 190, 192, 230n, 231n, 234n
 time, art of 102

nanotechnology 2, 129
natality 29, 66, *see also* having, been born

Native Americans   113f, 121f
natural resources, exploitation of   117f, 127
  human body tissue as   127f, 130f, 133f
navel   29
negativity   22, 51, 128, 131, 162, 172, 178
neo-fascism   88, 91
new age   4, 146, 166
newspaper   95, 107f, 149, 168, 206, 217, 275–6n
New Testament   93
nihilism   2, 12, 56, 74, 92, 96, 244n, 248n, 258n, 267n
non-violence   197f, 207f
nothingness   74, 96, 204, 224
  two   98, 128
now, the moment   105
nuclear armaments, destruction, preparedness   96, 101, 126
nuclear waste   274n
Nykia   42

Oldest System Programme   76
Only Fools Rush In   182, 191
opera   37, 90, 138, 191

painting   72, 84, 89, 106, 167
Palestinians   67, 112, 114
Pasteurization of France   92, 236n
peeping toms   148
Pelléas et Mélisande   181
perpetual motion machine   125
The Persistence of Memory   101
Phaedrus   58
phenomenology   x, 2f, 8f, 11, 23, 44f, 48, 111, 138, 181f
  hermeneutic   21, 44, 48f, 96, 98, 131, 153
  musical   71f, 111, 137
  sociology   148
philosophical anthropology, *see* anthropology
Philosophies for Sale   55
phonograph   145, 150, 176, 178
positivism, *see* analytic philosophy
posthumanism   4, 37, 42f, 102, 214, 242n, 243n, 275n, *see also* transhumanism

post-truth   39, 43, 139f, 218
prayer   44, 54, 76, 93, 124, 159, 165, 185, 243n
pre-emptive strike   88f, 108
pre-established harmony   86
priming   37, 39, 184, 200, *see also* programming
Princeton Radio Project   138, 200, 272n
privacy, private sphere   148f, 174f
profit   8, 72f, 118, 127, 187
programming   12f, 38ff, 61, 94, 184, 189f, 197ff, 272n
  self-   146f, 190
Prometheus   99, 214, 221f, 232n, 265n, 275n
  effect   143, 214
propaganda   107, 139, 141, 183f, 220
property   124, 150f, 175
pseudo-concreteness   42, 44, 51, 57, 63, 242n
psychoanalysis   88, 151, 203
psychology,   2, 6, 9, 41, 45ff, 48, 123, 138, 203, 229n, 264n, 275n
  psychologism   45f, 47, 246n
  social   124f, 203
psyop   61, 94, 203, *see also* priming; programming
public opinion   200, 206, 273n
public world, sphere   13, 36, 61, 148f, 179
Pygmalion   29

questioning   16, 21, 36, 44f, 48, 54, 73, 77, 128, 153f, 155f, 158, 189, 220f, 223, 240n, 244n, 246n, 277n

radio   15, 34f, 39f, 137ff, 147, 165, 261fn
  broadcast, transmission   138f, 142, 146f, 181ff, 192f, 199f
  face   137, 140, 188
  ghosts   137ff, 141f, 156f, 176f
  phenomenology of   137ff, 140ff, 188
  plugging   15, 273n
  spookiness   137ff (*see also* acoustic, double)
  theory   165, 178, 261n, 262n, 272n
radioactivity, radiation   193, 197f, 269n, 276n
rape, genocidal   115
rational choice   39, 248n

Subject Index

reactionary   35f, 38
reading, unlearning the art of   108
reception, cell phone, WLAN   36f, 182, 203
recording   28, 97, 179, 184, 188f, 191ff, 261n, 270n
   consciousness   190f
religion   2, 50, 74f, 76, 79, 96
Republic   67
rhythm   16, 30, 32, 37, 110, 165, 186, 257n, 270n
Ride of the Valkyries   199
robot   29f, 46, 228n, 240n, 263n
   sex   30, 32, 240n, 263n
rock intelligence   46
romanticism   24f, 192
R.U.R.   29

schizoid-topic, schizo-topia   141ff, 145, 160, 195f
Schlaraffenland   105
science   7, 17, 27f, 33f, 42, 45f, 48, 51ff, 70, 88, 130, 170, 172, 212, 223, 228n
   aesthetic   48, 185, 194
   capitalist   33
   experimentation   70, 130, 134, 205
   fiction   4, 228n
   German   130
   Jewish   130
   Nazi   130f
   philosophy of   88, 91, 219, 236n, 238n
   question of   48
   techno-   26f, 33, 108, 239n
Science as Vocation   55
scotosis   51, 105, 227n
Second World War   3, 28, 82, 98f, 105, 107, 114, 141, 200, 212
seduction   78, 200
Seinsfrage (Being question)   44, 47f
sex   30, 51, 53, 58, 63, 65f, 70, 123f, 240n, 250n, 260n
   -less   123f, 252n
shame   2f, 4, 13, 28f, 30, 32f, 51, 99f, 155, 168, 174, 214, 222
   Promethean   9, 13, 33, 40, 99, 105, 208, 231n
shepherd of Being   93, 96, 116
   of products/things   116
Siegfried motif   152f

Sigmund Freud Prize   83
simultaneity   74, 142, 154, 158, 194
situation   5, 8, 12f, 15, 22, 42, 69, 71ff, 74, 77f, 80, 99, 112, 126, 132f, 138, 152ff, 155f, 158, 161, 177f, 189, 205, 209, 270n
   attunement   140
   musical   8, 138f, 153f, 155ff, 159, 160ff, 164, 167f, 178f, 182, 188
   question-worthy   155
   at the same time   158, 194
   technopolitical   177, 204f, 209f
slavery, Slave Trade   113ff, 118ff
social media   4, 39, 93, 143f, 174f, 176f, 198f, 202f, 217f, 223f, 228n, 273n
sociology, 'social theory'   1f, 6f, 12, 21ff, 26, 37, 48, 107, 152f, 172, 219, 221f, 246n
Sonic Warfare   171, 199, 272n
Sorcerer's Apprentice   5, 10, 90, 111, 172f, 215, 223, 256n
Sorge [care]   123, 125
soundtrack   147, 179
space, spatiality   10, 77f, 79, 82, 97f, 101f, 105, 137, 142f, 165ff, 139f, 146, 148, 160, 179, 182, 186, 188, 257n
   ubiquity   178, 266n
spectacle   148, 179
'speech without response'   143f, 148, 150, 198, 201, 207
spook, spookiness   137f, 139, 141, 199
'Spuk und Radio'   72, 138, 200, 261n, 262n, 272n
standardized ubiquity   39, 82, 168, 199
stereophonic control   146
stereoscope, audio/acoustic   145, 166f, 188, 266n
Star Trek   5, 102
streaming   4, 39, 144f, 179, 184f, 199f
Stundenbuch/Book of Hours   75
style   6, 22, 24, 26, 38, 40, 45, 51, 60, 66, 70, 77, 92, 101, 112, 140, 187, 227n
subjectivity   49, 185, 189
superhuman   131, 133
surveillance   12f, 91f, 101, 148f, 151, 168ff, 174f, 179f, 261n, 233n
symphony
   community building power   148

taste   38f, 120, 179, 183, 185, 189f, 192, 244n, 253n, 266n
technification   95
technological reproducibility   3, 73
technological sabbath   177
technology, nothing technological   184
technophobia   88, 91
Teenage Mutant Ninja Turtles   109
telephone   61f, 102, 125, 175, 179
   cell phone, smart phone   13, 33f, 37, 39, 116, 126, 133, 148f, 150, 168, 174, 180, 182l, 197, 199, 200, 203, 217, 240n, 263n
   tapping   175f, 179
television   x, 2f, 4, 12f, 22, 30, 37, 39f, 56, 100, 101, 140f, 144, 147f, 150, 159, 168, 174, 177, 179, 184, 187, 189f, 198ff, 202, 233n, 247n, 269n
terror, terrorism   98
   consumer   98
   war on   169
Terror from the Air   94
theatre masks, in antiquity   140
theology, theological   2f, 23, 49, 78f, *see also* religion
   astronomy   85
   turn   2, 53f
theremin   137
thinking   16, 111f
Thomism, Thomists   38, 55
time   1f, 92ff, 95, 98, 100f, 127, 257n
   travellers   101f
Titans   99
To Be or Not To Be   90, 101, 185
tonal *ictus*   186
tone   78, 142, 162f, 231n, 265n
   data   163
tone-variator   9, 142, 231n
totalitarianism   150, 169, 174f
   soft   270n
total mobilization   46f
transhumanism   37, 42, 47, 126, 131f, 214, 239n, 243n, 268n, 275n
translator   84f
Trinity atomic bomb   103f

True Blood   56
Twilight Zone   101
Twitter   5, 40, 148f, 150, 174, 176f, 198f, 201

vaccine, vaccination   88, 91f, 98, 100f, 120f, 127, 131ff, 171f
Vatican   56, 86
Vietnam   3, 82, 105, 108, 113, 127, 181, 199, 212, 215, 259n
Viewmaster   167, 266n
violence   3, 15, 22, 57, 88f, 91f, 100, 114, 129, 199, 204f, 207f, 210, 214, 224, 273n, 275n
Of Vision and the Riddle   92
vocoder   146
voice   38f, 61, 71, 78, 125, 137f, 142, 145, 147, 159, 165f, 170, 178, 191, 244n, 266n
voluminosity   142, 166
voluptas, see desire

The Walking Dead   56
'The Walrus and the Carpenter'   122
war   3, 29f, 121, 126ff, 129, 131, *see also* atomic bomb; nuclear
   on terrorism   108, 129
weather control   3, 57, 109, 214f, 217, *see also* geoengineering
Weibermacht   59f
Whaling   119f
When I am Gone   188
Whig history   276n
Wirkungsgeschichte   1, 23, 36f, 82
Wormwood   56

xenotransplantation   126, 130, 277n

YouTube   5, 38, 144, 178, 184, 189f, 199f, 270n
You Want It Darker   193

Zarathustra   16, 43, 84, 92, 202
Zionism   67
Zoom instruction   94, 180, 198, 224, 249n

www.ingramcontent.com/pod-product-compliance
Lightning Source LLC
Chambersburg PA
CBHW052146300426
44115CB00011B/1538